Advance in Energy Harvesters/Nanogenerators and Self-Powered Sensors II

Advance in Energy Harvesters/Nanogenerators and Self-Powered Sensors II

Editors

Qiongfeng Shi
Jianxiong Zhu

MDPI

Basel • Beijing • Wuhan • Barcelona • Belgrade • Novi Sad • Cluj • Manchester

Editors

Qiongfeng Shi
School of Electronic Science and Engineering
Southeast University
Nanjing
China

Jianxiong Zhu
School of Mechanical Engineering
Southeast University
Nanjing
China

Editorial Office
MDPI
St. Alban-Anlage 66
4052 Basel, Switzerland

This is a reprint of articles from the Special Issue published online in the open access journal *Nanomaterials* (ISSN 2079-4991) (available at: www.mdpi.com/journal/nanomaterials/special_issues/T781775246).

For citation purposes, cite each article independently as indicated on the article page online and as indicated below:

Lastname, A.A.; Lastname, B.B. Article Title. *Journal Name* **Year**, *Volume Number*, Page Range.

ISBN 978-3-7258-0092-6 (Hbk)
ISBN 978-3-7258-0091-9 (PDF)
doi.org/10.3390/books978-3-7258-0091-9

Contents

About the Editors

Qiongfeng Shi

Qiongfeng Shi is currently a Professor in the School of Electronic Science and Engineering, Southeast University, Nanjing, China. He received his B.Eng. degree from the Department of Electronic Engineering and Information Science at the University of Science and Technology of China (USTC) in 2012, and received his Ph.D. from the Department of Electrical and Computer Engineering at the National University of Singapore (NUS) in 2018. He then worked as a Research Fellow in the same department at NUS from 2018 to 2022. His research interests include flexible/wearable electronics, energy harvesting, multimodal sensing, human-machine interfaces and intelligent systems. He has authored and co-authored more than 100 top-notch journal papers and conference papers in his research fields, which have received more than 8000 citations with an H-index of 50 and an i10-index of 78, according to Google Scholar. Up to now, he has served as the Young Advisory Board Member for InfoMat and InfoScience, the Guest Editor for several international journals including JMM, Nanomaterials, and Micromachines, as well as the Reviewer for more than 40 peer-reviewed journals.

Jianxiong Zhu

Jianxiong Zhu is currently an Associate Professor at Southeast University, China. He received his B.Eng. degree in Electrical and Automation Engineering from the Hubei University of Technology in 2006. After that, he obtained his Master's degree from the University of Science and Technology of China and Ph.D. from the University of Missouri Columbia in 2015. His research interests are focused on triboelectric sensing, MEMS zero-power sensors, wearable flexible sensors and gas sensors.

Preface

The rapid evolution of modern technologies in the new era has significantly promoted the development of cutting-edge research in the Internet of Things (IoT), wearable electronics, artificial intelligence, smart manufacturing, bionic robots, smart cities, etc. The realization of these applications is greatly dependent on numerous advanced sensors with multimodal and high-quality data acquisition capabilities. To power such a huge number of widely distributed sensors, energy supplying mode has gradually migrated from centralized energy supply to in situ and sustainable energy supply.

In this regard, various energy harvesting and self-powered sensing devices based on the photovoltaic, piezoelectric, triboelectric, electromagnetic, thermoelectric, pyroelectric, hydroelectric or their hybrid mechanisms, have been proposed as supplements to traditional batteries for higher sustainability and broader adaptability of the functional networks and systems. In the last year, we have witnessed the continuous and strong momentum in the field of energy harvesters/nanogenerators and self-powered sensors. Therefore, to showcase the latest technological advancement, we bring together ten research articles and one review article from related fields in this reprint. These articles are good examples for readers who want to learn more about the design, manufacturing, characterization, integration and application of energy harvesters, nanogenerators, self-powered sensors and their systems.

We hope that the articles included here are interesting and inspiring, from which new ideas and designs can be forged, contributing to solving the current energy crisis and enabling sustainable development. Last but not least, we would like to express our sincere gratitude to all authors for their remarkable efforts in the article preparation, and all reviewers for their valuable suggestions to maintain the high standard of this reprint.

Qiongfeng Shi and Jianxiong Zhu
Editors

nanomaterials

Editorial

Advances in Energy Harvesters/Nanogenerators and Self-Powered Sensors II

Jianxiong Zhu [1,2,*] and Qiongfeng Shi [3,*]

1 School of Mechanical Engineering, Southeast University, Nanjing 211189, China
2 Engineering Research Center of New Light Sources Technology and Equipment, Ministry of Education, Nanjing 211189, China
3 Interdisciplinary Research Center, School of Electronic Science and Engineering, Southeast University, Nanjing 211189, China
* Correspondence: mezhujx@seu.edu.cn (J.Z.); qiongfeng@seu.edu.cn (Q.S.)

The Internet of Things (IoT) has become a focal point in the realm of information technology and has facilitated the interconnectedness and communication of various objects, such as devices and sensors in smart cities, intelligent transportation, industrial automation, agriculture, healthcare, etc. To meet the extensive and widely dispersed power requirements essential for IoT nodes, self-sustained systems powered by energy harvesting technologies, including but not limited to piezoelectric, triboelectric, electromagnetic, thermoelectric, pyroelectric, and photovoltaic methods, offer the potential for substantial benefits. Research groups around the globe have extensively explored these technologies, resulting in innovative energy harvesters and self-powered devices that enable the realization of self-sustained and functional systems and open up a myriad of promising applications in the new era. For example, these applications span wearable sensors, medical devices, home systems, and entertainment. The ultimate goal is to enhance the convenience of human life through the integration of these technologies in various aspects of daily living. Meanwhile, the microfabrication techniques employed in creating these advanced sensing and smart systems have progressively evolved to become more sustainable and robust.

This Special Issue, "Advances in Energy Harvesters/Nanogenerators and Self-Powered Sensors II", aims to present the latest breakthroughs and contributions in the field of micro/nanogenerators, energy harvesters, self-powered sensors, and their system applications. It features research papers and review articles that introduce the design, manufacture, integration, characterization, and use of energy harvesters, self-powered sensors, nanogenerators, and systems.

It also includes ten research articles and one review article from researchers across the world that explore technological innovations in the field of energy harvesting and self-powered sensing. Zhao et al. [1] introduce a versatile, self-sustaining displacement sensor using the triboelectric effect. It is designed for seamless integration in linear feed systems, enabling the precise detection of displacements within a broad range. It also possesses the capability for real-time velocity detection in linear feed systems, achieving an impressive accuracy rate of below 0.5%. To precisely measure the occurrence of minor tangential slips at contact interfaces during cyclic loadings in a machine system, Zhao et al. [2] developed a self-powered displacement sensor utilizing a triboelectric nanogenerator. This sensor incorporates a 100 nm thick $BaTiO_3$ thin film positioned between the Kapton triboelectric layer and the Cu electrode. Then, Yoon et al. [3] introduce the design of an Ag:AZO electrode for a self-powered solar-blind ultraviolet (UV) photodetector utilizing the $Ag_2O/\beta\text{-}Ga_2O_3$ heterojunction. The 20 nm Ag:AZO electrode with the lowest surface roughness had an on/off ratio of 2.01×10^8, and its responsivity and detectivity were 56 mA/W and 6.99×10^{11}, respectively, when the optoelectronic properties were gauged under 254 nm UV-C light. Kim et al. [4] present a polymethyl methacrylate passivation layer

check for updates

Citation: Zhu, J.; Shi, Q. Advances in Energy Harvesters/Nanogenerators and Self-Powered Sensors II. *Nanomaterials* **2024**, *14*, 121. https://doi.org/10.3390/nano14010121

Received: 18 December 2023
Accepted: 19 December 2023
Published: 4 January 2024

on the cathode side to enhance the efficiency and stability of a self-powered photodiode using $CH_3NH_3PbI_3$ perovskite. When compared with a normal $MAPbI_3$ photodiode without a PMMA layer, this device showed a much higher specific detectivity value (~1.07×10^{12}) in the absence of external bias.

In terms of energy harvesting from ambient sources, Li et al. [5] introduce an innovative solid–liquid triboelectric nanogenerator using vortex-induced resonance to efficiently harvest ocean energy over extended periods. This device captures energy through the resonance between vortex-induced vibrations of a cylinder and the relative movements of solid–liquid friction pairs within the cylinder. Then, Mao et al. [6] present a technology to enhance the triboelectric achievement of a PDMS-based triboelectric nanogenerator through adding nanostructured cadmium sulfide. The N-Cds/PDMS substrates demonstrated superior triboelectric performance with a high short-circuit current and open-circuit voltage. Meanwhile, Duque et al. [7] designed, manufactured, and electrically evaluated a piezoelectric resonant energy harvester utilizing 3D printing. This device demonstrated a maximum output power of 1.46 mW under optimal conditions, i.e., 4 MΩ load impedance and 1 g acceleration. To take advantage of acoustic waves of lower frequency in marine and industrial environments, Xiao et al. [8] present a novel acoustic triboelectric nanogenerator comprising a quarter-wavelength resonant tube, an FEP membrane, an evenly perforated aluminum film, and a conductive coating of carbon nanotube. To efficiently manufacture Sb_2Se_3/CdS-based solar cells for solar energy harvesting, Kumari et al. [9] developed a device designed for easy reproduction in any laboratory through the utilization of the thermal evaporation technique. Wong et al. [10] utilized a BN/GaN layered composite along with aluminum to fabricate Schottky diodes with the goal of harvesting high-frequency wireless energy. Through the adjustment of the induced electric field at the interface, they discovered that the dielectric constant of the boron nitride monolayer atop the aluminum monolayer was as low as 1.5, indicating its potential for capturing high-band 5G signals.

Furthermore, for readers interested in an overview of triboelectric and piezoelectric nanogenerators, the review article by Delgado-Alvarado et al. [11] report the recent advancements in piezoelectric and triboelectric nanogenerators designed for self-powered monitoring healthcare devices. The article covers the working principles, fabrication processes, and materials of various nanogenerators. Discussions on the electrical behavior of both piezoelectric and triboelectric nanogenerators are provided. Additionally, they explored signal processing components and diverse packaging types for these nanogenerators. This review thoroughly summarizes the design, materials, and electromechanical performance of nanogenerators designed for biomechanical energy harvesting, emphasizing their potential for various medical applications.

Finally, we want to extend our gratitude to all the authors and reviewers who contributed to this Special Issue. We hope that the articles featured in this Special Issue are interesting and helpful for the broad readerships of *Nanomaterials* and can inspire new innovations in the current state of the art regarding energy harvesting, advanced sensing, and smart systems.

Funding: This study was supported by "the Fundamental Research Funds for the Central Universities" with No. 2242023K30035; "University-Industry Collaborative Education Program by Ministry of Education" with No. 3502002207; the Ministry of Education Chunhui Plan International Cooperation Project with No. HZKY20220130; the National Natural Science Foundation of Jiangsu Province with No. BK20221456; the Basic Research Program of Suzhou City with No. SYG202314; the National Natural Science Foundation of China (62301150); the Start-up Research Fund of Southeast University (RF1028623164); and the Nanjing Science and Technology Innovation Project for Returned Overseas Talent (4206002302).

Acknowledgments: The Guest Editors would like to thank the authors for submitting their excellent contributions to this Special Issue. Furthermore, the present Special Issue would not have been possible without the expert reviewers who carefully evaluated the manuscripts and provided helpful comments and suggestions for improvements. A special thank you goes to the Editors and the MDPI team for their outstanding management of this Special Issue.

Conflicts of Interest: The authors declare no conflicts of interest.

References

1. Zhao, T.; Li, D.; Cui, P.; Zhang, Z.; Sun, Y.; Meng, X.; Hou, Z.; Zheng, Z.; Huang, Y.; Liu, H. A Self-Powered Flexible Displacement Sensor Based on Triboelectric Effect for Linear Feed System. *Nanomaterials* **2023**, *13*, 3100. [CrossRef]
2. Zhao, W.; Qin, W.; Ba, M.; Sun, Y. Measurement of Slips at Contact Interfaces Using a Self-Powered Sensor Based on Triboelectric Nanogenerators. *Nanomaterials* **2022**, *12*, 3510. [CrossRef] [PubMed]
3. Yoon, Y.; Park, S.; Park, T.; Kim, H.; Kim, K.; Hong, J. Enhanced Responsivity and Optoelectronic Properties of Self-Powered Solar-Blind $Ag_2O/\beta\text{-}Ga_2O_3$ Heterojunction-Based Photodetector with Ag:AZO Co-Sputtered Electrode. *Nanomaterials* **2023**, *13*, 1287. [CrossRef] [PubMed]
4. Kim, W.; Park, J.; Aggarwal, Y.; Sharma, S.; Choi, E.H.; Park, B. Highly Efficient and Stable Self-Powered Perovskite Photodiode by Cathode-Side Interfacial Passivation with Poly (Methyl Methacrylate). *Nanomaterials* **2023**, *13*, 619. [CrossRef] [PubMed]
5. Li, X.; Zhang, D.; Zhang, D.; Li, Z.; Wu, H.; Zhou, Y.; Wang, B.; Guo, H.; Peng, Y. Solid-Liquid Triboelectric Nanogenerator Based on Vortex-Induced Resonance. *Nanomaterials* **2023**, *13*, 1036. [CrossRef] [PubMed]
6. Mao, J.; Seo, S. Improving the Performance of Polydimethylsiloxane-Based Triboelectric Nanogenerators by Introducing CdS Particles into the Polydimethylsiloxane Layer. *Nanomaterials* **2023**, *13*, 2943. [CrossRef] [PubMed]
7. Duque, M.; Murillo, G. Low-Cost Manufacturing of Monolithic Resonant Piezoelectric Devices for Energy Harvesting Using 3D Printing. *Nanomaterials* **2023**, *13*, 2334. [CrossRef] [PubMed]
8. Xiao, X.; Liu, L.; Xi, Z.; Yu, H.; Li, W.; Wang, Q.; Zhao, C.; Huang, Y.; Xu, M. Research on an Optimized Quarter-Wavelength Resonator-Based Triboelectric Nanogenerator for Efficient Low-Frequency Acoustic Energy Harvesting. *Nanomaterials* **2023**, *13*, 1676. [CrossRef] [PubMed]
9. Mamta; Kumari, R.; Yadav, C.; Kumar, R.; Maurya, K.K.; Singh, V.N. Thermally Deposited Sb_2Se_3/CdS-Based Solar Cell: Experimental and Theoretical Analysis. *Nanomaterials* **2023**, *13*, 1135. [CrossRef] [PubMed]
10. Wong, C.-H.; Lam, L.-Y.F.; Hu, X.; Tsui, C.-P.; Zatsepin, A.F. Schottky-Diode Design for Future High-Speed Telecommunications. *Nanomaterials* **2023**, *13*, 1448. [CrossRef] [PubMed]
11. Delgado-Alvarado, E.; Martínez-Castillo, J.; Zamora-Peredo, L.; Gonzalez-Calderon, J.A.; López-Esparza, R.; Ashraf, M.W.; Tayyaba, S.; Herrera-May, A.L. Triboelectric and Piezoelectric Nanogenerators for Self-Powered Healthcare Monitoring Devices: Operating Principles, Challenges, and Perspectives. *Nanomaterials* **2022**, *12*, 4403. [CrossRef] [PubMed]

nanomaterials

Article

A Self-Powered Flexible Displacement Sensor Based on Triboelectric Effect for Linear Feed System

Tingting Zhao [1,†], Dongsheng Li [1,†], Peijuan Cui [2], Zhongbin Zhang [1], Yuyang Sun [1], Xingyou Meng [1], Zhanlin Hou [2], Zaiping Zheng [2], Yuping Huang [2] and Huicong Liu [1,*]

[1] Jiangsu Provincial Key Laboratory of Advanced Robotics, Robotics and Microsystems Center, School of Mechanical and Electrical Engineering, Soochow University, Suzhou 215123, China; 20214229013@stu.suda.edu.cn (T.Z.); dsli@suda.edu.cn (D.L.)
[2] Laboratory of Aerospace Servo Actuation and Transmission, Beijing Institute of Precision Mechatronics and Controls, Beijing 100076, China
* Correspondence: hcliu078@suda.edu.cn
† These authors contributed equally to this work.

Abstract: The detection and feedback of displacement and velocity significantly impact the control accuracy of the linear feed system. In this study, we propose a flexible and self-powered displacement sensor based on the triboelectric effect, designed for seamless integration into linear feed systems. The displacement sensor comprises two parts, the mover and stator, operating in a sliding mode. This sensor can precisely detect the displacement of the linear feed system with a large detection range. Additionally, the sensor is capable of real-time velocity detection of linear feed systems, with an error rate below 0.5%. It also offers advantages, such as excellent flexibility, compact size, stability, easy fabrication, and seamless integration, with linear feed systems. These results highlight the potential of the self-powered displacement sensor for various applications in linear feed systems.

Keywords: flexible displacement sensor; self-powered; linear feed system; triboelectric effect

check for updates

Citation: Zhao, T.; Li, D.; Cui, P.; Zhang, Z.; Sun, Y.; Meng, X.; Hou, Z.; Zheng, Z.; Huang, Y.; Liu, H. A Self-Powered Flexible Displacement Sensor Based on Triboelectric Effect for Linear Feed System. *Nanomaterials* **2023**, *13*, 3100. https://doi.org/10.3390/nano13243100

Academic Editors: Antonios Kelarakis and Jung Woo Lee

Received: 26 October 2023
Revised: 22 November 2023
Accepted: 28 November 2023
Published: 7 December 2023

1. Introduction

Linear feed systems play a pivotal role in diverse fields, encompassing intelligent manufacturing, aerospace, and medical equipment [1–4]. The precision of machine tool machining and the accuracy of servo actuator control are significantly influenced by the detection and feedback of displacement and velocity for typical linear feed systems, such as linear motors and ball screws [5–7]. Consequently, it becomes imperative to incorporate a displacement sensor with high accuracy and a large detection range. Additionally, adherence to other requisites, such as compact size, ease of integration, robust stability, and cost-effectiveness, should be tailored to specific application scenarios [8–13].

At present, the displacement sensors employed in linear feed systems primarily include magnetic grating displacement sensors, laser displacement sensors, and grating displacement sensors [14–16]. Magnetic grating displacement sensors offer advantages, such as low cost and ease of operation, but they are susceptible to demagnetization due to external magnetic field interference, resulting in relatively poor stability [14]. Laser displacement sensors have gained widespread attention due to their high accuracy, rapid response, and capability to perform various detections, including displacement, velocity, and angular velocity [15]. Nevertheless, the large volume of their transmitters hinders direct integration into narrow spaces. Grating displacement sensors offer a substantial detection range and high accuracy but come with a higher cost, low tolerance to vibration, and the need for precise installation [16]. Despite the widespread use of these displacement sensors, their relatively large size renders them unsuitable for integration into narrow spaces and curved surfaces.

In recent years, flexible sensors have garnered significant attention due to their compact size, flexibility, and applicability to curved surfaces [17–21]. The triboelectric effect, rooted in the coupling of contact charging and electrostatic induction, has demonstrated substantial potential in the realm of flexible displacement detection. For instance, Zhu et al. proposed an angular displacement sensor utilizing the triboelectric effect to monitor the bending motion of human joints with low power consumption [22]. Zhou et al. showcased a self-powered, one-dimensional displacement and speed-sensing technology based on the coupling of the triboelectric effect and electrostatic induction, achieving high resolution, a broad dynamic range, and extensive detection capabilities [23]. Li et al. designed an innovative self-powered angle/displacement sensor, leveraging the coupling effect between triboelectrification and electrostatic induction. This sensor not only detects angle/displacement but also distinguishes the real-time direction of the freestanding electrode displacement [24]. These studies offer valuable insights for the displacement detection of linear feed systems.

In this study, we developed a flexible and self-powered displacement sensor based on the triboelectric effect. This sensor is designed for real-time displacement and velocity detection in linear feed systems, offering ease of fabrication and integration into narrow spaces. In a typical linear feed system, as illustrated in Figure 1a, the controller is responsible for managing the motor's operation, thereby driving the rotation of the screw and the linear feed of the nut. Our proposed flexible displacement sensor can be seamlessly installed within the confined space of the linear feed system, enabling efficient detection of displacement and velocity. The controller receives feedback signals from the flexible displacement sensor to regulate the motor's operation in a closed-loop system. As depicted in Figure 1b, the flexible displacement sensor comprises a mover attached to the moving feed platform and a stator fixed to the shell. The mover consists of grid electrodes on a PI film, while the stator incorporates interdigital electrodes (IDEs) sandwiched between two PI layers. The electrode of the mover is affixed to the surface of the stator, and the top PI film of the stator, in contact with the mover electrode, acts as the triboelectric layer. Operating in a sliding mode, the displacement sensor, based on the triboelectric effect, generates an electrical signal when the feed platform moves forth and back. This signal is then fed back to the controller. Figure 1c provides a photograph of the sensor seamlessly integrated into a linear feed system. The stator is attached to the shell, while the mover is affixed to the feed platform. This integration enables the real-time detection of displacement and velocity for the feed nut at different feed velocities.

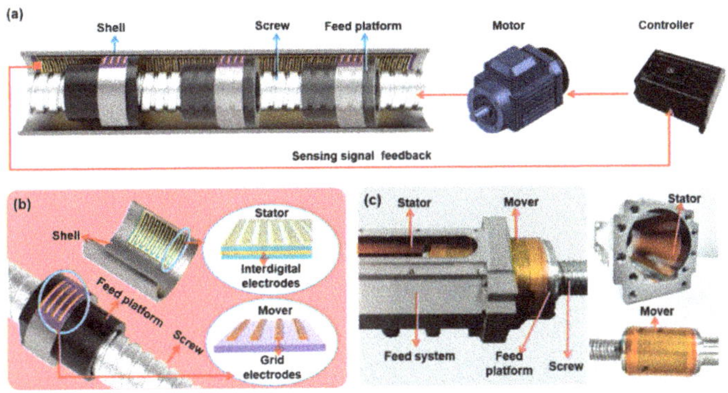

Figure 1. (**a**) Diagram of the displacement sensor integrated in a linear feed system; (**b**) structure diagram of the displacement sensor; (**c**) a photograph of the sensor integrated in a linear feed system.

2. Design and Fabrication

2.1. Working Principle of the Displacement Sensor

Figure 2a illustrates the operational diagram of the displacement sensor and the process of generating the triboelectric output signal (voltage difference between the stator electrodes, V_{AB}). The displacement sensor comprises two components, the mover and the stator, and its working principle is associated with contact electrification and in-plane sliding-induced charge transfer [23,25,26]. When the mover's electrode makes contact with the surface of the stator and slides from left to right, the triboelectric output signal is generated. Due to the difference in triboelectric polarities, as the mover starts to slide, its electrode becomes positively charged, while the PI film becomes negatively charged. When the electrodes of the mover (α) align with electrode A of the IDE in the stator (step i), there is no charge transfer between electrodes A and B due to electrostatic balance [23]. As the mover continues to slide to step ii, the original electrostatic equilibrium is disrupted, leading to electrons flowing from electrode A to B, establishing a new equilibrium state. As the mover further slides to step iii, electrons continue to flow from electrode A to B, resulting in a continuous increase in V_{AB} [27]. This completes half of the cycle for generating the triboelectric output signal. Similarly, as the mover continues to slide to step v, electrons flow back from electrode B to electrode A_1 until the mover's electrode (α) coincides with electrode A_1 of the stator. The mover returns to its original state, completing a full cycle of output signal generation. Figure 2a(vi) illustrates the corresponding output signal (V_{AB}) of the triboelectric sensor during a sliding period.

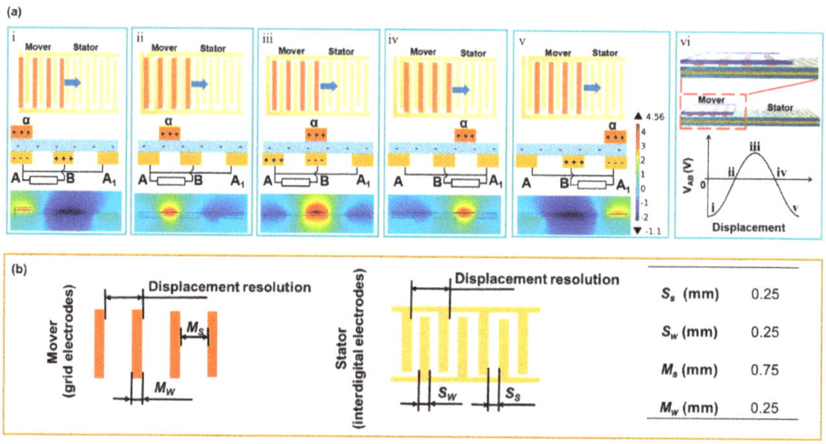

Figure 2. Working principle and structure of the displacement sensor: (**a**) working principle of the displacement sensor; (**b**) electrode structures of the mover and stator.

Resolution is the minimum detectable displacement, which is determined by the period distance of the sensor electrode. Figure 2b shows the electrode width and spacing of the mover and stator (M_w, M_s, S_w, and S_s), respectively. The resolution of the displacement sensor designed in this work is 1 mm, and M_w = 0.25 mm, M_s = 0.75 mm, and $S_s = S_w$ = 0.25 mm. The periods for electrodes of the mover and stator are consistent, that is, $2S_s + 2S_w = M_s + M_w$. This ensures a well-matched configuration between the movers and stators.

2.2. Fabrication Process of the Displacement Sensor

Figure 3a depicts the fabrication process of the displacement sensor. Copper was chosen as the electrode material, and PI was selected as the triboelectric material for its excellent abrasion resistance [28]. The sensor's preparation began with a PI film (i). Initially, a layer of copper was sputter-deposited onto the PI film surface (ii). Subsequently, the electrode with a grid structure was obtained through photolithography and an etching process (iii and iv), representing the preparation steps for the mover. For the stator, an electrode with an interdigital structure was prepared. Building on the first four steps, a PI insulating film was affixed to the electrode surface to shape the stator. Optical images of the fabricated stator and mover are presented in Figures 3b and 3c, respectively. The mover in this design measures 50 mm in length, 50 mm in width, and 0.06 mm in thickness, with a volume of only 150 mm^3. Similarly, the stator has dimensions of 150 mm in length, 50 mm in width, and 0.09 mm in thickness, with a volume of only 675 mm^3. The flexibility of the displacement sensor is demonstrated in Figure 3d, revealing its potential to integrate with a curved surface in a narrow space. It is seen that even on an extremely complex and irregular surface between the nut and shell, the flexible displacement sensor shows great adaptability for integration with the linear feed system.

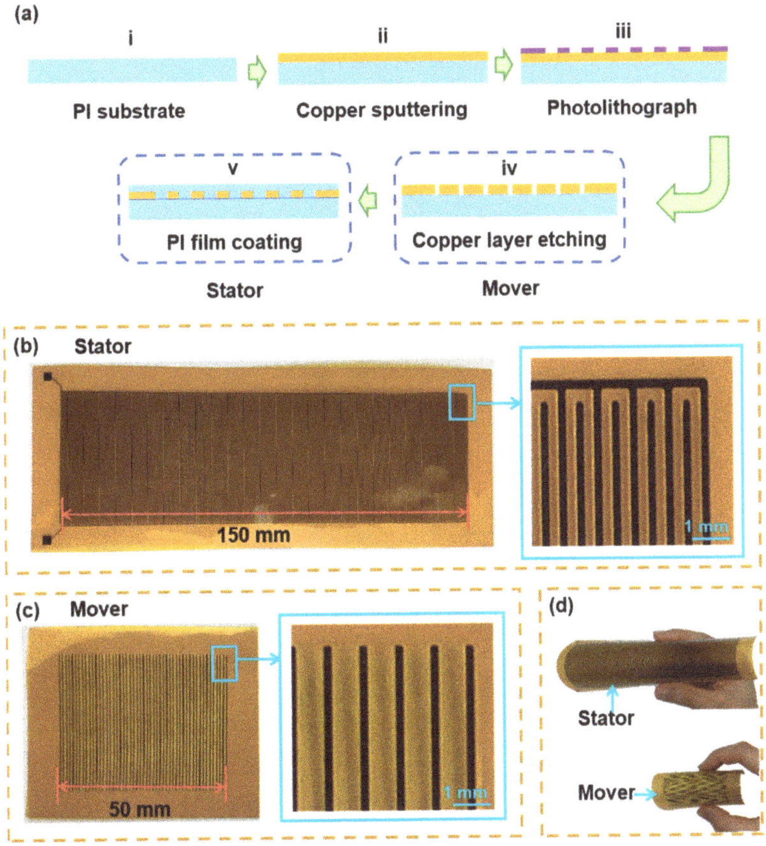

Figure 3. (**a**) Fabrication of the displacement sensor; (**b**) optical image of the stator; (**c**) optical image of the mover; (**d**) the flexibility of the displacement sensor.

2.3. Experimental Setup

Figure 4 illustrates the experimental setup for real-time displacement and velocity sensing calibration. The setup primarily comprises a linear motor (LinMot K05-Y/C-4), a data acquisition card (NI USB-6210), and a LabVIEW real-time monitoring interface. The mover of the displacement sensor was affixed to the linear motor, while the stator was attached to a tube using a 3D-printed bracket. The pair of electrodes extending from the stator is non-specific polarity and is connected to the AI0 and AI8 ports of the NI data acquisition card. This configuration enables precise acquisition and monitoring of experimental data. As the linear motor operates at different velocities, the mover on the motor slides along the surface of the stator, generating a self-powered sensing signal. The NI data acquisition card collects and transmits this signal to the LabVIEW monitoring interface. Consequently, the output voltage signal of the flexible displacement sensor is displayed in real time.

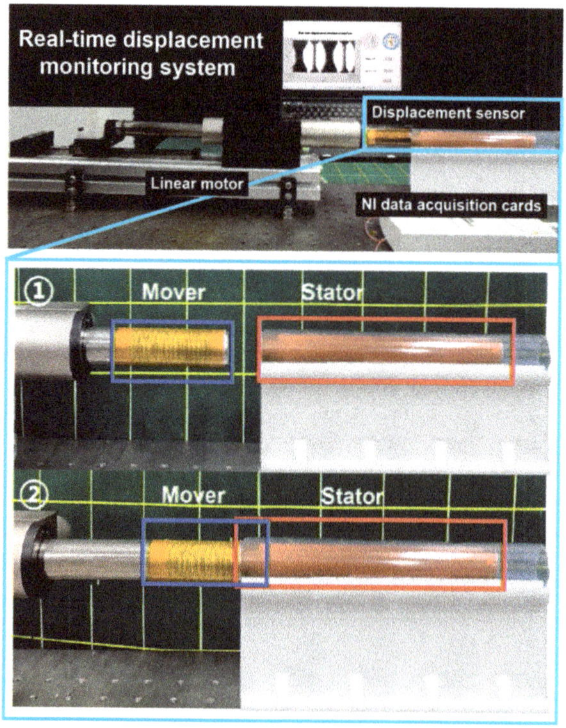

Figure 4. Experimental setup diagram of displacement sensor.

3. Results and Discussion

In this study, we assessed the performance of the displacement sensor in reciprocating motion. To evaluate its capabilities across various velocities, the feed velocities of the linear motors were set at 50 mm/s, 100 mm/s, 150 mm/s, and 200 mm/s, respectively. The corresponding voltage outputs of the flexible displacement sensor are depicted in Figure 5a–d. The maximum voltage outputs were recorded as 1.40 V, 2.12 V, 2.13 V, and 2.22 V at velocities of 50 mm/s, 100 mm/s, 150 mm/s, and 200 mm/s, respectively. Figure 5a(ii)–d(ii) illustrate the sensor's output voltage during a single motion cycle at different velocities. The output voltage of the sensor increases with the increase in velocity, but the trend of the increase in the output voltage decreases when the velocity exceeds a certain value.

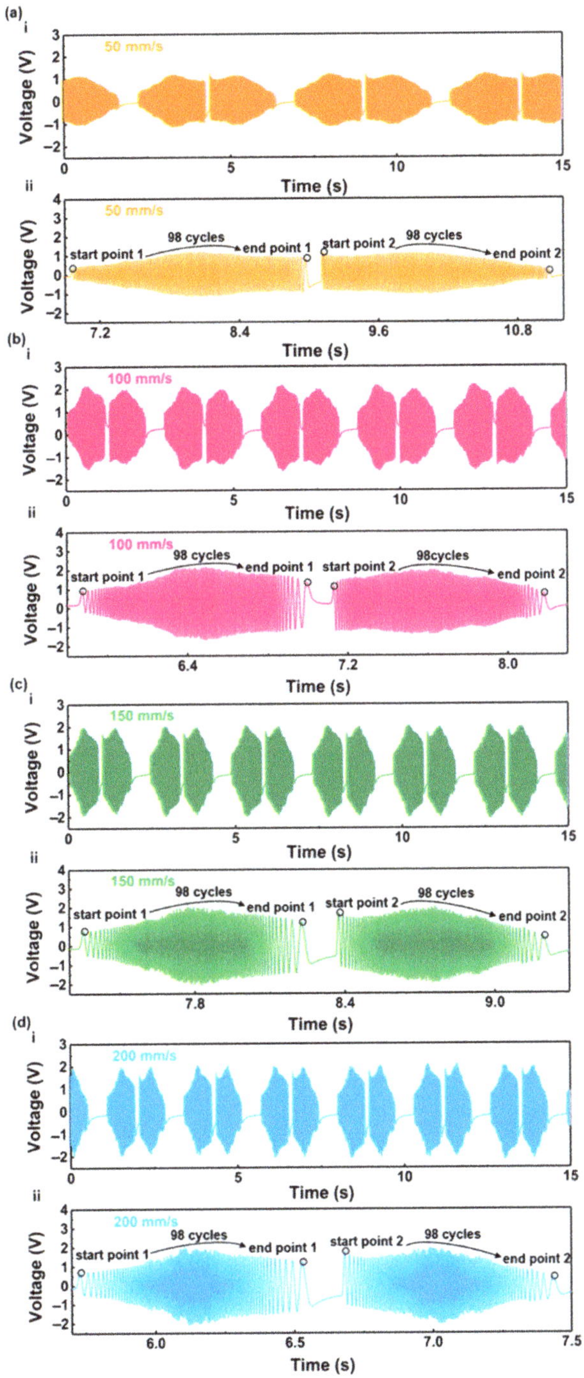

Figure 5. Output voltage of the displacement sensor at different velocities: (**a**) 50 mm/s; (**b**) 100 mm/s; (**c**) 150 mm/s; (**d**) 200 mm/s.

Based on the working principle of the designed triboelectric displacement sensor, the sensor's resolution is defined as 1 mm, indicating that the displacement sensor generates a sinusoidal cycle signal for every 1 mm it traverses. By calculating the cycle number of sine signals, the actual displacement of the feed system can be determined. This relationship is expressed in Equation (1), where D represents displacement, N is the cycle number of sine signals, and R is the sensor's resolution.

$$D = NR \tag{1}$$

As illustrated in Figure 5a(ii)–d(ii), there are 98 sine cycles from the starting point (start point 1) to the ending point (end point 1) during the forward stroke, and 98 sine signals from the starting point (start point 2) to the ending point of the reverse stroke (end point 2). This observation implies that the reset displacement of the linear feed system is 98 mm. Despite variations in the voltage peak generated by the sensor at different velocities, the displacements detected by the sensor remain consistent, affirming that the accuracy of displacement detection is unaffected by velocity.

The real-time displacement of the linear feed system can be determined based on the output of the displacement sensor, as illustrated in Figure 5 and Equation (1). The displacement curves are presented in Figure 6. The detected displacements during a single reciprocating motion at different velocities are compared with the actual displacements of the linear motor, as shown in Figure 6a(ii)–d(ii). The displacements detected by the displacement sensor exhibit a high degree of consistency with the actual displacements of the linear motor, indicating the accurate detection capability of the displacement sensor.

Furthermore, velocity can also be calculated based on the output of the sensor. The velocity is the distance traveled per unit of time, i.e., $V = \Delta S/\Delta t$. Because the resolution of the sensor is 1 mm, the time for the linear feed system with a 1 mm feed is the duration of the signal of a sinusoid cycle. Therefore, the velocity of the linear feed system in terms of the resolution (1 mm) of the sensor is shown in Equation (2), where V is velocity, Δt is the duration of each sinusoid cycle of the sensor output, and R is the resolution of the sensor.

$$V = R/\Delta \mathrm{t} \tag{2}$$

The real-time velocity of the sensor can be calculated based on the voltage output of the displacement sensor shown in Figure 5 and Equation (2). The real-time velocity curves are shown in Figure 7. Figure 7a(i)–d(i) illustrate the velocity detected by the displacement sensor. The detected velocity and the actual velocity in a single reciprocating motion of the linear motor are shown in Figure 7a(ii)–d(ii). The average velocities detected by the sensor during the constant velocity periods were 49.95 mm/s, 100.12 mm/s, 150.09 mm/s, and 200.79 mm/s for velocities of 50 mm/s, 100 mm/s, 150 mm/s, and 200 mm/s, respectively. Therefore, the errors of the velocity detections were 0.1%, 0.12%, 0.06%, and 0.40%, respectively. The velocities detected by the displacement sensor are consistent with the actual velocities of the linear motor, indicating that the sensor also has the capability for accurate velocity detection.

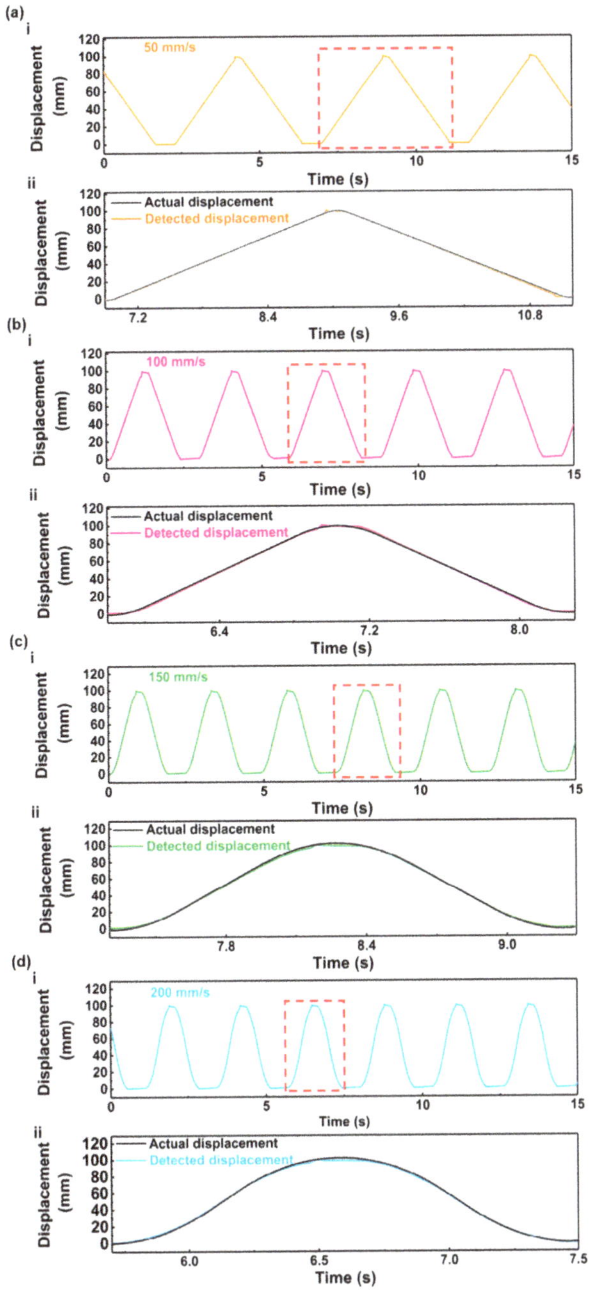

Figure 6. The displacements detected by the sensor and actual displacements at different velocities: (**a**) 50 mm/s; (**b**) 100 mm/s; (**c**) 150 mm/s; (**d**) 200 mm/s.

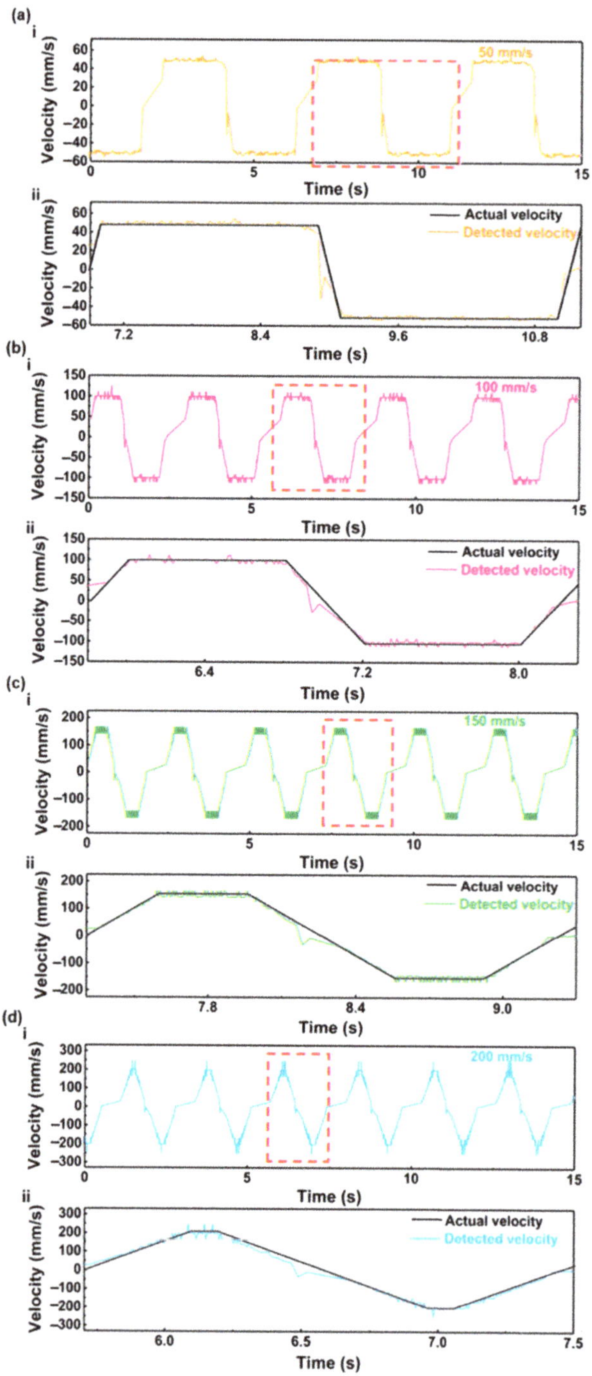

Figure 7. The velocities detected by the flexible sensor and actual velocities: (**a**) 50 mm/s; (**b**) 100 mm/s; (**c**) 150 mm/s; (**d**) 200 mm/s.

Stability is a crucial parameter for the displacement sensor, ensuring the accuracy and consistency of measurement data. The stability test of the displacement sensor was conducted at a velocity of 100 mm/s for over five hours, as depicted in Figure 8a. The sensor's output voltage remains stable after 8000 reciprocating movements, demonstrating its excellent stability. Figures 8b and 8c display the output voltage during the 70th and 7990th cycles of the reciprocating motion, respectively. The sensor's output voltages consistently align throughout the 8000 cycles of reciprocating movements. These results affirm that the displacement sensor exhibits exceptional stability for practical applications.

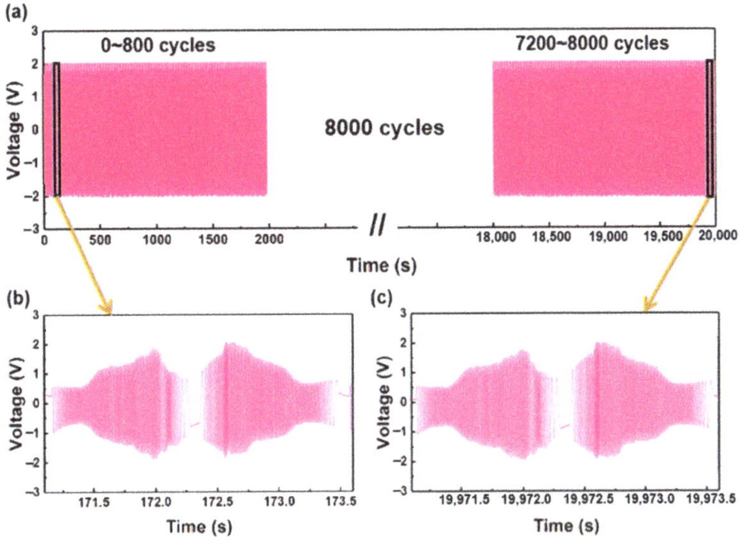

Figure 8. Stability test for the displacement sensor: (**a**) 0 s–20,000 s; (**b**) 70 cycles; (**c**) 7990 cycles.

4. Conclusions

In summary, this manuscript introduces a triboelectric-based, flexible, and self-powered sensor that finds application in real-time displacement detection within linear feed systems. Comprising a mover and a stator and operating in sliding mode, the sensor produces periodic output voltage corresponding to displacement and velocity. The performance of the displacement sensor underwent a comprehensive systematic investigation, validating its precise detection capabilities for linear feed system displacements. Furthermore, the sensor demonstrates the ability for real-time velocity detection, with an accuracy exceeding 99.5%. Rigorous testing also substantiated the exceptional stability of the sensor. The sensor boasts additional advantages, including flexibility, cost-effectiveness, robust stability, compact form factor, and straightforward fabrication, thereby showcasing its potential for integration into linear feed systems across diverse fields, including intelligent manufacturing, aerospace, and medical equipment.

Author Contributions: Conceptualization, T.Z., D.L. and X.M.; data curation, T.Z., Z.Z. (Zhongbin Zhang), D.L., P.C., Z.H. and Y.S.; supervision, H.L.; validation, P.C., Z.H., Y.H., Z.Z. (Zaiping Zheng) and H.L.; writing—original draft, T.Z. and D.L.; writing—review and editing, T.Z., D.L., Y.H., Z.Z. (Zaiping Zheng) and H.L. All authors have read and agreed to the published version of the manuscript.

Funding: This research received no external funding.

Data Availability Statement: Please contact the corresponding author for related data.

Conflicts of Interest: The authors declare no conflict of interest.

References

1. Yang, Z.Q.; Cui, W.; Zhang, W.B.; Wang, Z.H.; Zhang, B.Y.; Chen, Y.S.; Hu, N.; Bi, X.Y.; Hu, W. A New Performance Optimization Method for Linear Motor Feeding System. *Actuators* **2023**, *12*, 6. [CrossRef]
2. Liu, C.; Zhao, C.Y.; Wen, B.C. Dynamics analysis on the MDOF model of ball screw feed system considering the assembly error of guide rails. *Mech. Syst. Signal Pract.* **2022**, *178*, 109290. [CrossRef]
3. Shi, H.; Zhang, D.S.; Yang, J.; Ma, C.; Mei, X.S.; Gong, G.F. Experiment-based thermal error modeling method for dual ball screw feed system of precision machine tool. *Int. J. Adv. Manuf. Technol.* **2015**, *82*, 1693–1705. [CrossRef]
4. Altintas, Y.; Verl, A.; Brecher, C.; Uriarte, L.; Pritschow, G. Machine tool feed drives. *Cirp. Annals* **2011**, *60*, 779–796. [CrossRef]
5. Chen, J.S.; Huang, Y.K.; Cheng, C.C. Mechanical model and contouring analysis of high-speed ball-screw drive systems with compliance effect. *Int. J. Adv. Manuf. Technol.* **2004**, *24*, 241–250. [CrossRef]
6. Hu, H.L.; Zhang, F. Rational design of self-powered sensors with polymer nanocomposites for human–machine interaction. *Chinese J. Aeronaut.* **2022**, *35*, 155–177. [CrossRef]
7. Qiu, H.; Nisitani, H.; Kubo, A.; Yue, Y. Autonomous form measurement on machining centers for free-form surfaces. *Int. J. Mach. Tool. Manu.* **2004**, *44*, 961–969. [CrossRef]
8. Li, Y.; Zhang, J.; Su, D.X.; Zhou, C.X.; Zhao, W.H. Experiment-based thermal behavior research about the feed drive system with linear scale. *Adv. Mech. Eng.* **2018**, *10*, 11. [CrossRef]
9. Du, F.; Zhang, M.Y.; Wang, Z.G.; Yu, C.; Feng, X.Y.; Li, P.G. Identification and compensation of friction for a novel two-axis differential micro-feed system. *Mech. Syst. Signal Pract.* **2018**, *106*, 453–465. [CrossRef]
10. Erkorkmaz, K.; Wong, W. Rapid identification technique for virtual CNC drives. *Int. J. Mach. Tool. Manu.* **2007**, *47*, 1381–1392. [CrossRef]
11. Rajabi, N.; Abolmasoumi, A.H.; Soleymani, M. Sliding mode trajectory tracking control of a ball-screw-driven shake table based on online state estimations using EKF/UKF. *Struct. Control Health* **2018**, *25*, 4. [CrossRef]
12. Milić, P.; Marinković, D.; Klinge, S.; Ćojbašić, Ž. Reissner-Mindlin Based Isogeometric Finite Element Formulation for Piezoelectric Active Laminated Shells. *Teh. Vjesn.* **2023**, *30*, 416–425.
13. Milić, P.; Marinković, D.; Klinge, S.; Ćojbašić, Ž. Geometrically Nonlinear Analysis of Piezoelectric Active Laminated Shells by Means of Isogeometric FE Formulation. *Facta Univ. Ser. Mech. Eng.* **2023**. [CrossRef]
14. Zhang, Z.; Wang, L.; Cao, B.; Zhang, H.; Liu, J. A Moving Magnetic Grid-Type Long-Range Linear Absolute Displacement Sensor. *Sensors* **2023**, *23*, 2. [CrossRef] [PubMed]
15. Sabatini, R.; Richardson, M.A.; Gardi, A.; Ramasamy, S. Airborne laser sensors and integrated systems. *Prog. Aerosp. Sci.* **2015**, *79*, 15–63. [CrossRef]
16. Albert, J.; Shao, L.-Y.; Caucheteur, C. Tilted fiber Bragg grating sensors. *Laser Photonics Rev.* **2013**, *7*, 83–108. [CrossRef]
17. Trung, T.Q.; Lee, N.E. Flexible and Stretchable Physical Sensor Integrated Platforms for Wearable Human-Activity Monitoring and Personal Healthcare. *Adv. Mater.* **2016**, *28*, 4338–4372. [CrossRef]
18. Xu, F.; Li, X.; Shi, Y.; Li, L.; Wang, W.; He, L.; Liu, R. Recent Developments for Flexible Pressure Sensors: A Review. *Micromachines* **2018**, *9*, 11. [CrossRef]
19. Nag, A.; Mukhopadhyay, S.C.; Kosel, J. Wearable Flexible Sensors: A Review. *IEEE Sens. J.* **2017**, *17*, 3949–3960. [CrossRef]
20. Zang, Y.P.; Zhang, F.J.; Di, C.A.; Zhu, D.B. Advances of flexible pressure sensors toward artificial intelligence and health care applications. *Mater. Horiz.* **2015**, *2*, 140–156. [CrossRef]
21. Ha, H.; Qaiser, N.; Yun, T.; Cheong, J.; Lim, S.; Hwang, B. Sensing Mechanism and Application of Mechanical Strain Sensor: A Mini-Review. *Facta Univ. Ser. Mech. Eng.* **2023**, *20*, 2–2276723. [CrossRef]
22. Zhu, M.; Sun, Z.; Chen, T.; Lee, C. Low cost exoskeleton manipulator using bidirectional triboelectric sensors enhanced multiple degree of freedom sensory system. *Nat. Commun.* **2021**, *12*, 2692. [CrossRef]
23. Zhou, Y.S.; Zhu, G.; Niu, S.; Liu, Y.; Bai, P.; Jing, Q.; Wang, Z.L. Nanometer resolution self-powered static and dynamic motion sensor based on micro-grated triboelectrification. *Adv. Mater.* **2014**, *26*, 1719–1724. [CrossRef] [PubMed]
24. Li, C.; Wang, Z.; Shu, S.; Tang, W. A Self-Powered Vector Angle/Displacement Sensor Based on Triboelectric Nanogenerator. *Micromachines* **2021**, *12*, 3. [CrossRef]
25. Zhang, A.Q.; Wu, Y.; Yao, B.C.; Gong, Y.A. Optimization study on graphene-coated microfiber Bragg grating structures for ammonia gas sensing. *Photonic Sens.* **2014**, *5*, 84–90. [CrossRef]
26. Xie, Y.; Wang, S.; Niu, S.; Lin, L.; Jing, Q.; Yang, J.; Wu, Z.; Wang, Z.L. Grating-structured freestanding triboelectric-layer nanogenerator for harvesting mechanical energy at 85% total conversion efficiency. *Adv. Mater.* **2014**, *26*, 6599–6607. [CrossRef]
27. Li, W.J.; Liu, G.X.; Jiang, D.D.; Wang, C.; Li, W.; Guo, T.; Zhao, J.Q.; Xi, F.B.; Liu, W.B.; Zhang, C. Interdigitated Electrode-Based Triboelectric Sliding Sensor for Security Monitoring. *Adv. Mater. Technol.* **2018**, *3*, 11. [CrossRef]
28. Zhu, G.; Chen, J.; Zhang, T.J.; Jing, Q.S.; Wang, Z.L. Radial-arrayed rotary electrification for high performance triboelectric generator. *Nat. Commun.* **2014**, *5*, 3426. [CrossRef]

nanomaterials

Article

Measurement of Slips at Contact Interfaces Using a Self-Powered Sensor Based on Triboelectric Nanogenerators

Wangjia Zhao, Wenjie Qin *, Mingsen Ba and Yu Sun

School of Mechanical Engineering, Beijing Institute of Technology, Beijing 100081, China
* Correspondence: qinwj@bit.edu.cn; Tel.: +86-010-6891-5926

Abstract: The accumulation of tangential small slips at contact interfaces may occur in mechanical assemblies when subjected to cyclic loadings, and cause failure of parts. However, the slip is difficult to measure directly. In this paper, a self-powered displacement sensor based on triboelectric nanogenerator (TENG) is fabricated, which is used to measure the microslip of the specimen in the designed test bench. When the specimen slips under the cyclic tangential load with a frequency of 10 Hz, the resulted average slip displacement of the specimen under a single load cycle is 9.01 μm, which is measured by the Keithley DAQ 6510 multimeter (Tektronix, Shanghai, China). This value is close to that measured by the Keithley 6514 electrometer. In addition, adding a $BaTiO_3$ thin film with a thickness of 100 nm between the Kapton triboelectric layer and the Cu electrode, the output voltage of the sensor is increased by 51.5%. Thus, the sensitivity of the self-powered sensor is improved effectively.

Keywords: TENG; self-powered sensor; slip; contact

check for updates

Citation: Zhao, W.; Qin, W.; Ba, M.; Sun, Y. Measurement of Slips at Contact Interfaces Using a Self-Powered Sensor Based on Triboelectric Nanogenerators. *Nanomaterials* **2022**, *12*, 3510. https://doi.org/10.3390/nano12193510

Academic Editor: Dong-Joo Kim

Received: 25 August 2022
Accepted: 27 September 2022
Published: 7 October 2022

1. Introduction

There are many fastening structures in mechanical assemblies connected by frictional contacts. When they are subjected to cyclic tangential loads, frictional slips may occur along the interfaces, and the accumulation of slips may further lead to a significant global relative displacement between parts, such as the rotation of a bush in the small end of a connecting rod and loosening of bolted joints [1,2]. Given that slips will cause loosening of fastening connections, and even system failures, measuring the slip displacement between contact surfaces is of great significance to the safety of mechanical assemblies.

Sensors commonly used to measure displacement include eddy current sensors, laser sensors, LVDT sensors and so on [3–5]. Among them, a LVDT sensor is often used in fretting experiments to measure the tangential displacement between contact specimens as their slip. However, this displacement is not the same as the slip in the contact, because there are components with compliance which physically separate the contact from the point of measurement, and hence, the measured displacement amplitude is always larger than the contact slip amplitude [5]. In fact, the slip at different positions at the contact interface is different, and it is difficult to measure the slip at a certain position directly with the above sensors. In addition, these sensors need external energy during testing.

A triboelectric nanogenerator (TENG) can convert mechanical energy into electrical energy, and its output electrical signal can be used to analyze the mechanical motion in turn [6], so it can be used as a self-powered sensor. At present, a variety of self-powered displacement sensors based on the principle of TENG have been developed [7–10]. For example, Zhou et al. designed a one-dimensional displacement sensor composed of a pair of micro grating structures [11]. With the sliding of the measured object, the grating will be separated and contacted continuously, and the displacement of the object is obtained by calculating the number of voltage peaks and valleys. Given that grating-structured TENGs based on interdigital electrode structures are unable to sense the motion direction of the freestanding electrode in real time, Li et al. designed a self-powered vector/displacement sensor (SPVS) composed of one group of grating freestanding electrodes and three groups of interdigitated introduction electrodes with the identical period, which can distinguish the real-time direction of the freestanding electrode displacement [12]. Yang et al. fabricated a self-powered rotation speed and displacement sensor using the dual cylinder structure TENG to measure the displacement of the rotating mobile devices [13]. It is composed of two cylinders and rotates with the rolling of the measured object, generating voltage which is used to record the number of turns. The displacement of the measured object can be obtained by multiplying the number of turns by the length of the rotation object. However, these self-powered sensors based on the principle of TENG are often used to measure the macroscopic displacement of objects, and there are few reports on measuring the microslip between contact surfaces. We do not know whether this kind of self-powered sensor is sensitive enough to measure the micro contact slip and its accumulation. If it can, it can be applied to monitor the loosening failure of fastening connection structures.

In this paper, a self-powered sensor based on TENG is fabricated to measure the contact slips directly in the test bench built in this study. Fixing the two triboelectric layers of the sensor on the upper and lower specimen respectively, the slip displacement of several microns can be measured. Unlike an electrometer, such as a Keithley 6514 electrometer, which is widely used to obtain the electrical signal output of a TENG in many studies, a multimeter (Keithley DAQ 6510) is adopted to measure the output voltage of the sensor, which can obtain the close slip values compared with the results measured by the Keithley 6514 electrometer. Furthermore, in order to improve the sensitivity of the self-powered sensor, a BaTiO3 thin film is added between the triboelectric layer and the electrode to enhance the output voltage of the self-powered sensor.

2. Materials and Methods

2.1. Principle of Self-Powered Sensor

There are four working modes of TENG: contact-separation mode [14], sliding mode [15], single-electrode mode [16], and freestanding triboelectric-layer-based mode [17]. In this study, the contact-separation mode and sliding mode are considered.

The principle of a contact-separation TENG is the coupling of contact electrification and electrostatic induction [18]. After the two triboelectric layers with different materials being forced to contact with each other, charges with the same density and opposite sign will be induced on the contact surfaces. When the two triboelectric layers are separated under the force, electric potential difference is then established between the two electrodes which are connected to the triboelectric layers. If the area size of the metal electrodes is much larger than their separation distance, the relationship between the open circuit voltage (V_{OC}) generated by the TENG and the separation distance ($x(t)$) is [19]:

$$V_{OC} = \frac{\sigma x(t)}{\varepsilon_0}, \tag{1}$$

where σ is the charge density and ε_0 is the permittivity of vacuum.

As for a sliding-mode TENG, when the two contacting triboelectric layers slide relatively with a distance $x(t) << l$ in which l is the length of the triboelectric layer, electric potential difference is also established between the two electrodes due to the reduction of contact area. The relationship between the open circuit voltage (V_{OC}) and the sliding distance ($x(t)$) is [20]:

$$V_{OC} = \frac{\sigma x(t) d_0}{\varepsilon_0 [l - x(t)]}, \tag{2}$$

where and d_0 is effective thickness constant, which is defined as:

$$d_0 = \sum_{i=1}^{n} \frac{d_i}{\varepsilon_{ri}}, \tag{3}$$

where d_i is the thickness of the triboelectric layer and ε_{ri} is the permittivity of the triboelectric layer.

By comparing Equations (1) and (2), it can be seen that when $x << l$, the open circuit voltage of the contact-separation TENG is much greater than that of the sliding-mode TENG, and the open circuit voltage of the contact-separation TENG is directly proportional to the separation distance between the two electrodes. In order to ensure that the sensor can obtain larger electrical signals and better linearity, the contact-separation TENG is used as the self-powered sensor in this study.

2.2. Test Bench

The test bench designed in this experiment is shown in Figure 1. The upper specimen is in contact with the lower specimen under the uniformly distributed pressure which is applied by the steel cube. The positive triboelectric layer and the negative triboelectric layer of the self-powered sensor are fixed on the upper specimen and the lower specimen respectively, and perpendicular to the slip direction. The tangential cyclic load generated by the vibration exciter is applied to the upper specimen through a rod. A force sensor is mounted on one end of the rod to measure the exciting force on the rod. The preload is measured by another force sensor in the vibration exciter. When the cyclic load is great enough, the upper specimen begins to slide, resulting a separating distance between the positive and negative triboelectric layers of the sensor. The output electrical signal of the sensor is collected by a Keithley DAQ 6510 multimeter (Tektronix, Shanghai, China) and the slip displacement can be calculated from the measured voltage.

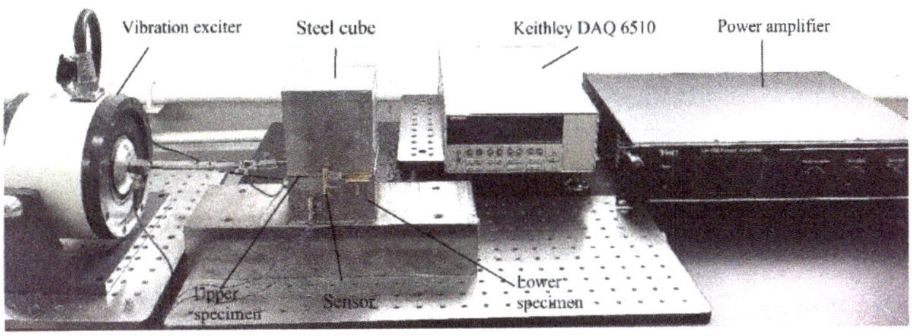

Figure 1. The test bench.

The self-powered sensor is a conductor-to-dielectric TENG, working in the contact-separation mode as shown in Figure 2. The size of each triboelectric layer is 25 mm × 10 mm. One triboelectric layer is a 0.06 mm copper film (also used as electrode), and the other triboelectric layer is a 0.06 mm Kapton film with a copper film as the other electrode. Both triboelectric layers are bonded to the acrylic substrates with a thickness of 2 mm. Two conducting wires connected with the two electrodes are accessed to the Keithley DAQ 6510 multimeter.

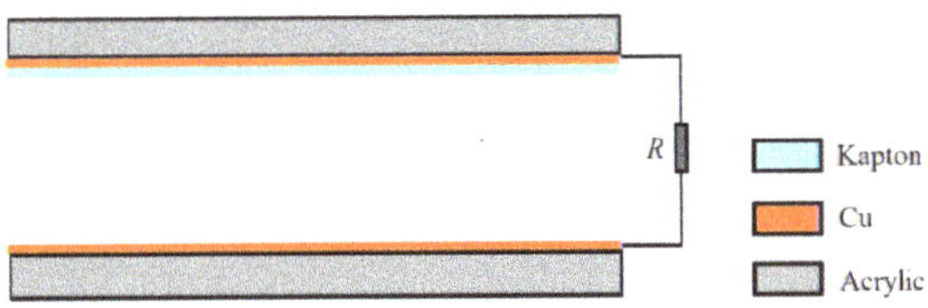

Figure 2. Structure of the self-powered sensor.

2.3. Signal Analysis and Test Instrument

At present, a Keithley 6514 electrometer is widely used to measure the voltage signal generated by the TENG [21,22]. In this study, the voltage signal of the self-powered sensor is measured by a multimeter, the Keithley DAQ 6510 multimeter, which can also achieve satisfactory accuracy. The equivalent circuit of the system including the TENG and the resistance load is shown in Figure 3 and its equation can be obtained by using Kirchhoff's law, as shown in Equation (4) [23].

$$R\frac{\mathrm{d}Q}{\mathrm{d}t} = -\frac{Q}{C} + V_{OC}, \tag{4}$$

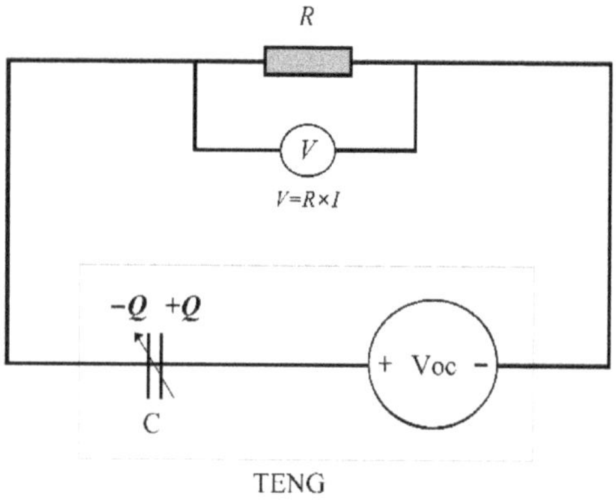

Figure 3. Equivalent circuit.

In Equation (4), Q is the transferred charge. Setting $Q(t = 0) = 0$, Equation (4) can be solved as and can be calculated as:

$$Q(t) = \frac{1}{R}exp\left[-\frac{1}{R}\int_0^t \frac{1}{C(x(t))}\mathrm{d}t\right]\int_0^t V_{OC}(x(t))exp\left[\frac{1}{R}\int_0^t \frac{1}{C(x(t))}\mathrm{d}t\right]\mathrm{d}t \tag{5}$$

As $I(t) = \frac{\mathrm{d}Q(t)}{\mathrm{d}t}$, the current in the system is:

$$I(t) = \frac{V_{OC}}{R} - \frac{1}{R^2C}exp\left[-\frac{1}{R}\int_0^t \frac{1}{C(x(t))}\mathrm{d}t\right]\int_0^t V_{OC}(x(t))exp\left[\frac{1}{R}\int_0^t \frac{1}{C(x(t))}\mathrm{d}t\right]\mathrm{d}t \tag{6}$$

Then, the voltage of the resistance R is:

$$V(t) = V_{OC} - \frac{1}{RC}exp\left[-\frac{1}{R}\int_0^t \frac{1}{C(x(t))}\mathrm{d}t\right]\int_0^t V_{OC}(x(t))exp\left[\frac{1}{R}\int_0^t \frac{1}{C(x(t))}\mathrm{d}t\right]\mathrm{d}t \tag{7}$$

If the separation distance of the electrodes is kept constant in the TENG, that is, $x(t)$ is a constant S, the output voltage is:

$$V(t) = V_{OC}e^{-\frac{t}{RC}}, \tag{8}$$

which decays in the form of negative exponential function.

In the Equations (8), the maximal output voltage V_{OC} is the open circuit voltage when the separation distance of the electrodes is S. Usually, the peak voltage measured by a multimeter (as shown in Figure 4) cannot reach the open circuit voltage of TENG. According to the study on the influence of the load resistance on the magnitude of the output voltage by Niu [23], it can be seen that if the resistance is greater than 1 GΩ, the output voltage is close to the open circuit voltage. Therefore, the resistance value of the multimeter is set to 10 GΩ in this study, and the measured peak voltage is close to the open circuit voltage V_{OC}.

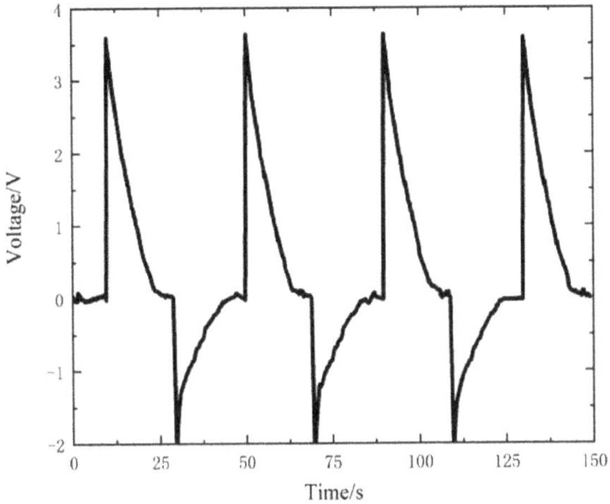

Figure 4. Voltage output of the sensor measured by the Keithley 6510 multimeter.

3. Results

3.1. Calibration of the Sensor

The two triboelectric layers are separated by a distance of 0.8 mm then kept for a period of time for several times, and the measured voltage results are shown in Figure 4. According to the analysis in Section 2.3, the peak voltage value is close to the open circuit voltage V_{OC} of the sensor, which is 3.59 V on average.

It can be seen that the measured voltage decreases exponentially with time after it gets to the peak. This will cause a serious distortion in voltage measuring at a low-frequency condition. As the excitation frequency of the cycle load in this study, 10 Hz, is concerned, the upper specimen slides every 0.1 s, and the measured voltage will decay for 0.1 s before the next slip. It can be deduced from Figure 4 that the attenuation is about 1/100 (from 3.59 V to 3.56 V), which can be ignored in the single slip measurement.

According to Equation (1), the open circuit voltage has a linear relationship with the separation distance of the self-powered sensor, so the relationship between the measured voltage V (V) and separation distance x (mm) is:

$$V = 4.49x \tag{9}$$

3.2. Dynamic Slip Test Results of Contact Surface under Cyclic Load

Setting the frequency of cycle load as 10 Hz, the time-varied cyclic load is applied to the upper specimen, and the change of the amplitude of the cyclic load is shown in Figure 5. When the time is about 81 s, the sensor starts to output voltage, that is, the upper specimen begins to slip. At this time, the amplitude of the cyclic load is 11.84 N, and the preload is 1.5 N. Therefore, the maximal force acting on the upper specimen is 13.34 N, which is the load to overcome the friction. The whole slip process lasts 7 s (from 81 s to 88 s). After that, the load is withdrawn.

According to the voltage value generated by self-powered sensor, and considering the relationship between the measured voltage and separation distance, the resulted slip curve of the upper specimen under the action of excitation force is shown in Figure 6. As the load is withdrawn, the measured voltage begins to drop. From the detail of the slip from the time of 81.15 s to 82.15 s, a total sliding distance of 90.13 μm for the 10 load cycles at the initial stage of slip is obtained, and the average slip value of each load cycle is 9.01 μm.

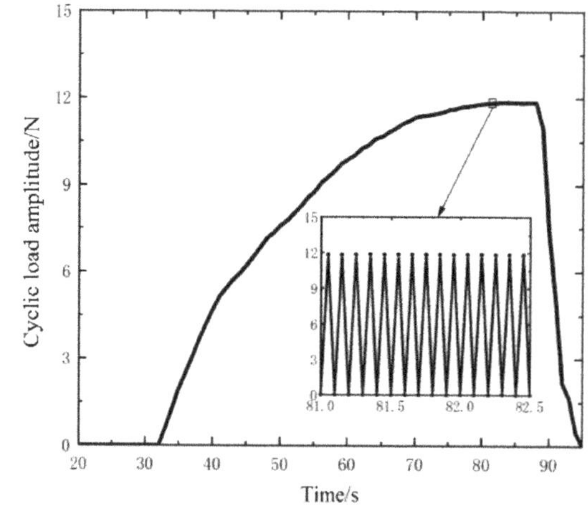

Figure 5. The amplitude of the cyclic load.

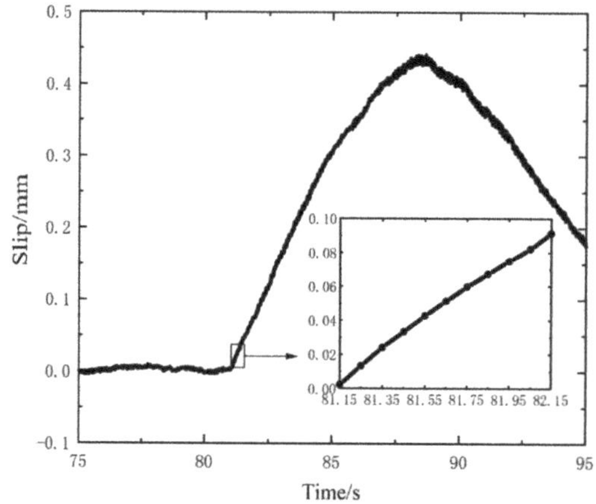

Figure 6. Slip measured by the sensor using the Keithley 6510 multimeter.

3.3. Verification of the Test Accuracy Using Keithley 6514 Electrometer

In order to investigate the accuracy of the above test results, the slip displacement as the same conditions is measured by the Keithley 6514 electrometer.

The two triboelectric layers are also separated by 0.8 mm then kept for a period of time for several times, and the measured voltage results are shown in Figure 7. Different from the voltage change measured by the Keithley 6510 multimeter, the voltage change measured by the Keithley 6514 electrometer is basically unchanged after reaching its peak value during the period when the triboelectric layers are separated by 0.8 mm and remains unchanged. The reason is that there is almost no charge transfer in the circuit during this period. The greatest value of the voltage is close to the open circuit voltage V_{OC} of the sensor. It can be seen from the figure that the maximum voltage value is 3.50 V on average. Taking this value as the open circuit voltage of the self-powered sensor, the relationship between the measured voltage V (V) and the separation distance x (mm) is:

$$V = 4.375x \quad (10)$$

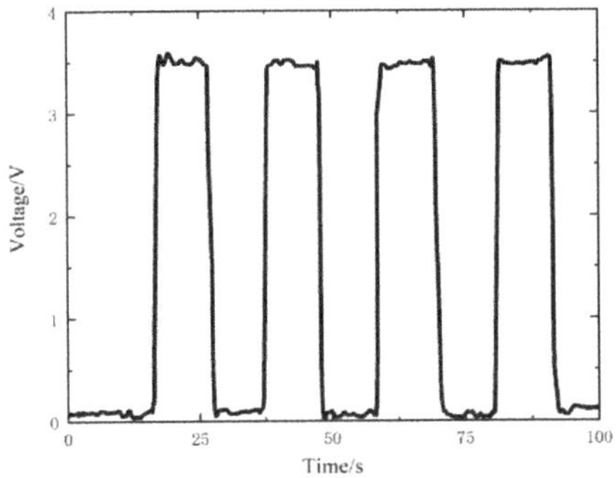

Figure 7. Voltage output of the sensor measured by the Keithley 6514 electrometer.

The frequency and amplitude of the cycle load is the same as in Section 3.2, and the slip curve of the upper specimen obtained by the dynamic slip test is shown in Figure 8. From the figure, it can be seen that the slip process lasts for 7 s (from 77.5 s to 84.5 s). After that, the load is withdrawn and the measured voltage remains constant. From the detail figure of the slip, a total sliding displacement of 94.50 μm for the 10 load cycles at the initial stage of slip is obtained, and the average slip value of each load cycle is 9.45 μm, which is close to the result measured by the Keithley 6510 multimeter.

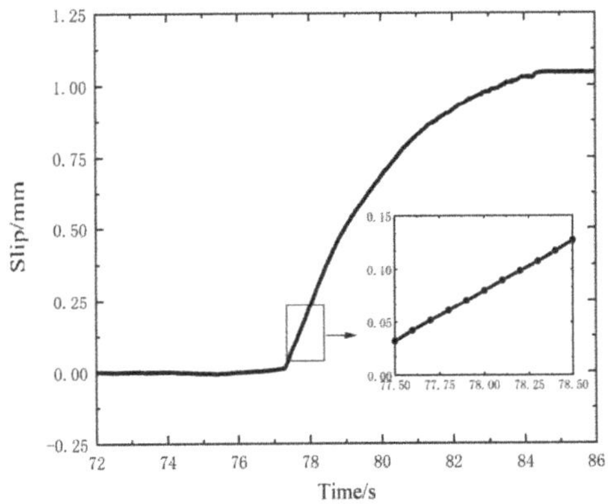

Figure 8. Slip measured by the sensor using the Keithley 6514 electrometer.

Comparing the measurement results of the Keithley DAQ 6510 multimeter and that of the Keithley 6514 electrometer, it is found that the test result of Keithley DAQ 6510 multimeter is close to that of Keithley 6514 electrometer, so it is feasible to obtain the voltage of the self-powered sensor used to measure the dynamic slip at this frequency with a multimeter such as the Keithley DAQ 6510 multimeter.

4. Performance Enhancement of Sensor by $BaTiO_3$ Thin Film

4.1. Role of $BaTiO_3$ Thin Film

According to Equation (1), the open circuit voltage of the contact-separation sensor is proportional to the charge density if the separation distance is constant. Therefore, the measured voltage of the self-powered sensor can be improved by increasing the charge density. Since $BaTiO_3$ is a ferroelectric ceramic material with dielectric polarization, it is known to have and residual dielectric polarization after being exposed to an electrical field. When the dielectric polarization and the surface polarization of the triboelectric layer are coupled, the residual polarization of $BaTiO_3$ can enhance the charge density, thereby increasing the open circuit voltage of the sensor [24]. Thus, a $BaTiO_3$ thin film is added between the triboelectric layer and the electrode in this study.

The $BaTiO_3$ thin film is deposited by an RF (radio frequency) magnetron sputtering on the copper electrode from the ceramic target. The sputtering time is 3 h and the obtained film thickness is 100 nm. It is placed between the Kapton layer and the copper electrode of the sensor in Section 2.2, as shown in Figure 9, and its SEM image is shown in Figure 10.

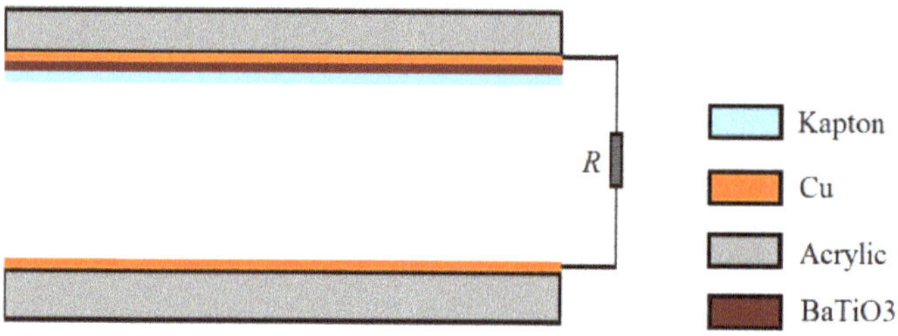

Figure 9. Structure of the sensor with a $BaTiO_3$ thin film.

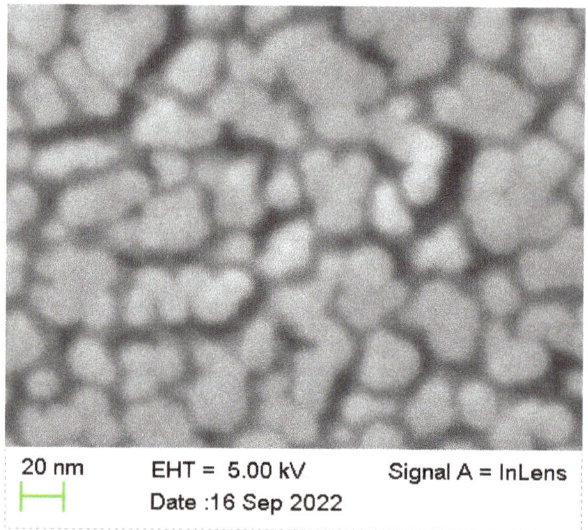

Figure 10. SEM of the $BaTiO_3$ thin film.

4.2. Measurement Results Using the Self-Powered Sensor with $BaTiO_3$ Thin Film

The self-powered sensor with a $BaTiO_3$ thin film is calibrated firstly by the same method in Section 3.1, and the measured voltage when the two triboelectric layers are separated by a distance of 0.8 mm is shown in Figure 11. The peak voltage is 5.44 V, so the relationship between the measured voltage V (V) and separation distance x (mm) is:

$$V = 6.8x \tag{11}$$

Setting the frequency and amplitude of the time-varied cyclic load is consistent with Section 3.2, and the slip curve of the upper specimen obtained by the dynamic slip test is shown in Figure 12. It can be seen that the slip process lasts for 7 s and the total displacement within 1 s at the initial stage of sliding is 89.41 μm from the slip details. Thus, the average slip displacement value of each cyclic load is 8.94 μm, which is close to that obtained in Section 3.

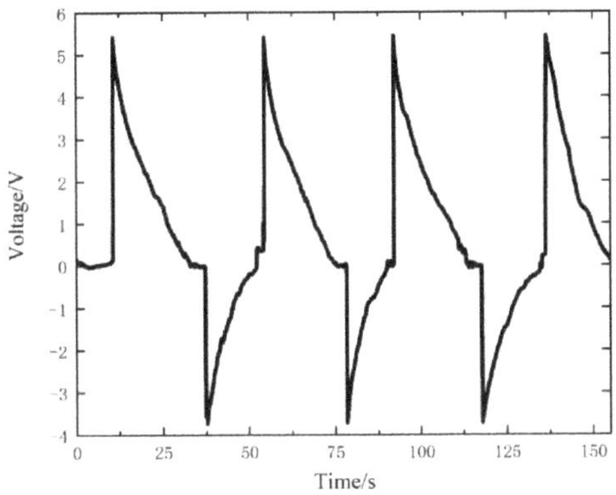

Figure 11. Voltage output of the sensor with a $BaTiO_3$ thin film.

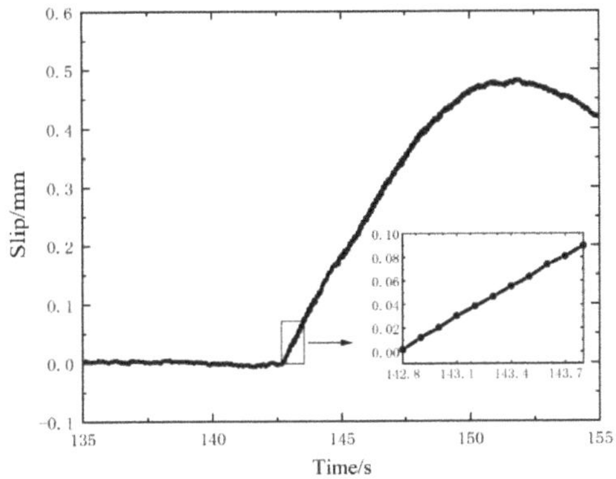

Figure 12. Slip measured by the sensor with a $BaTiO_3$ thin film.

Comparing the measurement results of the self-powered sensor with $BaTiO_3$ thin film and that of the self-powered sensor without the ferroelectric materials, the measured voltage value of the self-powered sensor with $BaTiO_3$ thin film is increased by 51.5%; thus, the sensitivity of the self-powered sensor is effectively improved.

5. Conclusions

A self-powered sensor is fabricated to measure the contact slips directly in the test bench built in this study, which is a conductor-to-dielectric TENG with one triboelectric layer is a 0.06 mm copper film and the other triboelectric layer is a 0.06 mm Kapton film. The sensor can directly measure the slip displacement of several microns of the contact surface in the test bench. When the specimen slips under the cyclic tangential load with a frequency of 10 Hz, the resulted average slip displacement of the specimen under a single load cycle measured by the Keithley DAQ 6510 multimeter is 9.01 μm, which is close to the slip value measured by the Keithley 6514 electrometer, that is 9.45 μm. Thus, a multimeter can be used to obtain the output voltage of the self-powered sensor, which is used to measure the dynamic slip at this frequency.

In addition, a $BaTiO_3$ thin film with a thickness of 100 nm is added between the Kapton layer and the Cu electrode of the sensor. The measured voltage is increased by 51.5% compared with the self-powered sensor without $BaTiO_3$ thin film. Thus, the sensitivity of the sensor can be effectively improved by adding a $BaTiO_3$ thin film under the triboelectric layer.

Author Contributions: Methodology, W.Z. and W.Q.; experiments, M.B. and Y.S.; data curation, Y.S. and M.B.; writing—original draft preparation, W.Z. and M.B.; writing—review and editing, W.Z. and W.Q. All authors have read and agreed to the published version of the manuscript.

Funding: This research was funded by the State Key Laboratory of Engine Reliability, grant number SKLER-202111.

Institutional Review Board Statement: Not applicable.

Informed Consent Statement: Informed consent was obtained from all subjects involved in the study.

Data Availability Statement: The data are available on reasonable request from the corresponding author.

Acknowledgments: The authors would like to thank the State Key Laboratory of Engine Reliability for funding this research.

Conflicts of Interest: The authors declare no conflict of interest.

References

1. Antoni, N.; Nguyen, Q.-S.; Ligier, J.-L.; Saffré, P.; Pastor, J. On the cumulative microslip phenomenon. *Eur. J. Mech. A-Solid* **2007**, *26*, 626–646. [CrossRef]
2. Gong, H.; Liu, J. Some factors affecting the loosening failure of bolted joints under vibration using finite element analysis. *Proc. Inst. Mech. Eng. Part C: J. Mech. Eng. Sci.* **2018**, *232*, 3942–3953. [CrossRef]
3. Li, W.; Wang, H.; Feng, Z. Non-contact online thickness measurement system for metal films based on eddy current sensing with distance tracking technique. *Rev. Sci. Instrum.* **2016**, *87*, 045005. [CrossRef]
4. Wang, X.; Gao, J.; Wang, L. Survey on the Laser Triangulation. *Yiqi Yibiao Xuebao* **2004**, *25*, 601–604+608.
5. Warmuth, A.R.; Shipway, P.H.; Sun, W. Fretting wear mapping: The influence of contact geometry and frequency on debris formation and ejection for a steel-on-steel pair. *Proc. R. Soc. A* **2015**, *471*, 20140291. [CrossRef]
6. Chen, M.; Li, X.; Lin, L.; Du, W.; Han, X.; Zhu, J.; Pan, C.; Wang, Z.L. Triboelectric nanogenerators as a self-powered motion tracking system. *Adv. Funct. Mater.* **2014**, *24*, 5059–5066. [CrossRef]
7. Jing, Q.; Xie, Y.; Zhu, G.; Han, R.P.; Wang, Z.L. Self-powered thin-film motion vector sensor. *Nat. Commun.* **2015**, *6*, 8031. [CrossRef]
8. Zhang, X.; Zheng, Y.; Wang, D.; Zhou, F. Solid-liquid triboelectrification in smart U-tube for multifunctional sensors. *Nano Energy* **2017**, *40*, 95–106. [CrossRef]
9. Liu, H.; Wang, H.; Lyu, Y. A novel triboelectric nanogenerator based on carbon fiber reinforced composite lamina and as a self-powered displacement sensor. *Nano Energy* **2020**, *224*, 111231. [CrossRef]
10. Wang, S.; Lin, L.; Wang, Z.L. Triboelectric nanogenerators as self-powered active sensors. *Nano Energy* **2015**, *11*, 436–462. [CrossRef]
11. Zhou, Y.S.; Zhu, G.; Niu, S.; Liu, Y.; Bai, P.; Jing, Q.; Wang, Z.L. Nanometer resolution self-powered static and dynamic motion sensor based on micro-grated triboelectrification. *Adv. Mater.* **2014**, *26*, 1719–1724. [CrossRef] [PubMed]
12. Li, C.; Wang, Z.; Shu, S.; Tang, W. A Self-Powered Vector Angle/Displacement Sensor Based on Triboelectric Nanogenerator. *Micromachines* **2021**, *12*, 231. [CrossRef]
13. Yang, H.; Liu, W.; Xi, Y.; Lai, M.; Guo, H.; Liu, G.; Wang, M.; Li, T.; Ji, X.; Li, X. Rolling friction contact-separation mode hybrid triboelectric nanogenerator for mechanical energy harvesting and self-powered multifunctional sensors. *Nano Energy* **2018**, *47*, 539–546. [CrossRef]
14. Wang, S.; Lin, L.; Wang, Z.L. Nanoscale triboelectric-effect-enabled energy conversion for sustainably powering portable electronics. *Nano Lett.* **2012**, *12*, 6339–6346. [CrossRef] [PubMed]
15. Wang, S.; Lin, L.; Xie, Y.; Jing, Q.; Niu, S.; Wang, Z.L. Sliding-triboelectric nanogenerators based on in-plane charge-separation mechanism. *Nano Lett.* **2013**, *13*, 2226–2233. [CrossRef] [PubMed]
16. Yang, Y.; Zhou, Y.S.; Zhang, H.; Liu, Y.; Lee, S.; Wang, Z.L. A single-electrode based triboelectric nanogenerator as self-powered tracking system. *Adv. Mater.* **2013**, *25*, 6594–6601. [CrossRef]
17. Wang, S.; Xie, Y.; Niu, S.; Lin, L.; Wang, Z.L. Freestanding triboelectric-layer-based nanogenerators for harvesting energy from a moving object or human motion in contact and non-contact modes. *Adv. Mater.* **2014**, *26*, 2818–2824. [CrossRef]
18. Zhu, G.; Pan, C.; Guo, W.; Chen, C.Y.; Zhou, Y.; Yu, R.; Wang, Z.L. Triboelectric-generator-driven pulse electrodeposition for micropatterning. *Nano Lett.* **2012**, *12*, 4960–4965. [CrossRef]
19. Niu, S.; Wang, S.; Lin, L.; Liu, Y.; Zhou, Y.S.; Hu, F.; Wang, Z.L.; Wang, Z.L. Theoretical study of contact-mode triboelectric nanogenerators as an effective power source. *Energy Environ. Sci.* **2013**, *6*, 3576–3583. [CrossRef]
20. Niu, S.; Liu, Y.; Wang, S.; Lin, L.; Zhou, Y.S.; Hu, Y.; Wang, Z.L. Theory of sliding-mode triboelectric nanogenerators. *Adv. Mater.* **2013**, *25*, 6184–6193. [CrossRef]
21. Bai, P.; Zhu, G.; Jing, Q.; Wu, Y.; Yang, J.; Chen, J.; Ma, J.; Zhang, G.; Wang, Z.L. Transparent and flexible barcode based on sliding electrification for self-powered identification systems. *Nano Energy* **2015**, *12*, 278–286. [CrossRef]

22. Chen, J.; Pu, X.; Guo, H.; Tang, Q.; Feng, L.; Wang, X.; Hu, C. A self-powered 2D barcode recognition system based on sliding mode triboelectric nanogenerator for personal identification. *Nana Energy* **2018**, *43*, 253–258. [CrossRef]
23. Niu, S.; Wang, Z.L. Theoretical systems of triboelectric nanogenerators. *Nano Energy* **2015**, *14*, 161–192. [CrossRef]
24. Wang, J.; Wu, C.; Dai, Y.; Zhao, Z.; Wang, A.; Zhang, T.; Wang, Z.L. Achieving ultrahigh triboelectric charge density for efficient energy harvesting. *Nat. Commun.* **2017**, *8*, 88. [CrossRef]

nanomaterials

MDPI

Article

Enhanced Responsivity and Optoelectronic Properties of Self-Powered Solar-Blind $Ag_2O/\beta\text{-}Ga_2O_3$ Heterojunction-Based Photodetector with Ag:AZO Co-Sputtered Electrode

Younghwa Yoon, Sangbin Park, Taejun Park, Hyungmin Kim, Kyunghwan Kim and Jeongsoo Hong *

Department of Electrical Engineering, Gachon University, 1342 Seongnamdaero, Seongnam 13120, Republic of Korea
* Correspondence: hongjs@gachon.ac.kr

Abstract: A Ag:AZO electrode was used as an electrode for a self-powered solar-blind ultraviolet photodetector based on a $Ag_2O/\beta\text{-}Ga_2O_3$ heterojunction. The Ag:AZO electrode was fabricated by co-sputtering Ag and AZO heterogeneous targets using the structural characteristics of a Facing Targets Sputtering (FTS) system with two-facing targets, and the electrical, crystallographic, structural, and optical properties of the fabricated thin film were evaluated. A photodetector was fabricated and evaluated based on the research results that the surface roughness of the electrode can reduce the light energy loss by reducing the scattering and reflectance of incident light energy and improving the trapping phenomenon between interfaces. The thickness of the electrodes was varied from 20 nm to 50 nm depending on the sputtering time. The optoelectronic properties were measured under 254 nm UV-C light, the on/off ratio of the 20 nm Ag:AZO electrode with the lowest surface roughness was 2.01×10^8, and the responsivity and detectivity were 56 mA/W and 6.99×10^{11} Jones, respectively. The $Ag_2O/\beta\text{-}Ga_2O_3$-based solar-blind photodetector with a newly fabricated top electrode exhibited improved response with self-powered characteristics.

Keywords: $\beta\text{-}Ga_2O_3$; DUV; photodetector; Co-sputtered electrode; Ag:AZO

Citation: Yoon, Y.; Park, S.; Park, T.; Kim, H.; Kim, K.; Hong, J. Enhanced Responsivity and Optoelectronic Properties of Self-Powered Solar-Blind $Ag_2O/\beta\text{-}Ga_2O_3$ Heterojunction-Based Photodetector with Ag:AZO Co-Sputtered Electrode. *Nanomaterials* **2023**, *13*, 1287. https://doi.org/10.3390/nano13071287

Academic Editors: Qiongfeng Shi, Jianxiong Zhu and Antonio Di Bartolomeo

Received: 21 February 2023
Revised: 3 April 2023
Accepted: 3 April 2023
Published: 6 April 2023

1. Introduction

Ultraviolet (UV) radiation plays an important role in various fields such as practical life, the military industry and environment, security applications, the chemical industry, and biology [1,2]. Of the total solar ultraviolet photons, deep-UV (DUV) light with a wavelength of less than 280 nm is absorbed by the ozone layer, which is the main region that absorbs the corresponding photons before they reach the surface of the earth. It is called a solar-blind UV light (200–280 nm) and has extremely low background noise and high responsibility [3]. Therefore, solar-blind UV photodetectors using these unique properties were developed and used extensively in medical imaging, chemical and biological analysis, flame sensor, and military-related industries, such as secure fields and missile detection [4–8].

Attempts were made to find suitable materials for the fabrication of a solar-blind photodetector, and a wide bandgap material of 4.42 eV or more to detect and measure UV light with a wavelength of <280 nm was required as a top priority. Studies have shown that Ga_2O_3 has several advantages as a material for the fabrication of solar-blind UV light photodetectors.

Ga_2O_3 has an ultra-wide band gap of 4.7–5.2 eV in bare state, excellent thermal and chemical stability, and high selectivity to UV light [9–11]. Owing to its great potential, many studies of the Ga_2O_3-based solar-blind UV photodetector have been conducted recently [12–14].

Photovoltaic Ga_2O_3 solar-blind photodetectors are being developed in the form of Schottky, PN, and PIN photodetectors [15]. However, solar-blind photodetectors are applied in extreme environments that require continuous and long-term use without a separate power supply and stability against corrosion and temperature. For this reason, a p-n junction photodetector is advantageous for self-powered photodetectors that use the built-in electric field of photovoltaic devices by transferring the photo-excited electron-hole pairs to generate an electrical signal with a fast response speed [16–21].

Recent research has shown that various materials and structures can be used to enhance photodetection performance. For instance, Ga_2O_3/zero-dimensional graphene quantum dot (GQD) heterojunction, Ga_2O_3/MXene nanowire networks, MSM structured ZnO/Au/ Ga_2O_3, and multi-layer thin film have all been applied to improve photodetection performance [22–26].

In a previous study, our group fabricated a $Ag_2O/\beta\text{-}Ga_2O_3$ p-n junction-based photodetector to increase photo-excited electron-hole pair transfer efficiency, which led to an improvement in key parameters such as responsivity [27]. Our study also observed differences in optical performance based on the thickness and roughness of the electrode. From these results, we identified the importance of the upper electrode used in the photodetector and the need for further research in this area.

Therefore, we designed a new electrode fabrication with low roughness from representative thin film materials, such as silver (Ag), aluminum-doped zinc oxide (AZO), copper (Cu), gold (Au), and graphene [28–30]. We fabricated a novel electrode material that combines the properties of Ag and AZO, which was expected to show a uniform thin film and low surface roughness, based on research on AZO/Ag/AZO multilayer thin film [31]. Although the effect of AZO on the formation of low roughness has not yet been studied, it is expected that the use of AZO with Ag will result in a low roughness surface of the electrode, just as AZO has been shown to assist in the formation of ultra-thin Ag films in studies on AZO/Ag/AZO thin films. Therefore, a thin film was deposited by the co-sputtering of Ag and AZO to create a Ag:AZO thin film, where the composition ratio of Ag and AZO was adjusted.

A co-sputtering technique using the Facing Targets Sputtering (FTS) system was applied to the deposition of a Ag:AZO thin film, which enabled the production of material compositions that are technically impossible to produce as a single target [32–34]. Moreover, a low-resistance, high-conductivity, and transparent Ag:AZO thin film was fabricated by controlling the co-sputtering conditions. Ag:AZO, Ag, and AZO electrodes were deposited on the Ag_2O layer of the $Ag_2O/\beta\text{-}Ga_2O_3$ heterojunction photodetector, and the photo responses of each photodetector were compared and evaluated. The thickness of each electrode was varied between 20 and 50 based on the sputtering time. The optoelectronic properties were measured under a 254 nm UVC light, and the Ag:AZO electrode exhibited a photoresponsivity of 56 mA/W without any additional voltage. The $Ag_2O/\beta\text{-}Ga_2O_3$-based solar-blind photodetector using a Ag:AZO upper electrode exhibited enhanced responsivity with the self-powered property.

2. Experiment

2.1. Materials

The fabrication of the p-n photodetector was accomplished using an n-type Sn-doped $\beta\text{-}Ga_2O_3$ wafer ($N_d–N_a$ = 6.3 × 10^{18} cm^{-3} with a thickness of 641 μm, Novel Crystal Technology, Inc., Saitama, Japan). For the epitaxial layer, Si-doped $\beta\text{-}Ga_2O_3$ ($N_d–N_a$ = 2.8 × 10^{16} cm^{-3} with a thickness of 8.6 μm, Novel Crystal Technology, Inc., Japan) was grown using halide vapor phase epitaxy. For the ohmic contact of the device, Ti/Au electrodes (10/40 nm) were deposited by electron-beam evaporation on the backside of the Sn-doped $\beta\text{-}Ga_2O_3$ wafer. Targets of 4-inch Ag and ZnO:Al_2O_3 (98:2 wt%) (RND Korea, Seoul, Republic of Korea) were purchased for the deposition of p-type Ag_2O and upper electrodes. The soda-lime glass substrate (size of 75 × 25 × 1 mm^3, Marienfeld, Lauda-Königshofen, Germany) was used in the experiment for measuring the characteristics of electrodes.

2.2. Fabrication and Characterization of the Ag:AZO Thin Film

As presented in Scheme 1, the co-sputtering method was applied with an FTS system to prepare Ag:AZO samples on soda-lime glass. The thin film was deposited under the sputtering conditions presented in Table 1, and the thickness of the thin film was measured by a KLA-Tencor α-Step D-500 Stylus Profiler and then deposited according to the calculated deposition rate shown in Table 2. The properties of Ag:AZO as an electrode material were evaluated, and the Ag and AZO samples were also evaluated after deposition. The X-ray diffraction (XRD) patterns were measured on a diffractometer (PANalytical; X'Pert Pro MPD, Almelo, The Netherlands). The surface morphologies of the films were observed using scanning electron microscopy (SEM; S-4700, Hitachi, Tokyo, Japan) and atomic force microscopy (AFM; Park NX10, Park Systems, Suwon, Republic of Korea). The optical properties were evaluated by a UV-visible spectrophotometer in the range of 190–800 nm (Lambda 750 UV-vis-near-infrared, PerkinElmer, Waltham, MA, USA).

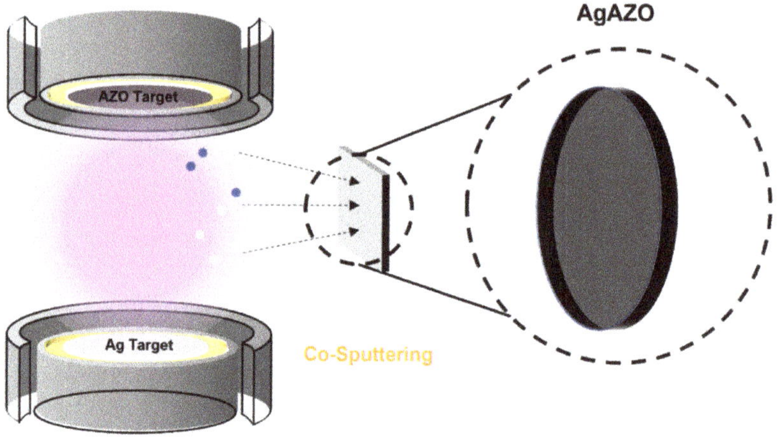

Scheme 1. The deposition of Ag:AZO thin film by co-sputtering method.

Table 1. Sputtering conditions of each electrode.

Parameters	Sputtering Conditions Ag	AZO	Ag:AZO
Targets	Ag (99.99%)	AZO (98:2 wt%)	Ag, AZO
Base pressure (Torr)	3×10^{-5}	3×10^{-5}	3×10^{-5}
Working pressure (mTorr)	2	1	2
Gas flow (sccm)	Ar: 10	Ar: 10	Ar: 10
Input power (W)	50 (DC)	150 (DC)	50 (DC)
Thickness (nm)		20, 30, 40, 50	

Table 2. Deposition rate of each electrode by thickness.

Targets	20 nm	30 nm	40 nm	50 nm
Ag	2 min 8 s	3 min 12 s	4 min 16 s	5 min 20 s
AZO	1 min 30 s	2 min 15 s	3 min	3 min 45 s
Ag:AZO	1 min 49 s	2 min 43 s	3 min 38 s	4 min 32 s

2.3. Fabrication and Evaluation of Ag_2O/β-Ga_2O_3-Based Photodetector with Different Electrodes

Scheme 2 shows the as-fabricated p-n junction photodetector. The 4 inch Ag targets were used to deposit the p-type Ag_2O layer on the top of the β-Ga_2O_3 epitaxial layer [35]. As shown in Table 3, the sputtering conditions were controlled during the deposition with FTS in an oxygen-saturated atmosphere. Each electrode was deposited in a circular shape on the Ag_2O layer by using a shadow mask with a radius of 300 μm. To enhance crystallization and reduce the interfacial defect of Ag_2O, the as-fabricated samples were post-annealed at 300 °C for 1 min at 100 mTorr in an Ar atmosphere using rapid thermal annealing. Electrical characteristics such as current-voltage (I–V) ratio and photosensitivity were measured and analyzed with a semiconductor analyzer (4200A-SCS Parameter Analyzer, Keithley, Cleveland, OH, USA) in response to various light intensities using a UV-C lamp (TN-4LC, wavelength: 254 nm, Korea Ace Scientific, Seoul, Republic of Korea).

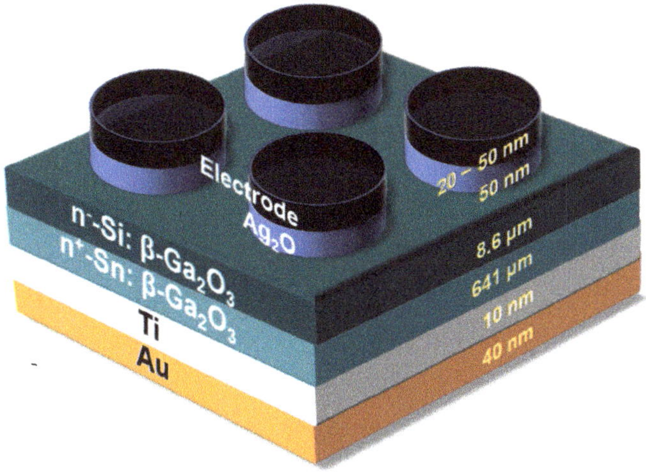

Scheme 2. Schematic illustration of the p-n junction photodetector.

Table 3. Sputtering conditions of p-type Ag_2O layer.

Parameters	Sputtering Conditions
Targets	Ag (99.99%)
Base pressure (Torr)	3×10^{-5}
Working pressure (mTorr)	2
Gas flow (sccm)	Ar: 10, O_2: 3
Input power (W)	50 (DC)
Thickness (nm)	50

3. Results and Discussion

3.1. Evaluation of As-Deposited Electrodes

In our previous study, Ag was used as the upper electrode material for the Ag_2O/β-Ga_2O_3-based p-n junction photodetector. This study confirmed that the surface condition of the electrode film has a significant influence on the performance of the photodetector. Therefore, as a further study, we tried to fabricate an electrode with an improved surface and apply it to the device. Therefore, a thin film was made by co-sputtering Ag and AZO, and the structural, optical, and electrical properties of the as-deposited thin film on the glass substrate were evaluated.

The properties of the electrodes were evaluated prior to deposition on the Ag_2O/β-Ga_2O_3 photodetector. Surface morphology in thin films analyzed by SEM are described in Figure 1a–e and Figures S1 and S2. The SEM images shown in Figure 1a–d are 20–50 nm thick Ag:AZO films deposited by co-sputtering with Ag and AZO. The films exhibited smooth and homogeneous surfaces, and grain boundaries increased with increasing thickness. The uniformity of the surface improves photosensitivity when used as a layer of a photodiode device by reducing the scattering and reflectivity of the irradiated light. Figure 1e is an energy-dispersive X-ray spectrometer (EDX) image of a 20 nm sample deposited at 50 W, showing the distribution and atomic percentages of Ag, Al, and ZnO present in the thin film. Since AZO is an oxide, it has a lower deposition rate than Ag, and it is inferred that more Ag is contained in the component table of the as-fabricated sample. Figures S1 and S2 show the surface SEM images according to the thickness of Ag and AZO thin films, respectively. It was observed to have a homogeneous surface like that of the Ag:AZO thin film.

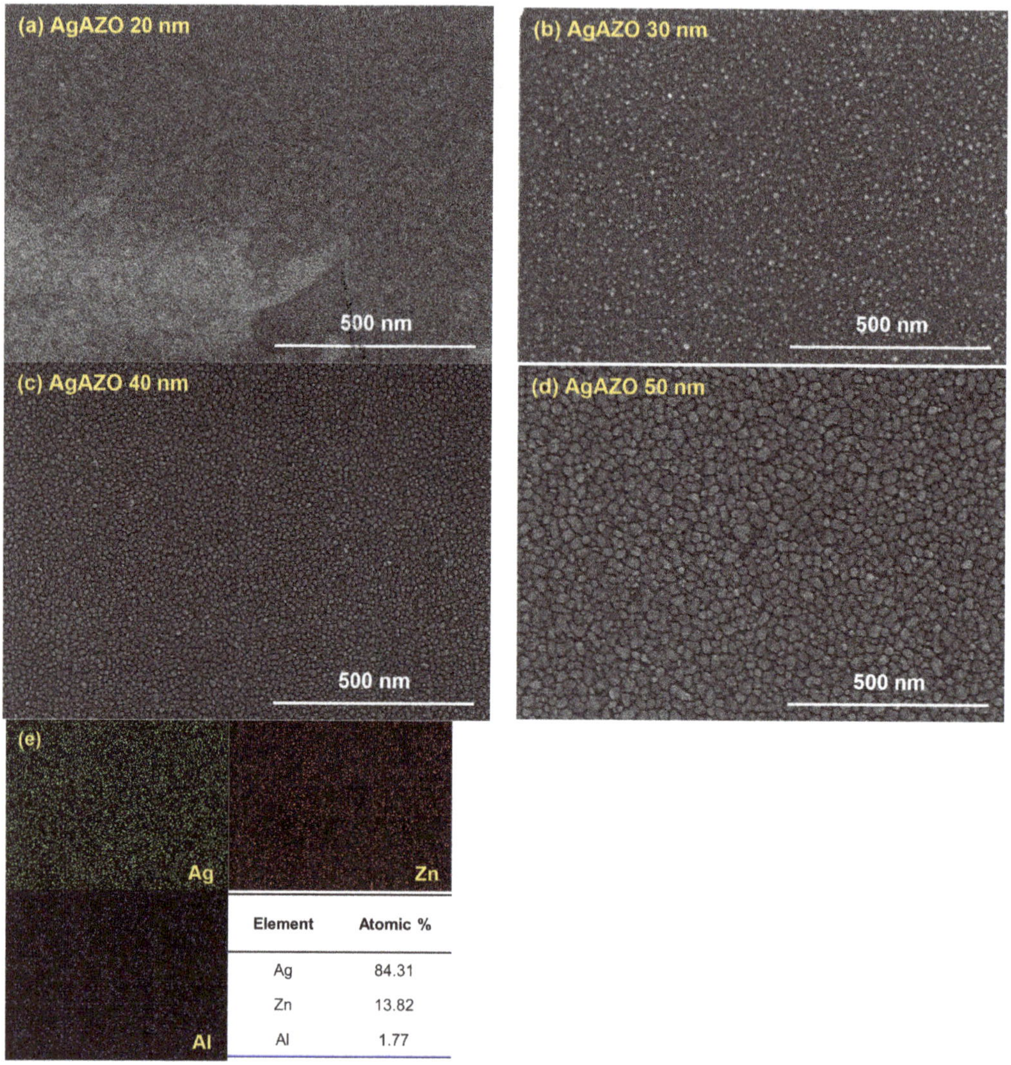

Figure 1. Thickness-dependent SEM images of Ag:AZO thin films. (**a**) 20 nm; (**b**) 30 nm; (**c**) 40 nm; (**d**) 50 nm; (**e**) EDX image of 20 nm Ag:AZO thin film.

The crystallographic characteristics of the as-fabricated Ag:AZO film for each thickness were evaluated by analyzing the XRD patterns in Figure 2a. The Ag:AZO samples had polycrystalline structures, and both Ag and AZO peaks were observed. Ag peaks at 2θ = 38.12, 64.4, and 77.5° corresponded to the (111), (220), and (311) planes, respectively, and the AZO peak at 2θ = 44.8° corresponded to the (400) plane (Ag; ICDD card 01-087-

0720, AZO; ICDD card 01-071-0968). There was no significant change based on thickness; however, compared with the peak intensity of the AZO electrode in Figure S3, the (400) peak intensity was lower because the AZO content in the Ag:AZO sample was 5.6 times lower than that of Ag, as we inferred from the EDX results.

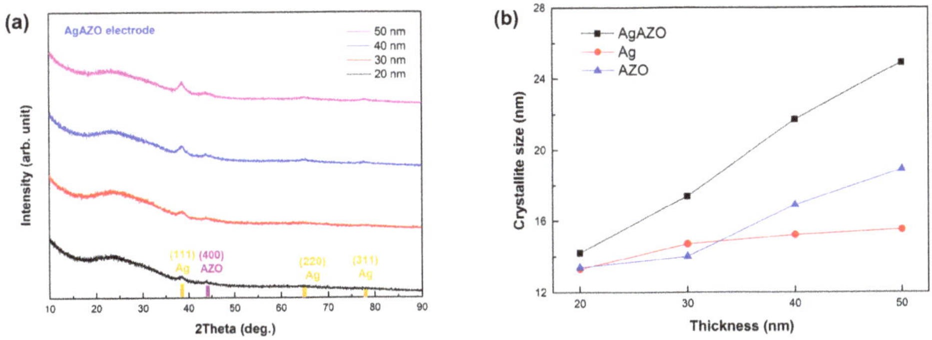

Figure 2. Structural characteristics of Ag:AZO thin film depending on thickness. (**a**) XRD measurement results. (**b**) Crystallite size according to the thickness of the electrodes.

The crystallite size was calculated using the Scherrer equation based on the (111) plane, which is the preferential growth plane of the XRD pattern of Ag:AZO, and it was inferred that the crystallite size depends on the thickness. The Scherrer equation is as follows [36,37]:

$$\tau = \frac{K\lambda}{\beta cos\theta} \quad (1)$$

In Equation (1), τ is the crystallite size, θ is the Bragg angle, K is the Scherrer constant, and β is the full width at half maximum (FWHM) inversely proportional to the crystallite size τ. Therefore, the larger the FWHM, that is, the wider the XRD pattern, the smaller the crystallite size. The crystallite size according to the thickness of the thin films is shown in Figure 2b. As the thickness increased, so did the crystallite size, and it was inferred that Ag:AZO, which had the smallest FWHM, had the largest crystallite size. Crystallite size increases as the density of a material increases; hence, a flat surface would be made of dense crystallites.

The low roughness of the thin film is an important factor for electrodes used in photodetectors, as it reduces loss such as scattering and reflection of incident light energy [38]. Therefore, the roughness of each thickness of the deposited samples was obtained using AFM in the range of 1×1 μm^2 images and is shown in Figures 3, S4 and S5. The RMS values representing the surface roughness of the Ag:AZO thin film ranged from 0.6 to 1.0 nm (20–50 nm Ag:AZO thin film). However, the range of RMS values of the Ag and AZO thin films were 1.0 to 1.3 nm and 0.9 to 1.2 nm, respectively, indicating that the Ag:AZO thin film had the lowest roughness compared with the same thickness. The roughness of the Ag:AZO 20 nm thin film was reduced by 25% compared with the previously used Ag 20 nm thin film. The roughness of the thin film is a similar feature to the crystallite size calculated by the Scherrer equation based on the XRD pattern, and by this, it can be inferred that a smoother surface is formed in the thin film with a large crystallite size. In addition, because the roughness of the thin film has a great effect on the performance of the photodetector, it is expected that a photodetector with high responsivity can be manufactured using the Ag:AZO electrode.

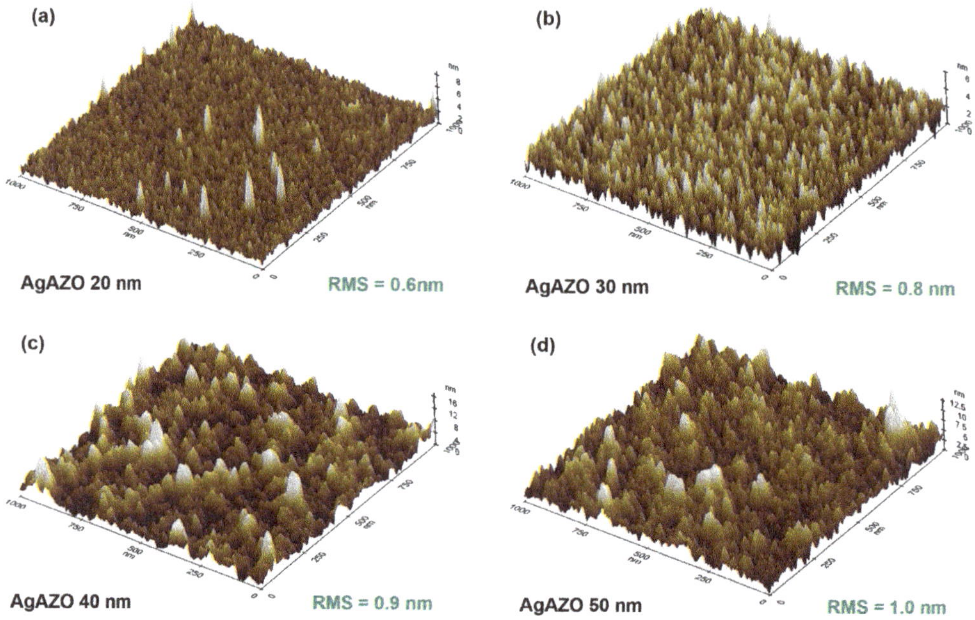

Figure 3. AFM images of the Ag:AZO thin film as a function of thickness. (**a**) 20; (**b**) 30; (**c**) 40; (**d**) 50 nm.

Figure 4 shows the transmittance of the fabricated thin films in the wavelength range of 190 to 800 nm using a spectrophotometer. AZO, a transparent electrode, exhibited a high transmittance in visible light and a transmittance of 89% at a thickness of 20 nm. However, Ag exhibited a peak transmittance of 57% in the 320 nm wavelength band, and a low transmittance in the other wavelength bands except for the corresponding region. The Ag:AZO thin film fabricated via the co-sputtering of Ag and AZO exhibited optical properties of both Ag and AZO, and had a peak at 320 nm. It also exhibited a transmittance of 55 to 70% at a thickness of 20 nm.

The Hall meter was used to evaluate the electrical properties of the fabricated thin film to be used as an electrode. As shown in Table 4, the sheet resistance, carrier concentration, and mobility were inferred, and after application to the device, changes in electrical properties were also observed after heat treatment of the electrode for post-heat treatment at 300 °C, to improve the crystallinity of Ag O [39–41].

The sheet resistance of various thicknesses of deposited thin films decreased as the thickness increased, and in the case of the Ag:AZO thin film, the sheet resistance was at least six times lower than that of the AZO thin film, which is an existing electrode material, at all thicknesses. In addition, after annealing at 300 °C, the sheet resistance of the 20 nm Ag:AZO thin film was reduced by half from 72.8 to 38.1 Ω/Sq. Regarding carrier concentration with respect to temperature change, all three electrodes exhibited an increasing change, but there was a difference in the increase rate depending on the electrode. This increase in carrier concentration is because defects generated during thin film deposition using sputtering are reduced via heat treatment, thereby preventing loss due to trapping during the movement of carriers [42]. However, owing to the increased electron concentration, the hole mobility decreased.

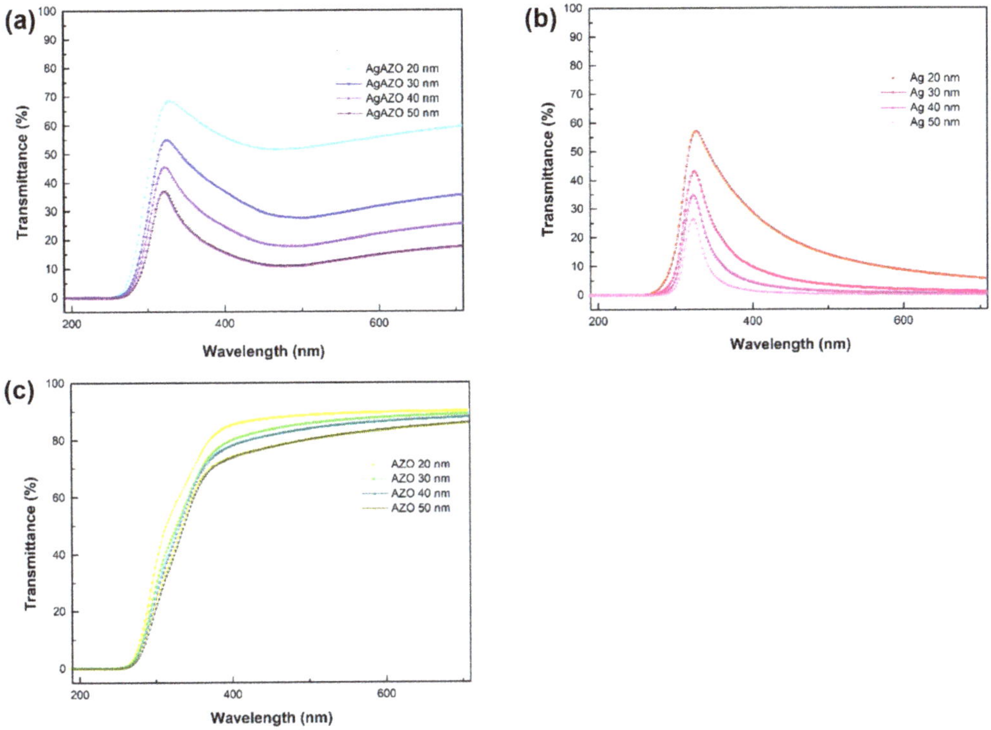

Figure 4. Optical characteristics of as-deposited electrodes by thickness. (**a**) Transmittance of Ag:AZO thin film (**b**) Transmittance of Ag thin film (**c**) Transmittance of AZO thin film.

Table 4. Electrical properties of as-deposited electrodes.

Ag:AZO	**R. T**	**300 °C**	**R. T**	**300 °C**	**R. T**	**300 °C**
Thickness	Sheet Res. (Ω/Sq.)		Mobility (cm^2/Vs)		Carrier Concentration (cm^{-3})	
20 nm	38.4	20.4	372	5.09	1.61×10^{20}	1.61×10^{23}
30 nm	44.2	26	152.7	3.3	3.08×10^{20}	2.43×10^{23}
40 nm	72.8	31.7	153.4	5.14	3.21×10^{20}	1.49×10^{23}
50 nm	120.6	142.3	496.9	1.97	2.08×10^{19}	4.46×10^{22}
Ag	R. T	300 °C	R. T	300 °C	R. T	300 °C
Thickness	Sheet Res. (Ω/Sq.)		Mobility (cm^2/Vs)		Carrier Concentration (cm^{-3})	
20 nm	0.952	0.966	4303.3	11.19	7.62×10^{20}	2.89×10^{23}
30 nm	0.572	0.515	5462.5	8.84	6.66×10^{20}	4.57×10^{23}
40 nm	0.394	0.3	6016.8	4.32	6.58×10^{20}	1.21×10^{24}
50 nm	0.289	0.236	6315.7	9.06	6.84×10^{20}	5.84×10^{23}
AZO	R. T	300 °C	R. T	300 °C	R. T	300 °C
Thickness	Sheet Res. (Ω/Sq.)		Mobility (cm^2/Vs)		Carrier Concentration (cm^{-3})	
20 nm	459.4	431.3	69.47	2.59	9.78×10^{19}	2.79×10^{21}
30 nm	483.4	383.9	749.1	8.69	5.75×10^{18}	6.24×10^{20}
40 nm	441.4	356.8	1405.3	3.63	2.52×10^{18}	1.21×10^{21}
50 nm	412.9	370.8	2614.9	3.02	1.16×10^{18}	1.11×10^{21}

As shown in Table 5, it was confirmed how the work function of Ag:AZO and co-sputtered Ag:AZO changed by measuring the work function of the deposited electrode. The work function of Ag:AZO is 4.66 eV, which is higher than Ag: 4.26 eV and AZO: 4.1 eV.

Table 5. Work function of electrodes measured by ultraviolet photoelectron spectroscopy.

Electrode	Work Function
Ag	4.26 eV
AZO	4.1 eV
Ag:AZO	4.66 eV

3.2. Characteristics of As-Deposited P-Type Ag_2O Thin Film

P-type Ag_2O thin films were fabricated using reactive sputtering in an oxygen atmosphere. To create an oxygen atmosphere in the chamber, gas was introduced at an Ar to O_2 ratio of 10:3 sccm, and sputtering was performed at an applied power of 50 W. Figure 5a is the XRD result of the deposited Ag_2O thin film. Peaks appearing in the XRD pattern are as follows. The Ag_2O peaks at 2θ = 33.66, 36.34, 63.36, and 66.74° corresponded to the (100), (011), (111), and (004) planes, respectively. (Ag_2O; ICDD card. 01-072-2108). After the heat treatment, the intensity of the (011) peak increased, indicating that the crystallinity of the preferentially grown surface of the Ag_2O thin film improved. In addition, it was inferred that the (004) plane, which was a fine peak around 66.74° after heat treatment, greatly increased [43]. The Ag_2O particle size calculated from the XRD results was 13.3 nm. The electrical properties of Ag_2O measured by a Hall meter were carrier concentration, mobility, and resistivity, measuring 6.35×10^{18} cm^{-3}, 71.4 cm^2/Vs, and 1.38×10^{-2} Ω·cm, respectively. In addition, the fabricated sample was measured by UV-vis and exhibited a transmittance of 40%, as shown in Figure 5b.

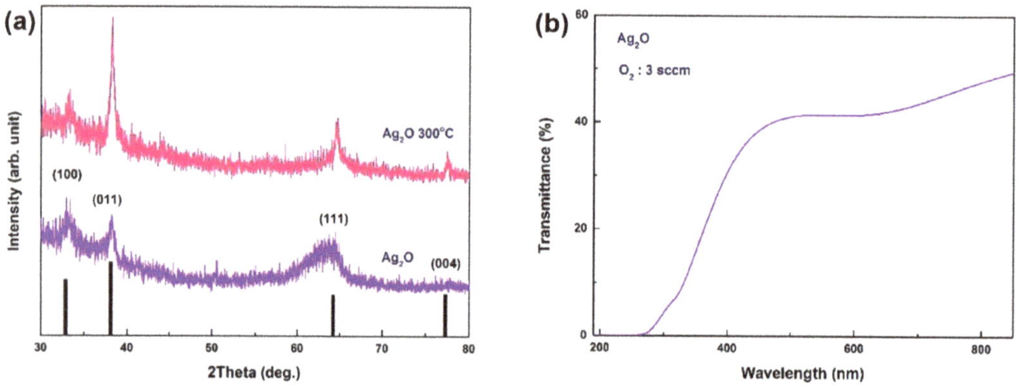

Figure 5. Characteristics of p-type Ag_2O thin film. (**a**) XRD pattern of Ag_2O thin film. (**b**) Transmittance of Ag_2O thin film.

3.3. Characteristic Evaluation of Fabricated Device

Scheme 3 is a photodetector fabricated by sequentially depositing the p-type Ag_2O and electrodes evaluated above on a β-Ga_2O_3 substrate. As shown in the figure, all three electrodes with the same thickness were deposited on one device to evaluate the semiconductor characteristics and photoreaction.

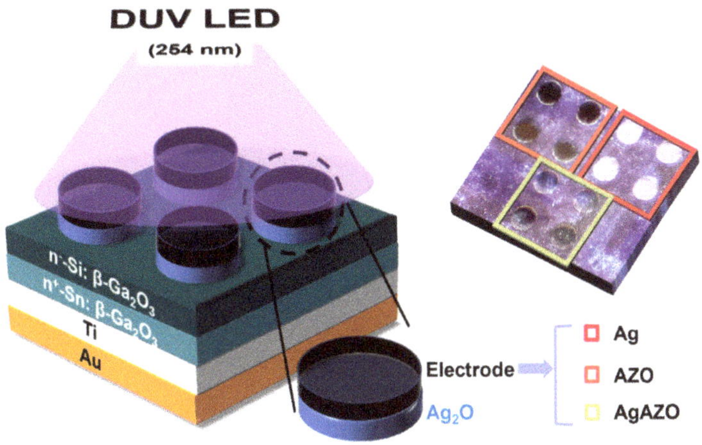

Scheme 3. Schematic diagram of a photodetector with three different electrodes applied.

Figure 6 is the I–V curve for each thickness of a Ag_2O/β-Ga_2O_3 photodetector based on the p-n junction deposited with a Ag:AZO electrode. Measurements were conducted in a dark room to eliminate the influence of light, and voltages from −2 to 4 V were applied to evaluate current changes according to voltage. The rectification characteristics of the p-n junction were observed at all thicknesses, and the off- and on-currents of the Ag:AZO 20 nm device were 2.56×10^{-11} A at 0 V and 5.16×10^{-3} A at 1.73 V, respectively. The on/off ratio was calculated as 2.01×10^{8}.

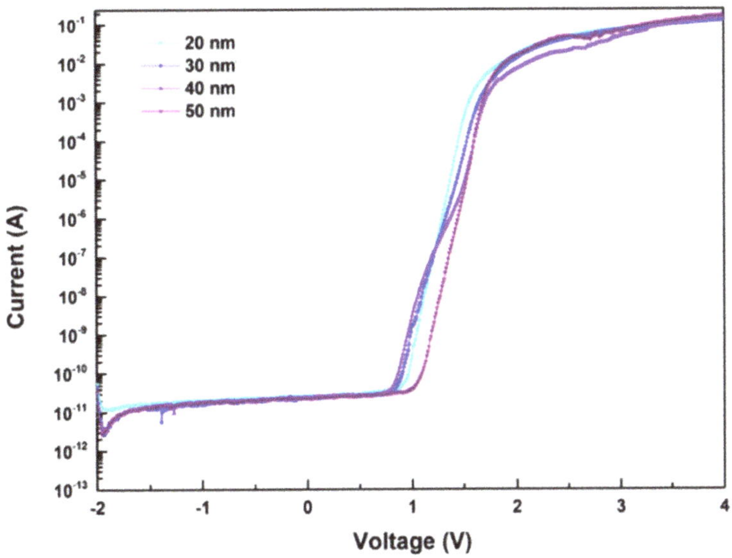

Figure 6. I-V curve of photodetector using Ag:AZO electrode.

Figure 7 shows the parameters calculated from the I–V curve of the photodetector shown in Figure 5 based on the thermionic emission model [44]. The parameters were on-resistance (R_{on}), barrier potential (φ_B), and ideality factor (n) based on thickness, and the thermionic emission model is as follows [45].

$$J = J_s \exp\left\{\frac{q(V - IR_s)}{nkT}\right\} \quad (2)$$

$$J_s = AA^*T^2 \exp\left(\frac{-q\varphi_B}{kT}\right) \quad (3)$$

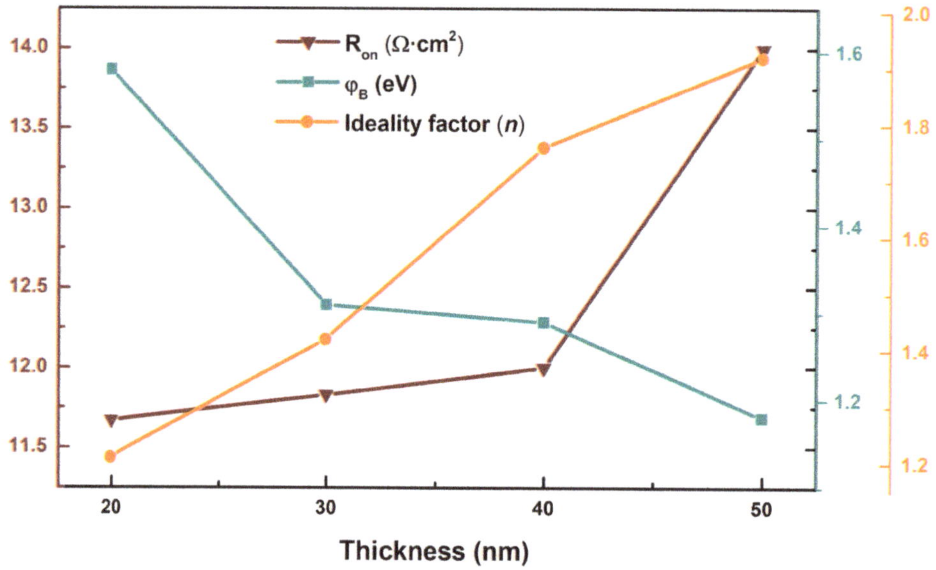

Figure 7. Electrical parameters of the photodetector as a function of Ag:AZO thickness.

In Equation (2), J_s is the saturation current density, q is the charge, V is the voltage across the diode, R_s is the series resistance, and n is the ideality coefficient representing the deviation between the ideal diode and the actual non-uniform barrier and tunneling diode. k is the Boltzmann constant and T is the Kelvin temperature. In Equation (3), A is the contact area, A^* is Richardson's constant, and β-Ga_2O_3 has a value of 41 A/(cm^2*K^2).

As the electrode thickness increased, the on-resistance increased from 11.67 to 14.0 Ω·cm^2, the ideality coefficient increased from 1.42 to 2.42, and the barrier potential decreased from 1.58 eV to 1.29 eV. The change in the ideality factor is related to the interface state between the electrode and the p-type layer. As the thickness of the electrode increased, the stress and roughness of the material increased, causing defects at the interface between the p-type and the electrode, which hindered the capture or movement of carriers, leading to an increase in resistance [46]. However, it was inferred that the barrier height increased as the thickness decreased because the roughness and interface defects decreased as the thickness of the electrode decreased. Defects present in the interfacial layer degrade the movement of carriers, so carriers move smoothly in samples with few defects, forming a wide depletion layer; hence, the barrier height increased [47]. A photodetector with a high ideality coefficient is affected by charge trapping and carrier recombination, so it was judged that the device with the 20 nm Ag:AZO electrode had the best electrical characteristics [48]. In addition, one of the causes of a high ideality coefficient is the hump phenomenon present in the I–V curve, which affects the current flow because of the high trap density due to the trapping of charged species in Ag and AZO combined crystals with different resistivity and conductivity. It was inferred that the hump phenomenon had occurred.

Figure 8 shows I–V curves of three electrodes with thicknesses of 20 nm. Rectification characteristics of the p-n junction were observed in all devices using each electrode, and the off- and on-currents of the Ag device were 1.62×10^{-11} A at 0 V and 6.23×10^{-4} A at 1.71 V, respectively, with an on/off ratio of 3.85×10^{7}. In addition, the off- and on-currents of the AZO device were found to be 2.58×10^{-11} A at 0 V and 7.03×10^{-4} A at 1.71 V, respectively, resulting in an on/off ratio of 2.72×10^{7}. Therefore, the Ag:AZO electrode was found to have the largest on/off ratio. Table 6 displays the photodetector's electrical parameters for each electrode. AZO, as an oxide, has a higher resistance than other electrodes, resulting in a relatively high R_{on} value compared with the other electrodes. However, it can be inferred that there are few interface defects with the p-type layer, as the ideality coefficient is closest to one and the potential barrier is high.

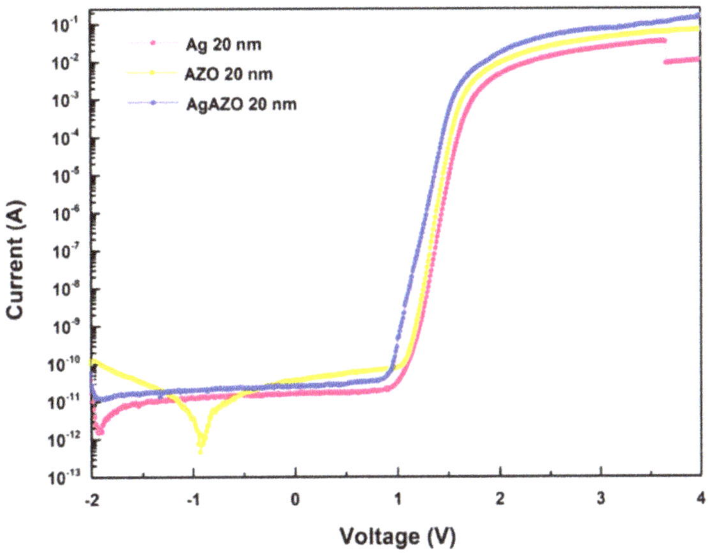

Figure 8. Photodetector I–V curves using 20 nm-thick electrodes.

Table 6. Electrical parameters of the photodetector for each electrode.

	Ag:AZO	Ag	AZO
R_{on} ($\Omega \cdot cm^2$)	11.67	43.33	2170.21
φ_B (eV)	1.58	1.68	1.93
n	1.21	1.29	1.12

Figure 9 depicts the I–V curve as a function of the intensity of UV irradiation on the Ag:AZO 20 nm device. Reverse bias is generated by applying irradiating light with a forward bias to the device, and reverse current is generated because of the generated electron-hole pair. Therefore, as shown in the figure, it was inferred that the stronger the light intensity, the greater the reverse current.

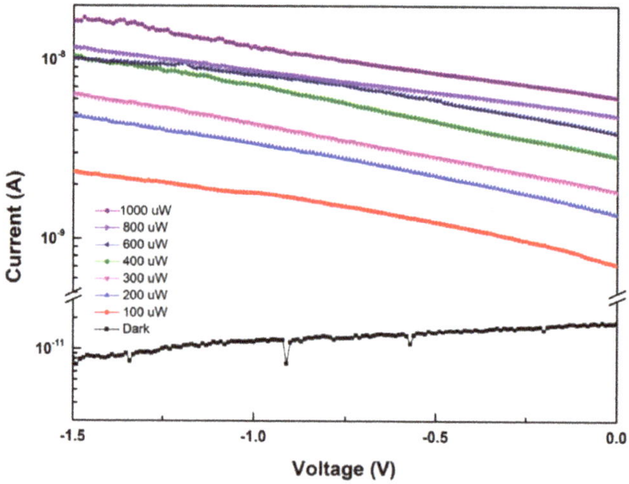

Figure 9. I–V curve of Ag:AZO 20 nm device at different UV light intensities.

Figure 10 shows the photocurrent densities of the photodetectors to which each electrode was applied over time, when irradiated with UV radiation with an intensity and wavelength of 1000 μW/cm^2 and 254 nm, respectively, in zero bias. Ag:AZO exhibited the highest photocurrent density of 31 μA/cm^2, compared with AZO and Ag, which had photocurrent densities of 27 μA/cm^2 and 18.6 μA/cm^2. There are two reasons for this phenomenon. First, because of the low surface roughness, loss due to scattering and the reflection of light energy incident on the surface of the thin film is relatively small. Second, because of the low roughness of the electrode, sufficient bonding between the interface layers is achieved, resulting in lower contact resistance. Therefore, it is expected that the largest photocurrent appears in the photodetector using the 20 nm Ag:AZO electrode with the lowest roughness.

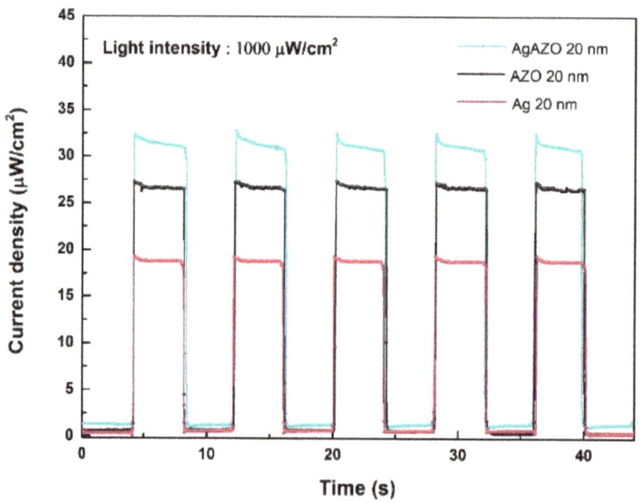

Figure 10. Photocurrent density of each device for 1000 μW/cm^2 UV light.

In Figure 11, the photocurrent density obtained by irradiating each element with UV from 100 to 1000 μW/cm^2 in zero bias is shown. As the light intensity increased, more electron-hole pairs were generated, and the photocurrent increased. The light intensity-photocurrent density of the devices exhibited a linear increase. Responsivity, which represents photosensitivity and is an important index for evaluating the performance of photodetectors, and detectivity, which is a performance index for the smallest detectable signal, were evaluated from the photocurrent density value based on the light intensity. Figure 12 shows the responsivity and detectivity as a function of the light intensity of each device. The equations for calculating responsivity and detectivity are as follows [49–52].

$$R = \left(\frac{J_{photo} - J_{dark}}{P}\right) \tag{4}$$

$$D = \frac{R}{(2eJ)^{1/2}} \tag{5}$$

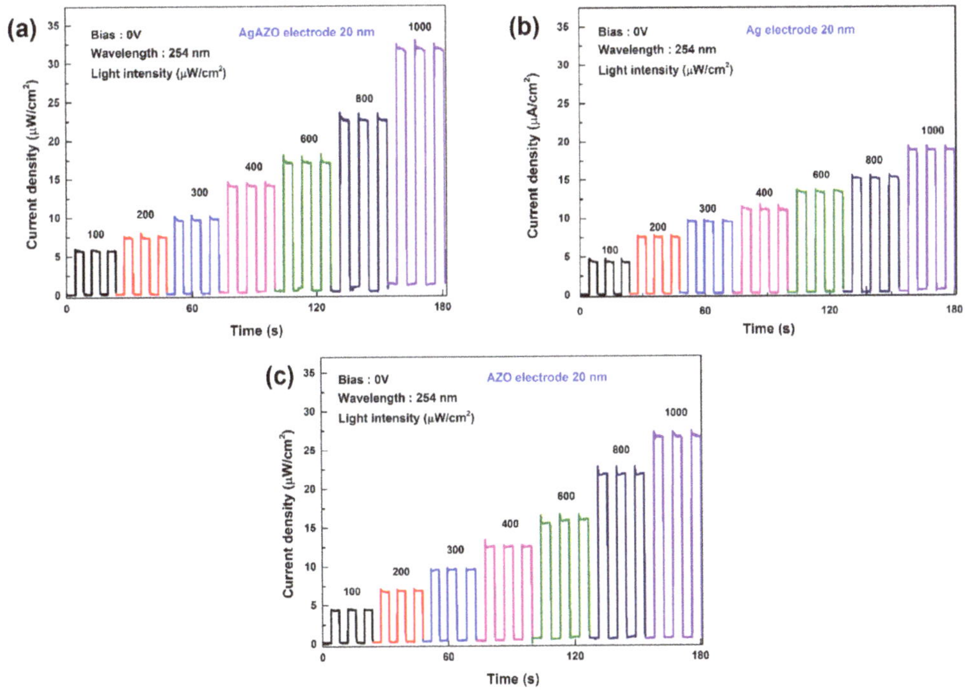

Figure 11. Photocurrent density change of device according to light intensity. (**a**) Ag:AZO 20 nm; (**b**) Ag 20 nm; (**c**) AZO 20 nm.

In Equation (4), J_{photo} is the photocurrent density, J_{dark} is the dark-current density, and P is the irradiated-light intensity. In Equation (5), e is the absolute value of charge, and J is the dark-current density. The responsivity and detectivity of each device calculated with the above formulas are shown in Figure 12. Responsivity and detectivity decreased as the intensity of light increased. At higher intensities, more electron-hole pairs were generated, which induced self-heating due to collisions and vibrations between electrons, increasing charge carriers, as well as increasing recombination rates. Therefore, the highest value was measured at 100 μW/cm^2, where the effect of self-heating was the least. The maximum values of responsivity and detectivity of each photodetector with 20 nm electrode applied were Ag:AZO: 56 mA/W, 6.99 × 10^{11} Jones; Ag: 42.5 mA/W, 5.3 × 10^{11} Jones; and AZO:

44.4 mA/W, 4.8 × 10^{11} Jones. In order to check the response spectra of the photodetector, the same measurement was performed under 365 nm light, but no photo-response was observed

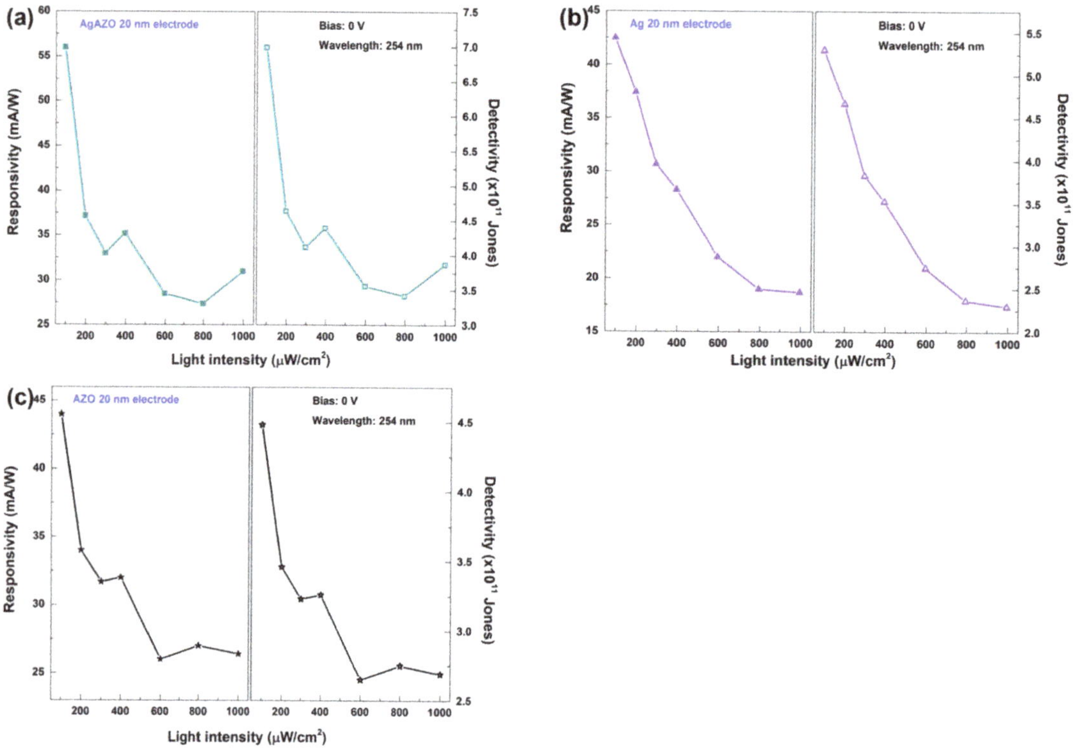

Figure 12. Device responsivity and detectivity change according to light intensity. (**a**) AgAZO 20 nm; (**b**) Ag 20 nm; (**c**) AZO 20 nm.

Rise and decay times, which evaluate the response speed of the device, are shown in Figure 13. Response speed, one of the important factors in evaluating photodetectors, is determined by the size of the internal potential caused by the difference in band gap between p-type Ag_2O and n-type β-Ga_2O_3 and measured using photocurrent characteristics over time [50,53–59]. When a UV radiation of 1000 μW/cm^2 was used in zero bias, the rise and fall times of the Ag:AZO, Ag, and AZO devices were 31.4 and 22 ms, 26 and 43 ms, and 63 and 63.8 ms, respectively. Based on the characteristics of the fabricated photodetector, the photodetector using the 20 nm-thick Ag:AZO electrode with the lowest surface roughness exhibited the highest photo response and responsivity. Furthermore, based on the current flowing characteristics, it was inferred that the p-n junction structure was driven without an external power supply owing to the built-in potential [60,61]. As shown in Figure S6, additional measurement with a light of 1000 μW/cm^2 was conducted to confirm the sustainability of the Ag:AZO photodetector after 6 months of fabrication.

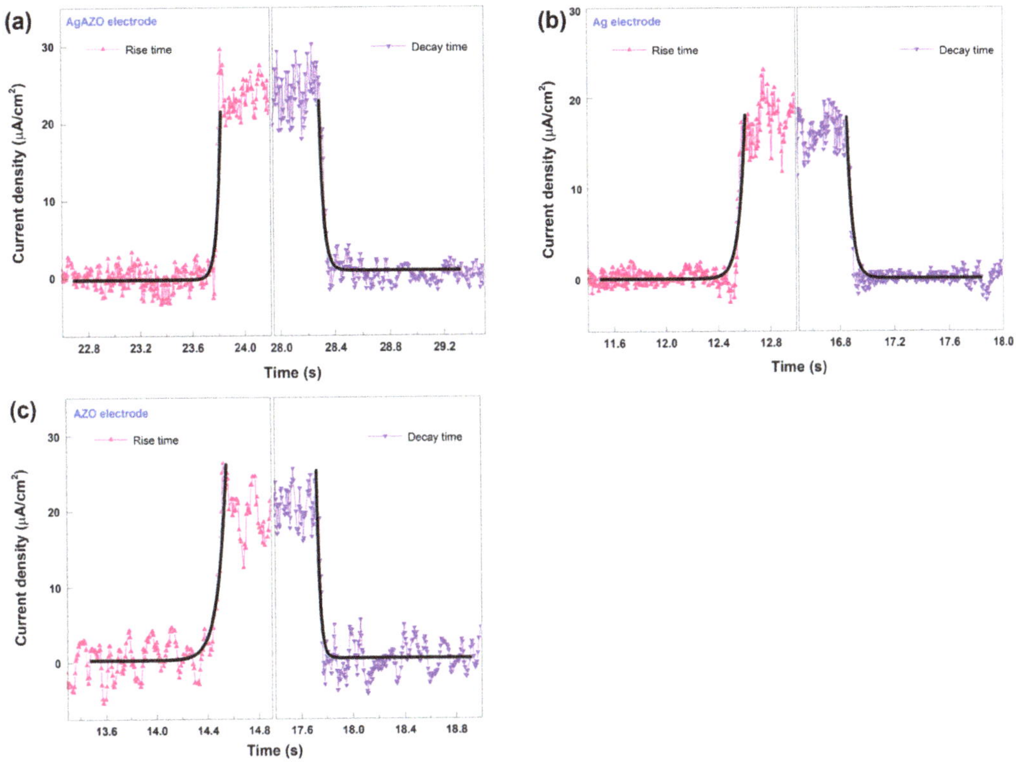

Figure 13. Rise time and decay time of time-dependent photocurrents of each photodetector irradiated with 1000 μW/cm^2 UV light. (**a**) Ag:AZO 20 nm; (**b**) Ag 20 nm; (**c**) AZO 20 nm.

4. Conclusions

The structural, optical, and crystallographic properties of the Ag:AZO electrode fabricated via co-sputtering using FTS were evaluated, and it was inferred that it had the lowest roughness at 20 nm. It was determined that the thin film with low roughness would improve photodetector performance by reducing light energy loss and the interfacial trap phenomenon. A $Ag_2O/\beta\text{-}Ga_2O_3$ photodetector was fabricated. The on/off ratio of the device was 2.01 $\times$ 10^8, and its responsivity and detectivity were 56 mA/W and 6.99 $\times$ 10^{11} Jones, respectively. The result was better than that of the previously used Ag and AZO electrodes, and similar features to the surface roughness value of each electrode were obtained. In addition, the rise and fall times of the Ag:AZO device under UV irradiation of 1000 μW/cm^2 in zero bias were 31.4 and 22 ms, respectively. Consequently, the performance of the photodetector was improved by using a thin film with low roughness to reduce the defect, which decreased the photocurrent owing to a large amount of light scattering and reflection on the surface of the film.

Supplementary Materials: The following supporting information can be downloaded at: https://www.mdpi.com/article/10.3390/nano13071287/s1, Figure S1: Thickness-dependent SEM images of Ag thin films; Figure S2: Thickness-dependent SEM images of AZO thin films; Figure S3: Structural characteristics of Ag:AZO thin film depending on thickness; Figure S4: AFM images of the Ag thin film as a function of thickness; Figure S5: AFM images of the AZO thin film as a function of thickness; Figure S6: Sustainability of Ag:AZO photodetector after 6 months.

Author Contributions: Conceptualization, Y.Y. and S.P.; methodology, Y.Y., S.P., T.P. and H.K.; software, J.H.; validation, K.K. and J.H.; formal analysis, Y.Y., S.P. and T.P.; investigation, J.H.; resources, J.H.; data curation, Y.Y. and H.K.; writing—original draft preparation, Y.Y.; writing—review and editing, Y.Y.; visualization, Y.Y.; supervision, K.K. and J.H.; project administration, J.H.; funding acquisition, J.H. All authors have read and agreed to the published version of the manuscript.

Funding: This research was supported by the Korea Institute for Advancement of Technology (KIAT) grant funded by the Korea Government (MOTIE) (P0012451, The Competency Development Program for Industry Specialist) and the Basic Science Research Capacity Enhancement Project through the Korea Basic Science Institute (National Research Facilities and Equipment Center) grant funded by the Ministry of Education (2019R1A6C1010016).

Data Availability Statement: The data is available on reasonable request from the corresponding author. The data is included in the main text and/or the Supplementary Materials.

Conflicts of Interest: The authors declare no conflict of interest.

References

1. Wolfe, R.L. Ultraviolet disinfection of potable water. *Environ. Sci. Technol.* **1990**, *24*, 768–773. [CrossRef]
2. Chen, H.; Liu, K.; Hu, L.; Al-Ghamdi, A.A.; Fang, X. New concept ultraviolet photodetectors. *Mater. Today* **2015**, *18*, 493–502. [CrossRef]
3. Xu, J.; Zheng, W.; Huang, F. Gallium oxide solar-blind ultraviolet photodetectors: A review. *J. Mater. Chem. C* **2019**, *7*, 8753–8770. [CrossRef]
4. Zhang, Z.-X.; Li, C.; Lu, Y.; Tong, X.-W.; Liang, F.-X.; Zhao, X.-Y.; Wu, D.; Xie, C.; Luo, L.-B. Sensitive deep ultraviolet photodetector and image sensor composed of inorganic lead-free $Cs_3Cu_2I_5$ perovskite with wide bandgap. *J. Phys. Chem. Lett.* **2019**, *10*, 5343–5350. [CrossRef] [PubMed]
5. Yoshikawa, A.; Ushida, S.; Nagase, K.; Iwaya, M.; Takeuchi, T.; Kamiyama, S.; Akasaki, I. High-performance solar-blind Al0.6Ga0.4N/Al0.5Ga0.5N MSM type photodetector. *Appl. Phys. Lett.* **2017**, *111*, 191103. [CrossRef]
6. Arora, K.; Goel, N.; Kumar, M.; Kumar, M. Ultrahigh performance of self-powered β-Ga_2O_3 thin film solar-blind photodetector grown on cost-effective Si substrate using high-temperature seed layer. *ACS Photonics* **2018**, *5*, 2391–2401. [CrossRef]
7. Qian, L.; Wang, Y.; Wu, Z.; Sheng, T.; Liu, X. β-Ga_2O_3 solar-blind deep-ultraviolet photodetector based on annealed sapphire substrate. *Vacuum* **2017**, *140*, 106–110. [CrossRef]
8. Park, T.; Hur, J. Self-Powered Low-Cost UVC Sensor Based on Organic-Inorganic Heterojunction for Partial Discharge Detection. *Small* **2021**, *17*, 2100695. [CrossRef]
9. Park, T.; Kim, K.; Hong, J. Effects of Drying Temperature and Molar Concentration on Structural, Optical, and Electrical Properties of β-Ga_2O_3 Thin Films Fabricated by Sol–Gel Method. *Coatings* **2021**, *11*, 1391. [CrossRef]
10. Pearton, S.; Yang, J.; Cary IV, P.H.; Ren, F.; Kim, J.; Tadjer, M.J.; Mastro, M.A. A review of Ga_2O_3 materials, processing, and devices. *Appl. Phys. Rev.* **2018**, *5*, 011301. [CrossRef]
11. Lin, T.-Y.; Han, S.-Y.; Huang, C.-Y.; Yang, C.-F.; Wei, S. Using Different Methods to Measure the Optical Energy Bandgap of Un-annealed and Annealed Ga_2O_3 Films. *Appl. Funct. Mater.* **2021**, *1*, 25–30. [CrossRef]
12. Shiojiri, D.; Fukuda, D.; Yamauchi, R.; Tsuchimine, N.; Koyama, K.; Kaneko, S.; Matsuda, A.; Yoshimoto, M. Room-temperature laser annealing for solid-phase epitaxial crystallization of beta-Ga_2O_3 thin films. *Appl. Phys. Express* **2016**, *9*, 105502. [CrossRef]
13. Jiao, S.; Lu, H.; Wang, X.; Nie, Y.; Wang, D.; Gao, S.; Wang, J. The Structural and Photoelectrical Properties of Gallium Oxide Thin Film Grown by Radio Frequency Magnetron Sputtering. *ECS J. Solid State Sci. Technol.* **2019**, *8*, Q3086–Q3090. [CrossRef]
14. Wu, C.; Guo, D.Y.; Zhang, L.Y.; Li, P.G.; Zhang, F.B.; Tan, C.K.; Wang, S.L.; Liu, A.P.; Wu, F.M.; Tang, W.H. Systematic Investigation of the Growth Kinetics of Ga_2O_3 Epilayer by Plasma Enhanced Chemical Vapor Deposition. *Appl. Phys. Lett.* **2020**, *116*, 072102. [CrossRef]
15. Park, T.; Park, S.; Park, J.H.; Min, J.Y.; Jung, Y.; Kyoung, S.; Kang, T.Y.; Kim, K.; Rim, Y.S.; Hong, J. Temperature-Dependent Self-Powered Solar-Blind Photodetector Based on Ag_2O/β-Ga_2O_3 Heterojunction. *Nanomaterials* **2022**, *12*, 2983. [CrossRef]
16. Mukhopadhyay, P.; Schoenfeld, W.V. High responsivity tin gallium oxide Schottky ultraviolet photodetectors. *J. Vac. Sci. Technol. A* **2020**, *38*, 013403. [CrossRef]
17. Bae, H.; Charnas, A.; Sun, X.; Noh, J.; Si, M.; Chung, W.; Qiu, G.; Lyu, X.; Alghamdi, S.; Wang, H. Solar-Blind UV Photodetector Based on Atomic Layer-Deposited Cu_2O and Nanomembrane β-Ga_2O_3 pn Oxide Heterojunction. *ACS Omega* **2019**, *4*, 20756–20761. [CrossRef]
18. Guo, D.; Su, Y.; Shi, H.; Li, P.; Zhao, N.; Ye, J.; Wang, S.; Liu, A.; Chen, Z.; Li, C.; et al. Self-Powered Ultraviolet Photodetector with Superhigh Photoresponsivity (3.05 A/W) Based on the GaN/Sn:Ga_2O_3 pn Junction. *ACS Nano* **2018**, *12*, 12827–12835. [CrossRef]
19. Chen, J.; Ouyang, W.; Yang, W.; He, J.; Fang, X. Recent Progress of Heterojunction Ultraviolet Photodetectors: Materials, Integrations, and Applications. *Adv. Funct. Mater.* **2020**, *30*, 1909909. [CrossRef]

20. Kim, H.; Seok, H.J.; Park, J.H.; Chung, K.B.; Kyoung, S.; Kim, H.K.; Rim, Y.S. Fully transparent InZnSnO/beta-Ga_2O_3/InSnO solar-blind photodetectors with high schottky barrier height and low-defect interfaces. *J. Alloy. Compd.* **2022**, *890*, 161931. [CrossRef]
21. Varshney, U.; Aggarwal, N.; Gupta, G. Current advances in solar-blind photodetection technology: Using Ga_2O_3 and AlGaN. *J. Mater. Chem. C* **2022**, *10*, 1573–1593. [CrossRef]
22. Zeng, G.; Li, X.X.; Li, Y.C.; Chen, D.B.; Chen, Y.C.; Zhao, X.F.; Chen, N.; Wang, T.Y.; Zhang, D.W.; Lu, H.L. A Heterostructured Graphene Quantum Dots/β-Ga2O3 Solar-Blind Photodetector with Enhanced Photoresponsivity. *ACS Appl. Mater. Interfaces* **2022**, *14*, 16846–16855. [CrossRef] [PubMed]
23. Zhu, X.; Wu, Y.; Li, G.; Zhang, K.; Feng, S.; Lu, W. Ga2O3–MXene Nanowire Networks with Enhanced Responsivity for Deep-UV Photodetection. *ACS Appl. Nano Mater.* **2023**, *6*, 2048–2062. [CrossRef]
24. Wang, H.; Ma, J.; Cong, L.; Song, D.; Fei, L.; Li, P.; Li, B.; Liu, Y. Solar-blind UV photodetector with low-dark current and high-gain based on ZnO/Au/Ga2O3 sandwich structure. *Mater. Today Phys.* **2022**, *24*, 100673. [CrossRef]
25. Chu, L.; Sui, M.A.O.; Chen, L.; Belfiore, L.A.; Tang, J. Depleting and hot-electron injection effect: Mechanism of boosting the photo-respond speed for ZnO nanorods MSM photodetectors by sandwiching Ag nanoparticles in AZO. *J. Alloy. Compd.* **2021**, *868*, 159125. [CrossRef]
26. Zhu, K.; Yang, K. Low-temperature fabrication of high-performance AlN/Ag/AlN thin films for transparent electrode applications. *Appl. Phys. A* **2022**, *128*, 1038. [CrossRef]
27. Park, S.; Park, T.; Park, J.H.; Min, J.Y.; Jung, Y.; Kyoung, S.; Kang, T.Y.; Kim, K.H.; Rim, Y.S.; Hong, J. Ag_2O/β-Ga_2O_3 Heterojunction-Based Self-Powered Solar Blind Photodetector with High Responsivity and Stability. *ACS Appl. Mater. Interfaces* **2022**, *14*, 25648–25658. [CrossRef]
28. Liu, Z.; Zou, Y.; Ji, C.; Chen, X.; Hou, G.; Zhang, C.; Wan, X.; Guo, L.J.; Zhao, Y.; Zhang, X. Broad-Spectrum Ultrathin-Metal-Based Oxide/Metal/Oxide Transparent Conductive Films for Optoelectronic Devices. *ACS Appl. Mater. Interfaces* **2021**, *13*, 58539–58551. [CrossRef] [PubMed]
29. Li, P.; Li, H.; Ma, J.; Zhou, Y.; Zhang, W.; Cong, L.; Xu, H.; Liu, Y. Structural Optimization of Oxide/Metal/Oxide Transparent Conductors for High-Performance Low-Emissivity Heaters. *Adv. Mater. Interfaces* **2018**, *5*, 1801287. [CrossRef]
30. Wang, Z.; Li, J.; Xu, J.; Huang, J.; Yang, Y.; Tan, R.; Chen, G.; Fang, X.; Zhao, Y.; Song, W. Robust Ultrathin and Transparent AZO/Ag-SnOx/AZO on Polyimide Substrate for Flexible Thin Film Heater with Temperature over 400 °C. *J. Mater. Sci. Technol.* **2020**, *48*, 156–162. [CrossRef]
31. Park, H.K.; Kang, J.W.; Na, S.I.; Kim, D.Y.; Kim, H.K. Characteristics of indium-free GZO/Ag/GZO and AZO/Ag/AZO multilayer electrode grown by dual target DC sputtering at room temperature for low-cost organic photovoltaics. *Sol. Energy Mater. Sol. Cells* **2009**, *93*, 1994–2002. [CrossRef]
32. Park, S.; Yoon, Y.; Lee, S.H.Y.; Park, T.; Kim, K.; Hong, J. Thermoinduced and Photoinduced Sustainable Hydrophilic Surface of Sputtered-TiO2 Thin Film. *Coatings* **2021**, *11*, 1360. [CrossRef]
33. Hong, J.S.; Matsushita, N.; Kim, K.H. Investigation of the effect of oxygen gas on properties of GAZO thin films fabricated by facing targets sputtering system. *Semicond. Sci. Technol.* **2014**, *29*, 075007. [CrossRef]
34. Hong, J.; Matsushita, N.; Kim, K. Effect of dopants and thermal treatment on properties of Ga-Al-ZnO thin films fabricated by hetero targets sputtering system. *Thin Solid Film.* **2013**, *531*, 238–242. [CrossRef]
35. Hong, J.S.; Jang, K.W.; Park, Y.S.; Choi, H.W.; Kim, K.H. Preparation of ZnO Based Thin Films for OLED Anode by Facing Targets Sputtering System. *Mol. Cryst. Liq. Cryst.* **2011**, *538*, 103–111. [CrossRef]
36. Fakhri, M.A. Annealing effects on opto-electronic properties of Ag_2O films growth using thermal evaporation techniques. *Int. J. Nanoelectron. Mater.* **2016**, *9*, 93–102.
37. Lee, M.; Park, Y.; Kim, K.; Hong, J. Influence of sputtering conditions on the properties of aluminum-doped zinc oxide thin film fabricated using a facing target sputtering system. *Thin Solid Film.* **2020**, *703*, 137980. [CrossRef]
38. Yoon, Y.; Katsumata, K.I.; Suzuki, N.; Nakata, K.; Terashima, C.; Kim, K.H.; Fujishima, A.; Hong, J. Rod-Shaped β-FeOOH Synthesis for Hydrogen Production under Light Irradiation. *ACS Omega* **2021**, *6*, 30562–30568. [CrossRef]
39. Zhang, C.; Zhao, D.; Gu, D.; Kim, H.; Ling, T.; Wu, Y.-K.R.; Guo, L.J. An Ultrathin, Smooth, and Low-Loss Al-Doped Ag Film and Its Application as a Transparent Electrode in Organic Photovoltaics. *Adv. Mater.* **2014**, *26*, 5696–5701. [CrossRef]
40. Zeng, G.; Zhang, M.R.; Chen, Y.C.; Li, X.X.; Chen, D.B.; Shi, C.Y.; Zhao, X.F.; Chen, N.; Zhang, D.W.; Lu, H.L. A solar-blind photodetector with ultrahigh rectification ratio and photoresponsivity based on the $MoTe_2$/Ta: β-Ga_2O_3 pn junction. *Mater. Today Phys.* **2023**, *33*, 101042. [CrossRef]
41. Wu, C.; Wu, F.; Hu, H.; Wang, S.; Liu, A.; Guo, D. Review of self-powered solar-blind photodetectors based on Ga_2O_3. *Mater. Today Phys.* **2022**, *28*, 100883. [CrossRef]
42. Waterhouse, G.I.N.; Bowmaker, G.A.; Metson, J.B. The Thermal Decomposition of Silver (I, III) oxide: A Combined XRD, FT-IR and Raman Spectroscopic Study. *Phys. Chem. Chem. Phys.* **2001**, *3*, 3838–3845. [CrossRef]
43. Zhong, W.; Li, G.; Lan, L.; Li, B.; Chen, R. Effects of annealing temperature on properties of InSnZnO thin film transistors prepared by Co-sputtering. *RSC Adv.* **2018**, *8*, 34817–34822. [CrossRef] [PubMed]
44. Zhang, Q.; Li, N.; Zhang, T.; Dong, D.; Yang, Y.; Wang, Y.; Dong, Z.; Shen, J.; Zhou, T.; Liang, Y.; et al. Enhanced gain and detectivity of unipolar barrier solar blind avalanche photodetector via lattice and band engineering. *Nat. Commun.* **2023**, *14*, 418. [CrossRef]

45. Michaelson, H.B. The work function of the elements and its periodicity. *J. Appl. Phys.* **1977**, *48*, 4729–4733. [CrossRef]
46. Venkata Krishna Rao, R.; Ranade, A.K.; Desai, P.; Kalita, G.; Suzuki, H.; Hayashi, Y. Temperature-dependent device properties of γ-CuI and β-Ga_2O_3 heterojunctions. *SN Appl. Sci.* **2021**, *3*, 796. [CrossRef]
47. Ohta, H.; Asai, N.; Horikiri, F.; Narita, Y.; Yoshida, T.; Mishima, T. Impact of Threading Dislocations in GaN p-n Diodes on Forward I-V Characteristics. *Jpn. J. Appl. Phys.* **2020**, *59*, 106503. [CrossRef]
48. Omar, S.U.; Sudarshan, T.S.; Rana, T.A.; Song, H.; Chandrashekhar, M. Interface Trap-Induced Nonideality in As-Deposited Ni/4H-SiC Schottky Barrier Diode. IEEE Trans. *Electron. Devices* **2015**, *62*, 615–621. [CrossRef]
49. Kim, Y.; Ha, T.-K.; Cho, Y.-J.; Kang, Y.-S.; Yu, S.; Kim, G.; Jeong, H.; Park, J.K.; Kim, O. Severe hump phenomenon induced by increased charge trapping and suppression of electron capture effect in amorphous In-Ga-Zn-O thin-film transistors under unipolar pulsed drain bias with static positive gate bias stress. *Solid-State Electron.* **2020**, *167*, 107785. [CrossRef]
50. Chen, Y.; Yang, X.; Zhang, Y.; Chen, X.; Sun, J.; Xu, Z.; Li, K.; Dong, L.; Shan, C. Ultra-sensitive flexible Ga_2O_3 solar-blind photodetector array realized via ultra-thin absorbing medium. *Nano Res.* **2021**, *15*, 3711–3719. [CrossRef]
51. Zhao, B.; Wang, F.; Chen, H.; Wang, Y.; Jiang, M.; Fang, X.; Zhao, D. Solar-blind avalanche photodetector based on single ZnO–Ga_2O_3 core–shell microwire. *Nano Lett.* **2015**, *15*, 3988–3993. [CrossRef]
52. Wu, C.; Wu, F.; Hu, H.; Ma, C.; Ye, J.; Wang, S.; Wu, H.; Wang, J.; Liu, A.; Guo, D. Work function tunable laser induced graphene electrodes for Schottky type solar-blind photodetectors. *Appl. Phys. Lett.* **2022**, *120*, 101102. [CrossRef]
53. Zheng, W.; Lin, R.; Ran, J.; Zhang, Z.; Ji, X.; Huang, F. Vacuum-Ultraviolet Photovoltaic Detector. *ACS Nano* **2018**, *12*, 425–431. [CrossRef]
54. Chen, M.; Ma, J.; Li, P.; Xu, H.; Liu, Y. Zero-Biased Deep Ultraviolet Photodetectors Based on Graphene/Cleaved (100) Ga_2O_3 Heterojunction. *Opt. Express* **2019**, *27*, 8717–8726. [CrossRef]
55. Yu, J.; Dong, L.; Peng, B.; Yuan, L.; Huang, Y.; Zhang, L.; Zhang, Y.; Jia, R. Self-Powered Photodetectors Based on β-Ga_2O_3/4H–SiC Heterojunction with Ultrahigh Current On/Off Ratio and Fast Response. *J. Alloy. Compd.* **2020**, *821*, 153532. [CrossRef]
56. Li, S.; Zhi, Y.; Lu, C.; Wu, C.; Yan, Z.; Liu, Z.; Yang, J.; Chu, X.; Guo, D.; Li, P. Broadband Ultraviolet Self-Powered Photodetector Constructed on Exfoliated β-Ga_2O_3/CuI Core–Shell Microwire Heterojunction with Superior Reliability. *J. Phys. Chem. Lett.* **2021**, *12*, 447–453. [CrossRef] [PubMed]
57. Wu, C.; Qiu, L.; Li, S.; Guo, D.; Li, P.; Wang, S.; Du, P.; Chen, Z.; Liu, A.; Wang, X. High Sensitive and Stable Self-Powered Solar-Blind Photodetector Based on Solution-Processed All Inorganic $CuMO_2$/Ga_2O_3 pn Heterojunction. *Mater. Today Phys.* **2021**, *17*, 100335. [CrossRef]
58. Zhuo, R.; Wu, D.; Wang, Y.; Wu, E.; Jia, C.; Shi, Z.; Xu, T.; Tian, Y.; Li, X. A Self-Powered Solar-Blind Photodetector Based on a MoS2/β-Ga_2O_3 Heterojunction. *J. Mater. Chem.* **2018**, *6*, 10982–10986. [CrossRef]
59. Guo, D.; Liu, H.; Li, P.; Wu, Z.; Wang, S.; Cui, C.; Li, C.; Tang, W. Zero-Power-Consumption Solar-Blind Photodetector Based on β-Ga_2O_3/NSTO Heterojunction. *ACS Appl. Mater. Interfaces* **2017**, *9*, 1619–1628. [CrossRef]
60. Kaur, D.; Debata, S.; Singh, D.P.; Kumar, M. Strain effects on the optoelectronic performance of ultra-wide band gap polycrystalline β-Ga_2O_3 thin film grown on differently-oriented Silicon substrates for solar blind photodetector. *Appl. Surf. Sci.* **2023**, *616*, 156446. [CrossRef]
61. Zhang, C.; Liu, K.; Ai, Q.; Sun, X.; Chen, X.; Yang, J.; Zhu, Y.; Cheng, Z.; Li, B.; Liu, L.; et al. High-performance fully transparent Ga_2O_3 solar-blind UV photodetector with the embedded indium–tin–oxide electrodes. *Mater. Today Phys.* **2023**, *33*, 101034. [CrossRef]

nanomaterials

Article

Highly Efficient and Stable Self-Powered Perovskite Photodiode by Cathode-Side Interfacial Passivation with Poly(Methyl Methacrylate)

Wonsun Kim [1], JaeWoo Park [2], Yushika Aggarwal [1], Shital Sharma [2], Eun Ha Choi [1,2] and Byoungchoo Park [1,2,*]

[1] Department of Electrical and Biological Physics, Kwangwoon University, Seoul 01897, Republic of Korea
[2] Department of Plasma-Bio Display, Kwangwoon University, Seoul 01897, Republic of Korea
* Correspondence: bcpark@kw.ac.kr; Tel.: +82-2-940-5237

Abstract: For several years now, organic–inorganic hybrid perovskite materials have shown remarkable progress in the field of opto-electronic devices. Herein, we introduce a cathode-side passivation layer of poly(methyl methacrylate) (PMMA) for a highly efficient and stable self-powered $CH_3NH_3PbI_3$ perovskite-based photodiode. For effective noise–current suppression, the PMMA passivation layer was employed between a light-absorbing layer of $CH_3NH_3PbI_3$ ($MAPbI_3$) perovskite and an electron transport layer of [6,6]-phenyl-C61-butyric acid methyl ester. Due to its passivation effect on defects in perovskite film, the PMMA passivation layer can effectively suppress interface recombination and reduce the leakage/noise current. Without external bias, the $MAPbI_3$ photodiode with the PMMA layer demonstrated a significantly high specific detectivity value (~1.07×10^{12} Jones) compared to that of a conventional $MAPbI_3$ photodiode without a PMMA layer. Along with the enhanced specific detectivity, a wide linear dynamic response (~127 dB) with rapid rise (~50 μs) and decay (~17 μs) response times was obtained. Furthermore, highly durable dynamic responses of the PMMA-passivated $MAPbI_3$ photodiode were observed even after a long storage time of 500 h. The results achieved with the cathode-side PMMA-passivated perovskite photodiodes represent a new means by which to realize highly sensitive and stable self-powered photodiodes for use in developing novel opto-electronic devices.

Keywords: photodetector; polymeric passivation layer; organic/inorganic hybrid perovskite

Citation: Kim, W.; Park, J.W.; Aggarwal, Y.; Sharma, S.; Choi, E.H.; Park, B. Highly Efficient and Stable Self-Powered Perovskite Photodiode by Cathode-Side Interfacial Passivation with Poly(Methyl Methacrylate). *Nanomaterials* **2023**, *13*, 619. https://doi.org/10.3390/nano13030619

Academic Editors: Qiongfeng Shi, Jianxiong Zhu and Julia Pérez-Prieto

Received: 19 January 2023
Revised: 29 January 2023
Accepted: 1 February 2023
Published: 3 February 2023

1. Introduction

Energy-harvesting photovoltaic (PV) technologies based on hybrid organic/inorganic perovskite materials have attracted the interest of many research groups due to the excellent optoelectronic characteristics, high charge carrier mobility, tunable bandgaps, and high diffusion ranges of these materials. Moreover, they are simple, operate at low temperatures, are solution processability, and have a low fabrication cost [1–6]. Given these characteristics, hybrid organic/inorganic perovskite materials are utilized in various optoelectronic devices such as PV or solar cells, light-emitting diodes, and especially in photodetectors [7–9]. In recent years, $CH_3NH_3PbI_3$ ($MAPbI_3$) organohalide perovskite photodetectors have shown significantly increased photo-detecting capabilities with rapid response times, with detectivity (D^*) values of ~7.8×10^{12} Jones (Jones = cm $Hz^{1/2}$/W) and responsivity (R) values close to 470 mA/W [10–13]. These results are slightly lower than or comparable to those of conventional silicon photodetectors (D^* ~1×10^{13} Jones) [14].

Despite the advent of these new types of device structures and novel materials for improving the performance capabilities of perovskite-based devices, deeply rooted problems including defect-rich and distorted lattice structures in the bulk and surfaces of the $MAPbI_3$ perovskite layers remain unsolved [15–20]. Specifically, defects in perovskite layers hamper the charge transport and extraction processes, causing interfacial recombination losses between the light-absorbing layer and adjacent charge transport layers and/or electrodes [17–20]. Due to these losses, much work remains for further improvements of the optoelectronic device performance and stability over conventional silicon-based devices [21,22].

Among the attempts to improve the interface quality of perovskite layers in PV cells, passivation layers consisting of functional organics or polymers such as poly(ethylene oxide) and poly(methyl methacrylate) (PMMA) have been suggested and successfully demonstrated. These were found to possess effective interface engineering based on their chemical and physical/electrical advantages [17–19,23,24]. Recently, the present authors also showed a self-powered $MAPbI_3$ perovskite-based photodiode device with an anode-side passivation layer of PMMA at the interface between a $MAPbI_3$ light-absorbing layer and a hole-transport layer (HTL) [25]. The anode-side PMMA passivation layer could reduce carrier recombination losses and noise currents. Thus, high specific detectivity (~0.81×10^{12} Jones) was achieved even without external bias [25].

Nonetheless, the introduction of a PMMA passivation layer between the light-absorbing layer and the electron-transport layer (ETL) for perovskite photodiodes has not been fully investigated and remains poorly understood [9,20]. Unlike the formation of the aforementioned anode-side passivation layer [25], a cathode-side passivation layer can be simultaneously formed during the antisolvent process used to crystallize the $MAPbI_3$ layer [26]. Even without an additional coating process, issues related to the grain boundary and the problem of defects at the interface between the perovskite light-absorbing layer and ETL can be solved to realize highly efficient perovskite photodiodes. Additionally, dark and/or noise currents, which directly affect the responsivity and specific detectivity of photodetectors, can also be controlled by the cathode-side passivation layer. Thus, it is crucial to select proper interface engineering techniques for the perovskite light-absorbing layer, as doing so can improve the photocurrent and reduce the leakage and/or noise currents.

Herein, a high-performance self-powered $MAPbI_3$ perovskite photodiode with a cathode-side passivation layer of PMMA was systematically investigated. To assess the optoelectronic characteristics, we studied the responsivity, detectivity, and linear dynamic response range of $MAPbI_3$ perovskite photodiodes under a self-powered condition. Along with these observations, we measured the response times (rise/decay times) and stability characteristics of the photodiodes. Furthermore, we evaluated the device characteristics of the reference perovskite photodiodes without a PMMA passivation layer to support our results. The obtained results highlight the superior device performance of the perovskite photodetector with cathode-side PMMA passivation compared to a conventional reference device as well as previous perovskite photodetectors that rely on anode-side PMMA passivation.

2. Materials and Methods

2.1. Materials

Nickel(II) nitrate hexahydrate ($Ni(NO_3)_2{\cdot}6H_2O$, 99.999%), anhydrous dimethyl sulfoxide ($(CH_3)_2SO$, DMSO, 99.9%), *N,N*-dimethylformamide ($HCON(CH_3)_2$, DMF, 99.8%), anhydrous mono chlorobenzene (C_6H_5Cl, CB, 99.8%), and anhydrous isopropyl alcohol ($(CH_3)_2CHOH$, IPA, 99.7%) were purchased from Sigma-Aldrich Korea (Seoul, Republic of Korea) PMMA ($(C_5O_2H_8)_n$, 950,000 M_w) was purchased from Kayaku Advanced Materials (Westborough, MA, USA). Methyl ammonium iodide (CH_3NH_3I, MAI) and lead(II) iodide (PbI_2, 99.9985%) for the perovskite precursor were purchased from Greatcell Solar (Queanbeyan, NSW, Australia) and Alfa Aesar (Haverhill, MA, USA), respectively. Ethylene glycol ($HO(CH_2)_2OH$, EG, 99%) and IPA for the nickel oxide (NiO_x) precursor were purchased from Daejung Chemicals and Metals (Siheung, Republic of Korea). Phenyl-C61-buthyric acid methyl ester ($C_{72}H_{14}O_2$, $PCBM_{60}$) was purchased from Nano-C (Westwood, MA, USA). Bathocuproine ($C_{26}H_{20}N_2$, BCP, 98%) and a colloidal suspension of ZnO nanoparticles were procured from Tokyo Chemical Industry Co. Ltd. (Tokyo, Japan) and Nanograde (Zurich, Switzerland), respectively. All materials were used as received without further purification.

2.2. Methods

Pre-patterned 80-nm-thick indium tin oxide (ITO, 20 Ω/sq) layers on glass substrates were used as transparent anodes for the photodiodes. The ITO substrates used here were ultrasonically cleaned with ethanol, a detergent, and deionized (DI) water, after which they underwent an ultraviolet ozone treatment in an oven for 10 min.

For the NiO_x HTL, a precursor solution with 29 mg of $Ni(NO_3)_2 \cdot 6H_2O$ dissolved in mixed solvents of EG and IPA at a 6:4 volume ratio was prepared by stirring for 3 h. The precursor solution was then spin-coated onto the ITO substrate at 2600 rpm for 40 s, followed by 60 min of annealing at 300 °C to form the a 10-nm-thick NiO_x HTL. Subsequently, the substrates coated with the NiO_x layer were placed in a nitrogen-filled glovebox.

To fabricate the perovskite light-absorbing active layers of $MAPbI_3$, an antisolvent-assisted spin-coating method was used [25,26]. The perovskite precursor solution was prepared by dissolving PbI_2 and MAI at a 1:1 molar ratio in a mixed polar solvent of DMF and DMSO at an 8:2 volume ratio, with the precursor solution then stirred overnight. In the nitrogen-filled glovebox, the precursor solution was spin-coated onto the NiO_x-coated substrate at 4400 rpm for 30 s. During the spinning process, anhydrous CB or CB containing 0.05 wt% of PMMA was dropped onto the perovskite precursor-coated substrate as an antisolvent at a dripping delay time of 5–8 s after the spinning process had started. When the dripping time was changed from 5–8 s, the film quality of the fabricated perovskite layer was degraded and the film became inhomogeneous and hazy. Hence, in this study, we set the dripping time to 5–8 s to fabricate a uniform and homogeneous perovskite precursor layer without or with a cathode-side PMMA passivation layer. The perovskite precursor layer was then dried at room temperature for 5 min and subsequently annealed at 100 °C for 20 min to crystallize the $MAPbI_3$. The thicknesses of the $MAPbI_3$ layers studied here were nearly identical to each other (~250 nm), and the thickness of the PMMA layer was estimated to be less than 5 nm.

Thereafter, $PCBM_{60}$ in CB and a colloidal suspension of ZnO nanoparticles were subsequently spin-coated above the coated layers, resulting in a uniform 50-nm-thick $PCBM_{60}$ layer and a 20-nm-thick layer of ZnO as ETLs. The substrates were then transferred to a vacuum chamber for thermal evaporation of the 12-nm-thick BCP layer and 70-nm-thick Al cathode layer, subsequently at a base pressure below 2.0×10^{-6} torr. Thus, a photodiode with the structure of [ITO/NiO_x/$MAPbI_3$/PMMA/$PCBM_{60}$/ZnO/BCP/Al] was fabricated. All fabricated photodiodes had an active area of 6 mm^2.

2.3. Characterization

The surface morphology of the $MAPbI_3$ perovskite layer was analyzed with a scanning electron microscope (Inspect F50, FEI Company, Hillsboro, OR, USA). ImageJ software (National Institutes of Health, Bethesda, MD, USA) was used to analyze the grain size distributions in the $MAPbI_3$ perovskite layers from the observed SEM images.

The UV–Visible optical adsorption spectra were studied using an UV–Vis spectrometer (Agilent 8453 Diode Array UV-VIS, Agilent Technologies, Inc., Santa Clara, CA, USA). The contact angles of the fabricated $MAPbI_3$ films were measured with a contact angle goniometer (Ossila, London, UK).

A monochromatic light source of a 637 nm diode laser (model COMPACT-100G-637-A, 100 mW, maximum internal modulation frequency: 50 kHz, World Star Tech, Markham, OT, Canada) was used to assess the performance of the photodetector. Keithley source meters (models 2400 and 2636, Tektronix, Inc. Beaverton, OR, USA) were used to assess the current-versus-voltage (*J–V*) characteristics. An incident photon-to-current conversion efficiency (IPCE) measurement system (ORIEL IQE-200, Newport, Irvine, CA, USA) was used to measure the spectral responsivity, R(λ), of the photodiode.

The noise current levels of the photodiode were obtained from the fast Fourier transform of the measured dark currents of the photodiode as a function of time using the 2636 Keithley source meter operated at a sampling rate of 1 kHz. The 3 dB cutoff bandwidths of the photodiodes were estimated from the logarithmic transform of their normalized photo responses, measured as a function of the modulation frequency of the irradiated light using the 637 nm laser system.

3. Results

3.1. Characteristics of $MAPbI_3$ Layers with a PMMA Passivation Layer

A schematic illustration of the passivation layer of the PMMA on a $MAPbI_3$ perovskite layer is shown in Figure 1a. As illustrated in the figure, PMMA dissolved in an antisolvent can simply be spin-coated during the formation of a 250-nm-thick $MAPbI_3$ perovskite light-absorbing layer. To verify the formation of the PMMA passivation layer on the perovskite layer, first, we observed the contact angles of DI water on the fabricated films (Figure 1b). As shown in the figure, the water contact angle on the $NiO_x/MAPbI_3$/PMMA perovskite film (sample) was 37.4°, significantly higher than that (~26.8°) of the $NiO_x/MAPbI_3$ perovskite film (reference). This increase in the water contact angle mainly stems from the intrinsic hydrophobicity of the PMMA layer in the sample film, providing evidence of the simple and reliable fabrication method of the PMMA passivation layer on the perovskite layer. Thus, such a hydrophobic PMMA passivation layer may affect the growth of the grains of the underlying $MAPbI_3$ perovskite by tailoring the nucleation of the perovskite crystal growth [3,13,27]. It is noteworthy that the coated PMMA polymers mainly existed at the perovskite/ETL interface, with some of the polymer possibly present inside the perovskite layer, similar to an earlier conjugated polymer that formed an interfacial passivation layer [20].

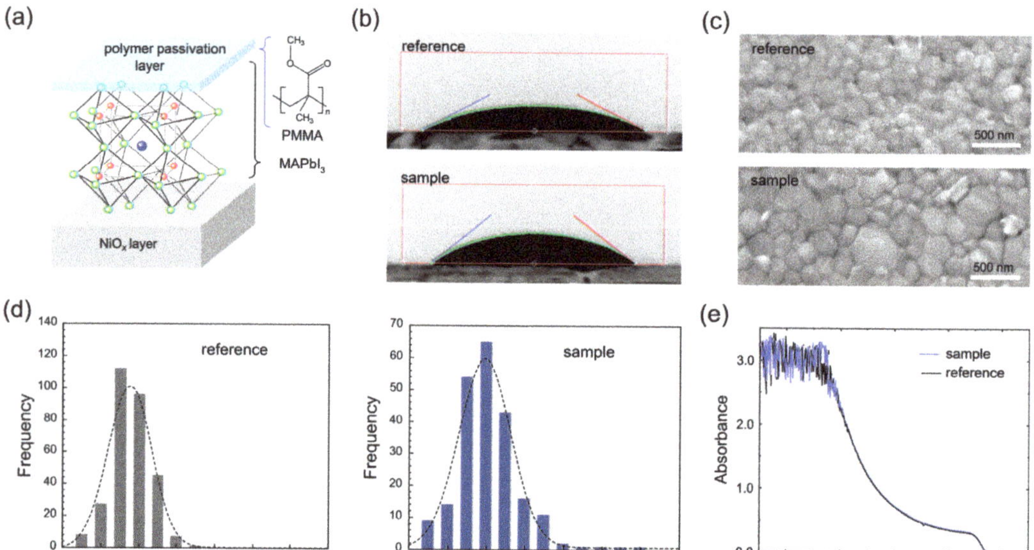

Figure 1. (**a**) Schematic illustration of the polymer passivation layer of PMMA on a $MAPbI_3$ perovskite layer and corresponding chemical structure. (**b**) Comparisons of the water contact angles on the $NiO_x/MAPbI_3$ perovskite (reference) and $NiO_x/MAPbI_3$ perovskite/PMMA (sample) layers, corresponding to (**c**) high-magnification top-view SEM images and (**d**) grain size distributions derived from the SEM images. (**e**) UV–Visible optical absorption spectra of the reference and sample layers.

To assess the effect of the PMMA passivation layer on the growth of grains for the $MAPbI_3$ perovskite films, scanning electron microscopy (SEM) was utilized. The surface morphologies obtained from the SEM observations are shown in Figure 1c. As shown in these SEM images, the surfaces of both perovskite films were fairly smooth and showed well-packaged grains. However, compared to the reference film without a passivation layer, it is clear that the introduction of the PMMA layer affected the increment of the grain size while also decreasing the number of grain sites in the sample perovskite film.

For an additional quantitative analysis, ImageJ software was used to adjust the contrast of the SEM images of the $MAPbI_3$ perovskite layers without/with a PMMA passivation layer (Figure 1c) and to identify the borders of the grains in the SEM images. Then, utilizing the black and white threshold of each adjusted SEM image, it was possible to estimate the range of the domain sizes of the surrounding grains. Histograms of the domain size distributions for $MAPbI_3$ perovskite layers without/with a PMMA passivation layer are presented in Figure 1d. As shown in this figure, the average grain size of the reference film was ~151 nm, while that of the sample film was ~196 nm, clearly demonstrating that the average grain size of the $MAPbI_3$ layer was increased significantly due to the presence of the PMMA passivation layer. From these SEM studies, it was confirmed that the generation of unnecessary $MAPbI_3$ nucleation could be significantly suppressed by the coordinate bonding of the carbonyl of the PMMA polymer with the uncoordinated Pb ions of $MAPbI_3$ [28–30]. Thus, the PMMA passivation layer enables an increase in the $MAPbI_3$ grain size while also mitigating film defects such as spikes and/or pinholes [31,32].

Apart from the increased grain size of the $MAPbI_3$ perovskite layer, the PMMA passivation layer can effectively cover defects and grain boundaries in the perovskite layer, possibly providing an effective passivation route for interface defects between the $MAPbI_3$ layer and the adjacent ETL [33,34]. Moreover, due to the improved interface quality of the $MAPbI_3$ layer stemming from the PMMA, the reduction in the recombination loss of charge carriers can improve the charge carrier extraction from the $MAPbI_3$ layer to the charge transport layers [31,32,34–36]. Thus, given these properties of the $MAPbI_3$ layer in the sample films, we hold that the introduction of a PMMA passivation layer can effectively improve the film and interface qualities of the $MAPbI_3$ layer, likely leading to an improvement in the optoelectronic performance of perovskite devices, even without any critical changes in the optical characteristics such as the optical absorption capabilities (Figure 1e) [25]. To confirm this, the optical absorption characteristics of the reference and sample perovskite films were also investigated, as shown in Figure 1e. In this figure, strong absorption in the visible range of 450~700 nm for the reference and sample films could be observed. Notably, the sample film showed an optical absorption nearly identical to that of the reference perovskite film, proving that the PMMA passivation layer did not significantly change or deteriorate the optical absorption properties of the perovskite layer. Thus, by introducing the PMMA passivation layer onto the perovskite layer, modification of the interface between the $MAPbI_3$ layer and the ETL can be anticipated.

3.2. Characterization of Perovskite Photodiodes without and with a PMMA Passivation Layer

Inspired by the desirable passivation effect of PMMA described above, we fabricated $MAPbI_3$ photodiodes with the device structure illustrated in Figure 2a. As shown in the figure, ITO acts as a transparent anode, with NiO_x used as a HTL, $MAPbI_3$ as a perovskite light-absorbing layer, and PMMA as the cathode-side passivation layer. $PCBM_{60}$, ZnO nanoparticles and BCP were utilized as the ETLs and Al was used as the top cathode electrode.

To evaluate the device performance of the fabricated $MAPbI_3$ photodiodes without (reference) and with a PMMA passivation layer (sample), the dark-current densities (J_{dark}) were investigated as a function of the applied voltage (J_{dark}-V). As indicated in Figure 2b, excellent diode characteristics with large rectifying ratios (*RR*s) and low leakage current densities were found in these photodiodes. Due to the suppression of the leakage currents by the PMMA passivation layer, a somewhat higher *RR* value, 6.1×10^5 (at 1.0 V), for the sample was obtained compared to that (5.3×10^5) of the reference, indicative of the useful film properties of the $MAPbI_3$ perovskite films, as above-mentioned. It is also noteworthy that the observed *RR* values were one order higher than the corresponding values in the literature [13]. Similarly, the PMMA passivation layer in the sample led to a reduced average J_{dark} value of 5.1×10^{-7} mA/cm^2 at zero applied voltage, lower than that (7.3×10^{-7} mA/cm^2) of the reference without a PMMA layer, providing evidence of the suppressed leakage currents caused by the improved interfacial quality of the perovskite light-absorbing layer.

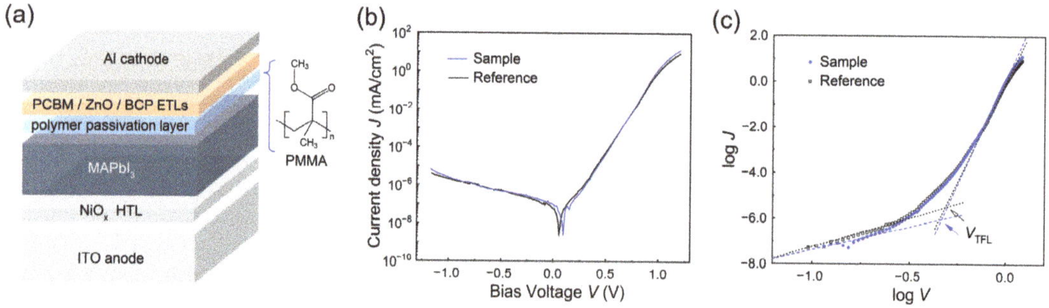

Figure 2. (**a**) Schematic device configuration of a $MAPbI_3$ photodiode with a cathode-side PMMA passivation layer. Dark-current density-voltage (J_{dark}-V) characteristics of $MAPbI_3$ photodiodes without (reference) and with a cathode-side PMMA passivation layer (sample) on (**b**) a semi-log scale and (**c**) a log–log scale.

Furthermore, for an in-depth examination of the perovskite photodiodes presented here, their trap-filled limit voltages (V_{TFL}s) were estimated from the J_{dark}–V curves on a log–log scale (Figure 2c) [13,20,32,37,38]. The obtained V_{TFL} values for the reference and sample were ~0.49 and 0.47 V, respectively. Based on the V_{TFL} values, the trap density states (n_{trap}) can also be estimated using the space–charge–limited current model, as follows [38,39]:

$$V_{TFL} = \frac{q n_{trap} L^2}{2 \varepsilon_0 \varepsilon_p} \quad (1)$$

In this equation, q is the elementary charge, L is the thickness of the perovskite layer, while ε_0 and ε_p are the dielectric constants of the vacuum and perovskite, respectively. The estimated n_{trap} values were around 1.30×10^{15} cm^{-3} for the reference and 1.25×10^{15} cm^{-3} for the sample, indicating that the PMMA layer in the sample could effectively passivate defects at the interface between the perovskite layer and the ETL.

Subsequently, we measured the photocurrents as a function of the bias voltage in the range of $\pm$1.2 V (I_{light} -V) for several input power levels (Ps) of incident laser light with a wavelength of λ = 637 nm, as shown in Figure 3a. In the figure, the open-circuit potential (V_{OC}) and the short-circuit current (I_{SC}) increased considerably as the input power of the incident light was increased. For a detailed comparison, the built-in potentials (V_{bi}s) of the reference and sample were obtained using the Shockley diode model equation, expressed as follows,

$$V_{bi} = -\left(\frac{n k_B T}{e}\right) \ln J_0, \quad (2)$$

where k_B is the Boltzmann's constant; T is the temperature; n is the ideality factor; and J_0 is the reverse-saturated current density [13,40–42]. The estimated values of V_{bi} and n were approximately 0.58 V and 1.49, respectively, for the sample, slightly higher than the V_{bi} value of approximately 0.57 V and the n value of about 1.47 for the reference. As the thickness of the PMMA layer was considered to be less than ~5 nm, it is clear that the PMMA passivation layer did not degrade the built-in potential despite its intrinsic insulating properties. It should be noted that the short-circuit current I_{SC} and built-in potential V_{bi} began to decrease noticeably as the thickness of each functional layer, especially the PMMA layer or the $MAPbI_3$ layer, was changed from its optimized thickness.

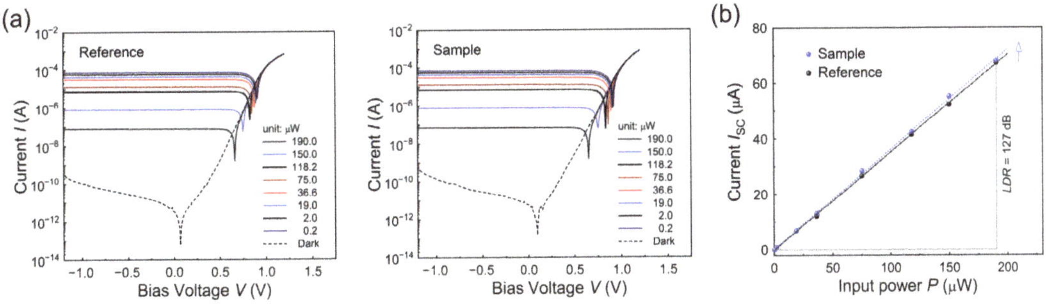

Figure 3. (**a**) Photocurrent characteristics of the $MAPbI_3$ photodiodes of the reference (**left**) and sample (**right**) as a function of the bias voltage for several different intensity levels of irradiating light with a wavelength of 637 nm. (**b**) I_{SC} curves as a function of the input power of incident light (637 nm) for the reference and sample photodiodes.

Next, the I_{SC} values of the reference and sample were determined using Figure 3a and are plotted as a function of the input power of the incident laser light (Figure 3b). It is clear from the figure that even under the self-powered condition (at zero bias voltage), large numbers of charge carriers become separated by the built-in potential, resulting in a considerable increment of I_{SC}. With the power law of

$$I_{SC} = \kappa \times P^{\theta}, \quad (3)$$

where κ is a proportional constant and θ is the power law index, the trap states existing in the $MAPbI_3$ perovskite light-absorbing layer studied here can be analyzed [13,20,43]. We obtained θ values for both the reference and sample using the power law above with the best fitted parameter values, determining an θ value of 0.999 for the sample, which is much closer to θ = 1.0 for an ideal photodiode compared to the value of θ = 0.984 for the reference without a PMMA layer. Hence, the cathode-side PMMA passivation layer can effectively decrease the number of trap states and reduce the second-order recombination loss under a short-circuit condition (i.e., self-powered condition).

Next, based on the photocurrent data in Figure 3b, the responsivity at the wavelength of λ (R_λ) for the reference and sample were estimated using the equation.

$$R_\lambda = \left(\frac{I_{PH}}{P}\right) \text{ and } I_{PH} = I_{light} - I_{dark} \quad (4)$$

where I_{PH} denotes the net photocurrent [7,9,12–14,20]. The responsivity R_{637} value at the wavelength λ = 637 nm of incident light under zero bias voltage as obtained here was approximated as 360 mA/W for the sample with the PMMA passivation layer, higher than that (~352 mA/W) of the reference at zero bias voltage.

To evaluate the photodiode performance over a wide light intensity range, we also calculated the linear dynamic range (*LDR*) of the photodiodes using the equation [7,9,12,14,20]:

$$LDR = 20\log\left(\frac{I_{PH}}{I_{dark}}\right) \quad (5)$$

Even at zero bias voltage, a notably higher *LDR* value was realized, ~127 dB, for the sample compared to that (~124 dB) of the reference. These high values of R_λ and *LDR* are clear evidence of the excellent photoelectric conversion ability and remarkably good linearity over a wide range of incident light intensity for the sample with the cathode-side PMMA passivation layer. It should also be noted that the sample device showed a relatively high *LDR* value compared to those of previous perovskite thin-film photodetectors and industrial silicon photodetectors (90~120 dB). Given the nearly ideal value of θ and the high values of R_λ and LDR, the cathode-side PMMA passivation layer can serve to realize trap-less perovskite photodiodes.

For a further evaluation of the photodiode performance, the R_λ spectra were also measured using an IPCE system in the self-powered condition, as presented in Figure 4a. This figure indicates that the peak R_λ value of ~401 mA/W for the sample was notably higher than that (386 mA/W) for the reference and higher than those of the $MAPbI_3$-based photodiodes in the literature [9,13,25]. This finding clearly shows that the improved interface of the $MAPbI_3$ layer for efficient charge extraction is crucial when attempting to realize high responsivity from perovskite photodiodes.

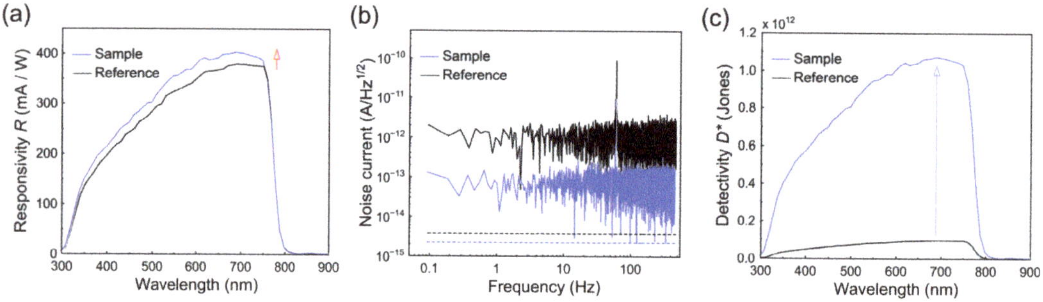

Figure 4. (**a**) Responsivity R_λ spectra at zero bias voltage, (**b**) noise currents vs. frequency characteristics, and (**c**) specific detectivity D^* spectra at zero bias voltage for the photodiodes of the reference and sample. The dotted lines in (**b**) show the shot noise levels for the reference (black) and sample (blue).

Subsequently, we analyzed the noise current (i_n) of the photodiodes. At an ambient temperature, the dark currents (I_{dark}) were measured over time with no voltage applied. Subsequently, the dark currents were fast Fourier transformed and plotted as a function of the frequency, as shown in Figure 4b [7,10,11,20,44], where the photodiodes exhibited white noise unrelated to the frequency in the ranges observed here. Additionally, as shown in the figure, the noise level of each photodiode was evaluated at a bandwidth of 1 Hz. The estimated noise level of the sample was ~0.09 pA $Hz^{-1/2}$ at a bandwidth of 1 Hz, somewhat lower than that (0.96 pA $Hz^{-1/2}$) of the reference. This indicates that the proposed cathode-side PMMA passivation layer clearly suppresses the noise level by passivating unwanted film defects at the interface between the $MAPbI_3$ layer and the $PCBM_{60}$ ETL.

Next, the obtained values of i_n and R_λ were applied to assess the noise equivalent power (NEP) in an effort to determine the minimum detectable incident optical power using the relationship of [7,9,11,12,14,45,46]:

$$NEP = \frac{i_n}{R_\lambda} \tag{6}$$

The obtained minimum NEP value for the sample was ~230 fW/$Hz^{1/2}$, notably less than that (~2520 fW/$Hz^{1/2}$) for the reference at zero bias voltage. This clearly indicates that the cathode-side PMMA passivation layer can enhance the low-power-detection ability of the $MAPbI_3$ photodiode compared to a non-passivated $MAPbI_3$ photodiode.

Based on the NEP values presented above, the specific detectivity D^* values of the photodiodes were also evaluated under the self-powered condition in order to evaluate their weak-signal detection capacities. This was carried out with the following equation,

$$D^* = \frac{\sqrt{A \cdot B}}{NEP}, \tag{7}$$

where A and B are the active area of the device and the 1 Hz specific bandwidth, respectively [7,9–14,20,44–49]. In the self-powered condition, the estimated specific detectivity D^* spectra for the reference and sample using Equation (7) are plotted in Figure 4c. As shown in this figure, while the peak value of D^* was ~0.97 × 10^{11} Jones for the reference, the peak value of D^* for the sample was ~1.07 × 10^{12} Jones, which is more than eleven times greater for weak-signal detection compared to the reference. Hence, the suppression of unnecessary leakage/noise currents by the cathode-side PMMA passivation layer clearly increased the photodetector performance outcomes. Additionally, the D^* value of 1.07 × 10^{12} Jones for the sample was conspicuously higher than those (D^* value ~0.81 × 10^{12} Jones) of the $MAPbI_3$ photodiodes in the literature with an anode-side PMMA passivation layer at the interface between the $MAPbI_3$ perovskite light-absorbing layer and the NiO_x HTL; this was also comparable to those of commercial silicon photodetectors [14,20,25,47].

As another comparison, we also estimated the conventional simplified specific detectivity D^* values of the reference and sample based on the shot noise ($i_{n,s}$), as estimated from the dark-current density (J_{dark}) with the simple relationship of $D^* \sim R_\lambda/\sqrt{2eJ_{dark}}$ [7,9,13,14,48,49]. The estimated peak value of the simplified D^* for the reference was ~2.5 × 10^{13} Jones. In contrast, the peak value of simplified D^* for the sample showed an increase to ~4.5 × 10^{13} Jones. It should also be noted that the simplified D^* value for the sample was significantly higher compared to earlier values for other $MAPbI_3$ thin-film-based photodiodes previously reported in the literature [25,48,49]. However, such simplified D^* values may be overestimated because the shot noise levels were lower than their measured noise current values, as indicated by the dotted lines in Figure 4b [49].

To assess the dynamic characteristics of the photodiodes studied here, the temporal responses were investigated at zero bias voltage by irradiating monochromatic light (λ = 637 nm, P = 190 μW) modulated at a frequency of 2 kHz (Figure 5a). In Figure 5a, the measured temporal responses of the sample clearly indicated higher photocurrents than those of the reference, as expected. The response times of the rise (τ_r) and decay (τ_d) times were determined as the time intervals required to change the photocurrent signal amplitudes between 10% (I_{10}) and 90% (I_{90}) up the rising and decay edges of the signal curves, respectively; τ_r and τ_d for the sample were ~50 and ~17 μs, respectively, similar to those (~49 and ~18 μs) of the reference. Moreover, the obtained response times for the sample and reference were considerably shorter than those of $MAPbI_3$-based photodiodes in the literature [13].

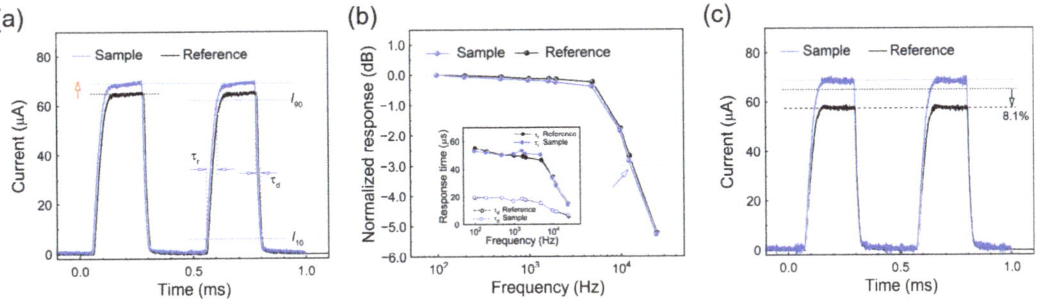

Figure 5. (**a**) Temporal photo responses of the reference and sample $MAPbI_3$ photodiodes at zero bias voltage when turning the incident light (λ = 637 nm, P = 190 μW) on and off (2 kHz). The rise (τ_r) and decay (τ_d) times were determined as the time intervals between the 10% (I_{10}) and 90% (I_{90}) amplitude levels of the signal, respectively. (**b**) Normalized photocurrent versus the modulation frequency of incident light (p = 190 μW, 637 nm). The inset shows the dependences of τ_r and τ_d on the modulation frequency for the photodiodes studied here. (**c**) Temporal photo responses of the reference and sample photodiodes when stored for 500 h.

Next, the 3 dB cutoff bandwidths ($f_{\text{-3dB}}$) for the photodiodes were measured. These values were $f_{\text{-3dB}}$ = ~1.40 × 10^4 and ~1.33 × 10^4 Hz for the reference and sample, respectively (Figure 5b). These dynamic photo responses of the photodiodes indicate that the limitation of the response bandwidths in the devices may stem mainly from the *RC* time constants of the devices, implying that additional improvements in the response speed can be achieved by greater optimization of the device architecture [45]. Additionally, it should be noted that as the modulation frequency applied to the irradiated laser light was increased, the observed rise and decay times began to decrease continuously in the frequency range investigated here, as shown in the inset of Figure 5b.

Finally, we investigated another important functionality of the PMMA passivation layer based on its excellent hydrophobicity, as shown in Figure 1b. Herein, the storage stability of the photodiodes was measured in terms of the photocurrent flowing through the photodiode devices. Given that the storage lifetime is closely linked to water permeation into the functional layers, particularly the $MAPbI_3$ active layer, measuring the temporal responses of the stored photodiodes can provide information pertaining to the degradation mechanisms in these devices [50].

To measure the storage stability, the devices in this study were stored in a nitrogen glovebox between successive measurements of their temporal responses at a regular time interval of approximately 48 h. Temporal response measurements were also taken at zero bias at room temperature. Figure 5c presents the representative temporal response characteristics for the photodiodes stored for a total storage time of ~500 h. As shown in the figure, the photocurrent of the reference was reduced significantly to ~91.9% of its original value after being stored for approximately 500 h. In contrast, the sample exhibited a relatively negligible decrement (~0%) of the photo-response, even after the long storage time used here. Thus, as clearly shown in the figure, the sample with the cathode-side PMMA passivation layer is much more stable than the reference without a PMMA layer, indicating once again that the PMMA passivation layer significantly enhances the storage stability of $MAPbI_3$ perovskite photodiodes. Additional details related to the long-term and/or continuous stability of the photodiode with the PMMA passivation layer, together with a more stable Ag, Cu, or Au cathode instead of the Al cathode used here, will be discussed elsewhere.

4. Discussion

The observations above highlight the effects of the cathode-side PMMA passivation layer; it not only improved the quality of the film but also decreased the leakage/noise currents through the $MAPbI_3$ perovskite layers, thus providing a new paradigm of a high, fast, and durable performance of $MAPbI_3$ perovskite-based photodiodes. Additional developments of organic/inorganic perovskite layers and/or the introduction of other novel functional layers will result in further improvements in the device performance for self-powered, highly sensitive, and stable solution-processable perovskite photodiodes with a cathode-side PMMA passivation layer.

5. Conclusions

In this study, we demonstrated that the introduction of a cathode-side PMMA passivation layer between an absorber layer of $MAPbI_3$ and the ETL effectively suppressed the leakage/noise currents and in turn improved the performance of the self-powered solution-processable $MAPbI_3$ photodiodes. With the cathode-side PMMA passivation layer, the interface quality of the $MAPbI_3$ perovskite layer was shown to be improved, and the film defects were reduced. Moreover, we noted an increase in the grain size and effective decrements in the recombination losses as well as the noise currents. Consequently, even at zero bias voltage, the $MAPbI_3$ photodiode with the PMMA passivation layer exhibited a notably high specific detectivity D^* of ~1.07×10^{12} Jones. Moreover, a significantly low *NEP* of ~230 fW/Hz$^{1/2}$ and wide *LDR* of ~127 dB were achieved. The $MAPbI_3$ photodiode with the PMMA passivation layer revealed excellent device performance compared to that of a conventional $MAPbI_3$ photodetector without a passivation layer. A rapid rise response time of 50 μs and decay response times of 17 μs were also obtained, as were highly durable dynamic responses of the photodiode, even after a long storage time of nearly 500 h. These findings provide clear evidence of the remarkably improved device performance of a perovskite photodiode with a noise-reducing PMMA passivation layer at the interface between the $MAPbI_3$ perovskite light-absorbing layer and the ETL. Thus, a cathode-side PMMA passivation layer in a perovskite photodiode, as presented here, can enhance the applicability of interface-engineered perovskites in photodiodes, imaging sensors, and various low-energy-consuming light-detecting devices.

Author Contributions: Conceptualization, W.K. and Y.A.; Data curation, W.K., J.P. and S.S.; Formal analysis, W.K., J.P. and S.S.; Methodology, J.P.; Project administration, E.H.C. and B.P.; Writing—original draft, W.K. and Y.A.; Writing—review & editing, J.P., E.H.C. and B.P. All authors have read and agreed to the published version of the manuscript.

Funding: This research was funded by the National Research Foundation of Korea (2020R1A2B5B03097060, 2021R1A6A1A03038785) and by Kwangwoon University (2022).

Institutional Review Board Statement: Not applicable.

Informed Consent Statement: Not applicable.

Data Availability Statement: Data presented in this study are available on request from the corresponding author.

Conflicts of Interest: The authors declare no conflict of interest.

References

1. Kojima, A.; Teshima, K.; Shirai, Y.; Miyasaka, T. Organometal Halide Perovskites as Visible-Light Sensitizers for Photovoltaic Cells. *J. Am. Chem. Soc.* **2009**, *131*, 6050–6051. [CrossRef] [PubMed]
2. Lee, M.M.; Teuscher, J.; Miyasaka, T.; Murakami, T.N.; Snaith, H.J. Efficient Hybrid Solar Cells based on Meso-Superstructured Organometal Halide Perovskites. *Science* **2012**, *338*, 643–647. [CrossRef] [PubMed]
3. Bi, C.; Wang, Q.; Shao, Y.; Yuan, Y.; Xiao, Z.; Huang, J. Non-wetting Surface-Driven High-Aspect-Ratio Crystalline Grain Growth for Efficient Hybrid Perovskite Solar Cells. *Nat. Commun.* **2015**, *6*, 7747. [CrossRef] [PubMed]
4. Momblona, C.; Gil-Escrig, L.; Bandiello, E.; Hutter, E.M.; Sessolo, M.; Lederer, K.; Blochwitz-Nimoth, J.; Bolink, H.J. Efficient Vacuum Deposited p-i-n and n-i-p Perovskite Solar Cells Employing Doped Charge Transport Layers. *Energy Environ. Sci.* **2016**, *9*, 3456–3463. [CrossRef]
5. Eperon, G.E.; Leijtens, T.; Bush, K.A.; Prasanna, R.; Green, T.; Wang, J.T.-W.; McMeekin, D.P.; Volonakis, G.; Milot, R.L.; May, R.; et al. Perovskite-Perovskite Tandem Photovoltaics with Optimized Bandgaps. *Science* **2016**, *354*, 861–865. [CrossRef]
6. Yang, W.S.; Park, B.-W.; Jung, E.H.; Jeon, N.J.; Kim, Y.C.; Lee, D.U.; Shin, S.S.; Seo, J.; Kim, E.K.; Noh, J.H.; et al. Iodide management in formamidinium-lead-halide-based perovskite layers for efficient solar cells. *Science* **2017**, *356*, 1376–1379. [CrossRef]
7. Dou, L.; Yang, Y.M.; You, J.; Hong, Z.; Chang, W.-H.; Li, G.; Yang, Y. Solution-Processed Hybrid Perovskite Photodetectors with High Detectivity. *Nat. Commun.* **2014**, *5*, 5404. [CrossRef]
8. Kwon, K.C.; Hong, K.; Van Le, Q.; Lee, S.Y.; Choi, J.; Kim, K.-B.; Kim, S.Y.; Jang, H.W. Inhibition of Ion Migration for Reliable Operation of Organolead Halide Perovskite-Based Metal/Semiconductor/Metal Broadband Photodetectors. *Adv. Funct. Mater.* **2016**, *26*, 4213–4222. [CrossRef]
9. Veeramalai, C.P.; Feng, S.; Zhang, X.; Pammi, S.V.N.; Pecunia, V.; Li, C. Lead–halide Perovskites for Next-Generation Self-Powered Photodetectors: A Comprehensive Review. *Photonics Res.* **2021**, *9*, 968–991. [CrossRef]
10. Shen, L.; Fang, Y.; Wang, D.; Bai, Y.; Deng, Y.; Wang, M.; Lu, Y.; Huang, J. A Self-Powered, Sub-nanosecond-Response Solution-Processed Hybrid Perovskite Photodetector for Time-Resolved Photoluminescence-Lifetime Detection. *Adv. Mater.* **2016**, *28*, 10794–10800. [CrossRef]
11. Fang, Y.; Huang, J. Resolving Weak Light of Sub-picowatt per Square Centimeter by Hybrid Perovskite Photodetectors Enabled by Noise Reduction. *Adv. Mater.* **2015**, *27*, 2804–2810. [CrossRef]
12. Zhao, Y.; Li, C.; Shen, L. Recent Advances on Organic-Inorganic Hybrid Perovskite Photodetectors with Fast Response. *InfoMat* **2019**, *1*, 164–182. [CrossRef]
13. Afzal, A.M.; Bae, I.-G.; Aggarwal, Y.; Park, J.; Jeong, H.-R.; Choi, E.H.; Park, B. Highly Efficient Self-Powered Perovskite Photodiode with an Electron-Blocking Hole-Transport NiO_x Layer. *Sci. Rep.* **2021**, *11*, 169. [CrossRef]
14. Gong, X.; Tong, M.; Xia, Y.; Cai, W.; Moon, J.S.; Cao, Y.; Yu, G.; Shieh, C.-L.; Nilsson, B.; Heeger, A.J. High-Detectivity Polymer Photodetectors with Spectral Response from 300 nm to 1450 nm. *Science* **2009**, *325*, 1665–1667. [CrossRef]
15. Son, D.-Y.; Lee, J.-W.; Choi, Y.J.; Jang, I.-H.; Lee, S.; Yoo, P.J.; Shin, H.; Ahn, N.; Choi, M.; Kim, D.; et al. Self-Formed Grain Boundary Healing Layer for Highly Efficient $CH_3NH_3PbI_3$ Perovskite Solar Cells. *Nat. Energy* **2016**, *1*, 16081. [CrossRef]
16. Bi, D.; Tress, W.; Dar, M.I.; Gao, P.; Luo, J.; Renevier, C.; Schenk, K.; Abate, A.; Giordano, F.; Baena, J.-P.C.; et al. Efficient Luminescent Solar Cells based on Tailored Mixed-Cation Perovskites. *Sci. Adv.* **2016**, *2*, e1501170. [CrossRef]
17. Zheng, X.; Chen, B.; Dai, J.; Fang, Y.; Bai, Y.; Lin, Y.; Wei, H.; Zeng, X.C.; Huang, J. Defect Passivation in Hybrid Perovskite Solar Cells using Quaternary Ammonium Halide Anions and Cations. *Nat. Energy* **2017**, *2*, 17102. [CrossRef]

18. Aydin, E.; De Bastiani, M.; De Wolf, S. Defect and Contact Passivation for Perovskite Solar Cells. *Adv. Mater.* **2019**, *31*, 1900428. [CrossRef]
19. Cai, F.; Cai, J.; Yang, L.; Li, W.; Gurney, R.S.; Yi, H.; Iraqi, A.; Liu, D.; Wang, T. Molecular Engineering of Conjugated Polymers for Efficient Hole Transport and Defect Passivation in Perovskite Solar Cells. *Nano Energy* **2018**, *45*, 28–36. [CrossRef]
20. Wang, D.; Xu, W.; Min, L.; Tian, W.; Li, L. Interfacial Passivation and Energy Level Alignment Regulation for Self-Powered Perovskite Photodetectors with Enhanced Performance and Stability. *Adv. Mater. Interfaces* **2022**, *9*, 2101766. [CrossRef]
21. Wang, C.; Liu, M.; Rahman, S.; Pasanen, H.P.; Tian, J.; Li, J.; Deng, Z.; Zhang, H.; Vivo, P. Hydrogen Bonding Drives the self-Assembling of Carbazole-Based Hole-Transport Material for Enhanced Efficiency and Stability of Perovskite Solar Cells. *Nano Energy* **2022**, *101*, 107604. [CrossRef]
22. Li, R.; Liu, M.; Matta, S.K.; Hiltunen, A.; Deng, Z.; Wang, C.; Dai, Z.; Russo, S.P.; Vivo, P.; Zhang, H. Sulfonated Dopant-Free Hole-Transport Material Promotes Interfacial Charge Transfer Dynamics for Highly Stable Perovskite Solar Cells. *Adv. Sustain. Syst.* **2021**, *5*, 2100244. [CrossRef]
23. Piao, C.; Xi, J.; Choi, M. Directionally Selective Polyhalide Molecular Glue for Stable Inverted Perovskite Solar Cells. *Sol. RRL* **2020**, *4*, 2000244. [CrossRef]
24. Kim, M.; Motti, S.G.; Sorrentino, R.; Petrozza, A. Enhanced Solar Cell Stability by Hygroscopic Polymer Passivation of Metal Halide Perovskite Thin Film. *Energy Environ. Sci.* **2018**, *11*, 2609–2619. [CrossRef]
25. Park, J.; Aggarwal, Y.; Kim, W.; Sharma, S.; Choi, E.H.; Park, B. Self-powered $CH_3NH_3PbI_3$ Perovskite Photodiode with a Noise-Suppressible Passivation Layer of Poly(Methyl Methacrylate). *Opt. Express* **2023**, *31*, 1202–1213. [CrossRef]
26. Frolova, L.A.; Davlethanov, A.I.; Dremova, N.N.; Zhidkov, I.; Akbulatov, A.F.; Kurmaev, E.Z.; Aldoshin, S.M.; Stevenson, K.J.; Troshin, P.A. Efficient and Stable $MAPbI_3$-Based Perovskite Solar Cells using Polyvinylcarbazole Passivation. *J. Phys. Chem. Lett.* **2020**, *11*, 6772–6778. [CrossRef]
27. Du, Y.; Xin, C.; Huang, W.; Shi, B.; Ding, Y.; Wei, C.; Zhao, Y.; Li, Y.; Zhang, X. Polymeric Surface Modification of NiO_x-Based Inverted Planar Perovskite Solar Cells with Enhanced Performance. *ACS Sustain. Chem. Eng.* **2018**, *6*, 16806–16812. [CrossRef]
28. Shi, Y.-R.; Wang, K.-L.; Lou, Y.-H.; Zhang, D.-B.; Chen, C.-H.; Chen, J.; Ni, Y.-X.; Öz, S.; Wang, Z.-K.; Liao, L.-S. Unraveling the Role of Active Hydrogen Caused by Carbonyl Groups in Surface-Defect Passivation of Perovskite Photovoltaics. *Nano Energy* **2022**, *97*, 107200. [CrossRef]
29. Cai, Y.; Cui, J.; Chen, M.; Zhang, M.; Han, Y.; Qian, F.; Zhao, H.; Yang, S.; Yang, Z.; Bian, H.; et al. Multifunctional Enhancement for Highly Stable and Efficient Perovskite Solar Cells. *Adv. Funct. Mater.* **2021**, *31*, 2005776. [CrossRef]
30. Lin, Y.; Shen, L.; Dai, J.; Deng, Y.; Wu, Y.; Bai, Y.; Zheng, X.; Wang, J.; Fang, Y.; Wei, H. π-Conjugated Lewis Base: Efficient Trap-Passivation and Charge-Extraction for Hybrid Perovskite Solar Cells. *Adv. Mater.* **2017**, *29*, 1604545. [CrossRef]
31. Taguchi, M.; Suzuki, A.; Ueoka, N.; Oku, T. Effects of poly(Methyl Methacrylate) Addition to Perovskite Photovoltaic Devices. *AIP Conf. Proc.* **2019**, *2067*, 020018.
32. Jo, B.; Han, G.S.; Yu, H.M.; Choi, J.; Zhu, J.; Ahn, T.K.; Namkoong, G.; Jung, H.S. Composites of Cross-Linked Perovskite/Polymer with Sodium Borate for Efficient and Stable Perovskite Solar Cells. *J. Mater. Chem. A* **2022**, *10*, 14884–14893. [CrossRef]
33. Ochoa-Martinez, E.; Ochoa, M.; Ortuso, R.D.; Ferdowsi, P.; Carron, R.; Tiwari, A.N.; Steiner, U.; Saliba, M. Physical Passivation of Grain Boundaries and Defects in Perovskite Solar Cells by an Isolating Thin Polymer. *ACS Energy Lett.* **2021**, *6*, 2626–2634. [CrossRef]
34. Peng, J.; Khan, J.I.; Liu, W.; Ugur, E.; Duong, T.; Wu, Y.; Shen, H.; Wang, K.; Dang, H.; Aydin, E.; et al. A Universal Double-Side Passivation for High Open-Circuit Voltage in Perovskite Solar Cells: Role of Carbonyl Groups in Poly(methyl methacrylate). *Adv. Energy Mater.* **2018**, *8*, 1801208. [CrossRef]
35. Yang, F.; Lim, H.E.; Wang, F.; Ozaki, M.; Shimazaki, A.; Liu, J.; Mohamed, N.B.; Shinokita, K.; Miyauchi, Y.; Wakamiya, A.; et al. Roles of Polymer Layer in Enhanced Photovoltaic Performance of Perovskite Solar Cells via Interface Engineering. *Adv. Mater. Interfaces* **2018**, *5*, 1701256. [CrossRef]
36. Kim, H.; Lee, K.S.; Paik, M.J.; Lee, D.Y.; Lee, S.-U.; Choi, E.; Yun, J.S.; Seok, S.I. Polymethyl Methacrylate as an Interlayer Between the Halide Perovskite and Copper Phthalocyanine Layers for Stable and Efficient Perovskite Solar Cells. *Adv. Funct. Mater.* **2022**, *32*, 2110473. [CrossRef]
37. Choi, K.; Choi, H.; Min, J.; Kim, T.; Kim, D.; Son, S.Y.; Kim, G.-W.; Choi, J.; Park, T. A Short Review on Interface Engineering of Perovskite Solar Cells: A Self-Assembled Monolayer and Its Roles. *Sol. RRL* **2020**, *4*, 1900251. [CrossRef]
38. Bube, R.H. Trap Density Determination by Space-Charge-Limited Currents. *J. Appl. Phys.* **1962**, *33*, 1733–1737. [CrossRef]
39. Sworakowski, J.; Ferreira, G.F.L. Space-Charge-Limited Currents and Trap-Filled Limit in One-Dimensional Insulators. *J. Phys. D Appl. Phys.* **1984**, *17*, 135. [CrossRef]
40. Speirs, M.J.; Dirin, D.N.; Abdu-Aguye, M.; Balazs, D.M.; Kovalenko, M.V.; Loi, M.A. Temperature Dependent Behaviour of Lead Sulfide Quantum Dot Solar Cells and Films. *Energy Environ. Sci.* **2016**, *9*, 2916–2924. [CrossRef]
41. Ryu, S.; Nguyen, D.C.; Ha, N.Y.; Park, H.J.; Ahn, Y.H.; Park, J.-Y.; Lee, S. Light Intensity-dependent Variation in Defect Contributions to Charge Transport and Recombination in a Planar $MAPbI_3$ Perovskite Solar Cell. *Sci. Rep.* **2019**, *9*, 19846. [CrossRef] [PubMed]
42. Cao, Q.; Li, Y.; Zhang, H.; Yang, J.; Han, J.; Xu, T.; Wang, S.; Wang, Z.; Gao, B.; Zhao, J.; et al. Efficient and Stable Inverted Perovskite Solar Cells with Very High Fill Factors via Incorporation of Star-Shaped Polymer. *Sci. Adv.* **2021**, *7*, eabg0633. [CrossRef] [PubMed]

43. Lu, H.; Tian, W.; Cao, F.; Ma, Y.; Gu, B.; Li, L. A Self-Powered and Stable All-Perovskite Photodetector–Solar Cell Nanosystem. *Adv. Funct. Mater.* **2016**, *26*, 1296–1302. [CrossRef]
44. Xue, J.; Zhu, Z.; Xu, X.; Gu, Y.; Wang, S.; Xu, L.; Zou, Y.; Song, J.; Zeng, H.; Chen, Q. Narrowband Perovskite Photodetector-Based Image Array for Potential Application in Artificial Vision. *Nano Lett.* **2018**, *18*, 7628–7634. [CrossRef] [PubMed]
45. Li, C.; Wang, H.; Wang, F.; Li, T.; Xu, M.; Wang, H.; Wang, Z.; Zhan, X.; Hu, W.; Shen, L. Ultrafast and broadband photodetectors based on a perovskite/organic bulk heterojunction for large-dynamic-range imaging. *Light: Sci. Appl.* **2020**, *9*, 31. [CrossRef]
46. Wang, Y.; Liu, Y.; Cao, S.; Wang, J. A Review on Solution-Processed Perovskite/Organic Hybrid Photodetectors. *J. Mater. Chem. C* **2021**, *9*, 5302–5322. [CrossRef]
47. Guo, F.; Yang, B.; Yuan, Y.; Xiao, Z.; Dong, Q.; Bi, Y.; Huang, J. A Nanocomposite Ultraviolet Photodetector Based on Interfacial Trap-Controlled Charge Injection. *Nat. Nanotechnol.* **2012**, *7*, 798–802. [CrossRef]
48. Li, T.; Li, Q.; Tang, X.; Chen, Z.; Li, Y.; Zhao, H.; Wang, S.; Ding, X.; Zhang, Y.; Yao, J. Environment-Friendly Antisolvent Tert-Amyl Alcohol Modified Hybrid Perovskite Photodetector with High Responsivity. *Photonics Res.* **2021**, *9*, 781–791. [CrossRef]
49. Wang, S.; Li, T.; Li, Q.; Zhao, H.; Zheng, C.; Li, M.; Li, J.; Zhang, Y.; Yao, J. Inhibition of Buried Cavities and Defects in Metal Halide Perovskite Photodetectors via a Two-Step Spin-Coating Method. *J. Mater. Chem. C* **2022**, *10*, 7886–7895. [CrossRef]
50. Ma, N.; Jiang, J.; Zhao, Y.; He, L.; Ma, Y.; Wang, H.; Zhang, L.; Shan, C.; Shen, L.; Hu, W. Stable and Sensitive Tin-Lead Perovskite Photodetectors Enabled by Azobenzene Derivative for Near-Infrared Acousto-Optic Conversion Communications. *Nano Energy* **2021**, *86*, 106113. [CrossRef]

nanomaterials

Article

Solid-Liquid Triboelectric Nanogenerator Based on Vortex-Induced Resonance

Xiaowei Li [1], Di Zhang [1], Dan Zhang [1], Zhongjie Li [1,*], Hao Wu [1], Yuan Zhou [1], Biao Wang [2], Hengyu Guo [3] and Yan Peng [2,4]

1 School of Mechatronic Engineering and Automation, Shanghai University, Shanghai 200444, China
2 Institute of Artificial Intelligence, Shanghai University, Shanghai 200444, China
3 Department of Applied Physics, Chongqing University, Chongqing 400044, China
4 Shanghai Artificial Intelligence Laboratory, Shanghai 200232, China
* Correspondence: lizhongjie@shu.edu.cn

Abstract: Energy converters based on vortex-induced vibrations (VIV) have shown great potential for harvesting energy from low-velocity flows, which constitute a significant portion of ocean energy. However, solid-solid triboelectric nanogenerators (TENG) are not wear-resistant in corrosive environments. Therefore, to effectively harvest ocean energy over the long term, a novel solid-liquid triboelectric nanogenerator based on vortex-induced resonance (VIV-SL-TENG) is presented. The energy is harvested through the resonance between VIV of a cylinder and the relative motions of solid-liquid friction pairs inside the cylinder. The factors that affect the output performance of the system, including the liquid mass ratio and the deflection angle of the friction plates, are studied and optimized by establishing mathematical models and conducting computational fluid dynamics simulations. Furthermore, an experimental platform for the VIV-SL-TENG system is constructed to test and validate the performance of the harvester under different conditions. The experiments demonstrate that the energy harvester can successfully convert VIV energy into electrical energy and reach maximum output voltage in the resonance state. As a new type of energy harvester, the presented design shows a promising potential in the field of 'blue energy' harvesting.

Keywords: vortex-induced vibration; solid-liquid triboelectric nanogenerator; computational fluid dynamic

Citation: Li, X.; Zhang, D.; Zhang, D.; Li, Z.; Wu, H.; Zhou, Y.; Wang, B.; Guo, H.; Peng, Y. Solid-Liquid Triboelectric Nanogenerator Based on Vortex-Induced Resonance. *Nanomaterials* **2023**, *13*, 1036. https://doi.org/10.3390/nano13061036

Academic Editor: Sotirios Baskoutas

Received: 3 February 2023
Revised: 6 March 2023
Accepted: 10 March 2023
Published: 13 March 2023

1. Introduction

Recent decades have witnessed rapid development in integrated circuits, and MEMS technology has made electronic devices much smaller in size and considerably more energy-efficient. However, these electronic devices still rely on batteries for power supply, and the safe disposal of chemical waste poses a particular challenge [1,2]. Therefore, harvesting energy from the surrounding circumstance is a more reasonable solution. Blue energy contains tidal energy, ocean current energy, temperature difference energy, etc. [3,4]. Among them, ocean current energy is a widely recognized source of clean energy. The main method of harnessing ocean current energy is to utilize turbines, which work better in areas with high flow velocities. However, turbines struggle to function efficiently in low-flow speed environments, whereas vortex-induced vibration can produce periodic oscillations, and the vibration energy can effectively convert low-speed flow energy into electrical energy.

Vortex-induced vibration is a typical fluid-structure interaction phenomenon caused by the viscous effect of fluids, which generates alternating shedding vortices around the back of the spoiler cylinder and causes the cylinder to reciprocate motion perpendicular to the flow direction. Moreover, the structure can still vibrate, even at a low flow speeds. In the early days, research on vortex-induced vibration mainly focused on prevention and suppression to prevent damage to the structure. M.L. Facchinetti et al. [5] simulated a low-order model of vortex-induced vibration with three coupling terms and found that the acceleration coupling model performed better. Bearman [6] elaborated detailed conclusions on the response characteristics of rigid cylinders to one degree-of-freedom (1 DOF) VIV. Moe and Wu [7] conducted forced and free vibration experiments on the cylinder, and the results can predict the amplitude of a cylinder in a given flow. Zhou et al. [8] studied the elastic cylinders via the discrete vortex method incorporating the vortex-in-cell (VIC) and found the maximum amplitude of vibration can reach 0.57 diameter. Bishop and Hasson [9] expounded the experimental conclusions of lift and drag forces on the vibrating cylinder and proposed a wake oscillator model. Parkinson [10] qualitatively described major features of near-wake vortex shedding and VIV using analytical and numerical methods.

From another perspective, the VIV energy can convert kinetic energy into mechanical energy, which can be recovered and utilized in low-velocity water flow through energy harvesters. At present, many researchers have carried out related research on vortex-induced vibration energy harvesting. Junlei Wang et al. [11] proposed a bluff body with different interfaces that can couple VIV and gallop to improve energy harvesting performance. Iman Mehdipour et al. [12] conducted a comprehensive experimental study on the shape of a blunt body causing VIV and enhance the performance of low-speed wind energy harvester. Toma et al. [13] proposed a vertical moving cylinder piezoelectric energy harvesting device. Baoshou Zhang et al. [14] proposed a novel biomimetic V-shaped layout, increasing the energy recovery area of downstream cylinders. Ying Gong et al. [15] proposed a VIV energy harvester with a direction-adaptive wing, which is well adapted to the direction-varying flow environments. Zhaoyong Mao et al. [16] studied the effect of spacing between cylinders on VIV energy harvesting and optimized the layout of four cylinders. Junlei Wang et al. [17] introduced metasurface into VIV and found that certain surfaces can increase voltage and amplitude. Mingjie Zhang et al. [18] studied the effect of the Reynolds number on energy harvesting of cylindrical VIV through piezoelectric energy harvesters. Mengfan Gu et al. [19] studied the effect of submerged depth on VIV energy harvesting and found that the conversion efficiency reached a peak at 0.5 m. Yufei Mei et al. [20] greatly improved the net energy output of VIV by controlling jet flow through deep reinforcement learning. Peng Han et al. [21] studied the vibration and energy harvesting characteristics of low-mass square bluff bodies at different incident angles.

Energy harvesters that convert mechanical energy into electrical energy are predominantly divided into three categories according to the principle: electromagnetic, piezoelectric, and electrostatic [22,23]. Among them, the electromagnetic energy harvesters are large in size and high in cost, and the piezoelectric energy harvesters can only be effective under high-frequency vibrations [24,25]. As an emerging energy harvesting technology, TENG utilizes triboelectric electrification and electrostatic induction to convert mechanical energy into electrical energy. It exhibits the advantages of low cost, light weight, and, particularly for low-frequency vibration energy harvesting, presents better performance [26,27]. Two main types of TENG have been invented, namely, solid-solid TENGs and solid-liquid TENGs.

Recent decades have witnessed various prototypes and structures of solid–solid TENGs. Wang et al. [28] were the first to demonstrate the conversion of mechanical energy into electrical energy via triboelectric electrification and electrostatic inductive coupling. As the research progressed, nanoscale triboelectric-effect-enabled energy conversion was discovered, which was named the triboelectric nanogenerator due to the nano-scale surface roughness. Zize Liu et al. [29] proposed a temperature sensor by TENG to detect human body temperature without an external power supply. Jing Liu et al. [30] proposed a triboelectric hydrophone for monitoring underwater low-frequency sound signals and offering high-sensitivity. Yan Wu et al. [31] fabricated a generator that combines TENG and an electromagnetic generator (EMG) to collect low-frequency wave energy. Liang Xu et al. [32] studied different connection methods of the TENG network and improved the performance of TENG cluster. Huamie Wang et al. [33] improved the performance of TENGs by encapsulating the shuttle of charges in conductive domains and increasing the charge density. Tian Xiaoxiao et al. [34] used silicone rubber and carbon black as flexible electrodes, improving the performance of TENG. Bolang Cheng et al. [35] proposed a temperature differential TENG that can improve the performance in high-temperature circumstances. Nonetheless, the solid-solid TENGs faced several problems: first, the wear of the solid-solid friction pair diminishes the long-term working stability of TENGs. Additionally, high humidity normally exerts an adverse effect on the performance of solid–solid TENGs, while solid-liquid TENGs have drawn much attention for their stable output performance and durability.

Xiaoyi Li et al. [36] proposed a solid-liquid TENG of nanowires, which significantly increases the contact surface of fluorinated ethylene propylene (FEP) with water and improved the performance. Lun Pan et al. [37] designed a U-tube SL-TENG to study the effect of liquid properties on output performance. Wei Tang et al. [38] developed a mercury-based triboelectric nanogenerator (LM-TENG) with significantly improved output efficiency. Xiya Yang et al. [39] developed a polytetrafluoroethylene (PTFE)-based triboelectric nanogenerator for water wave energy harvesting. Jinhui Nie et al. [40] proposed a triboelectric nanogenerator based on the interaction of two pure liquids. Shunmin Zhu et al. [41] developed a triboelectric nanogenerator without solid moving parts for waste heat recovery. Jing You et al. [42] proposed the equivalent circuit model of solid-liquid TENG and solved the electrical phenomena on the liquid-solid interface. Cun Xin Lu et al. [43] studied the effect of ambient temperature on the output performance of triboelectric nanogenerators. Liqiang Zhang et al. [44] found that temperature, ionic concentration, and pH affected the output of SL-TENGs via experimental research. Liyun Ma et al. [45] developed smart protective clothing based on triboelectric nanogenerators, which collect motion to power the monitoring system. Xinkai Xie et al. [46] utilized TENG to design a self-powered gyroscope angle sensor to detect the relative rotation angle. Song Wang et al. [47] designed a self-powered tilt sensor based on an SL-TENG, which demonstrates stable output and high sensitivity. Due to the stability and insensitivity to humidity, SL-TENGs are more suitable for ocean flow energy harvesting.

In recent decades, numerous researchers have systematically studied VIV and summarized its characteristics. At the same time, many researchers have proposed energy harvesters based on VIV and optimized the power generation efficiency of the energy harvester via various methods. With the proposed triboelectric nanogenerators, this technology has been applied in many scenarios in recent years. However, no research on the SL-TENG was utilized to collect VIV energy.

In this paper, a VIV-SL-TENG is proposed for harvesting low-velocity flow energy in oceans and rivers. The SL-TENG is embedded in a cylinder to collect vortex-induced vibration energy. As shown in Figure 1, when the device is placed in a water environment, it generates vibrations under the action of water, and the internal SL-TENG unit converts the vibration energy into electric energy. More details are shown below. The main novelties are as follows. Firstly, we establish a theoretical model to explore the physical parameters that influence the state of the cylinder. In addition, we investigate how the cylinder diameter and the speed affect the vortex frequency by computational fluid dynamics (CFD). Then, we find that the VIV-SL-TENG performs best in the resonance state. At the same time, in the resonance state, to increase the efficiency of the internal liquid-solid friction pair, we investigate the performance of the VIV-SL-TENG under different deflection angles.

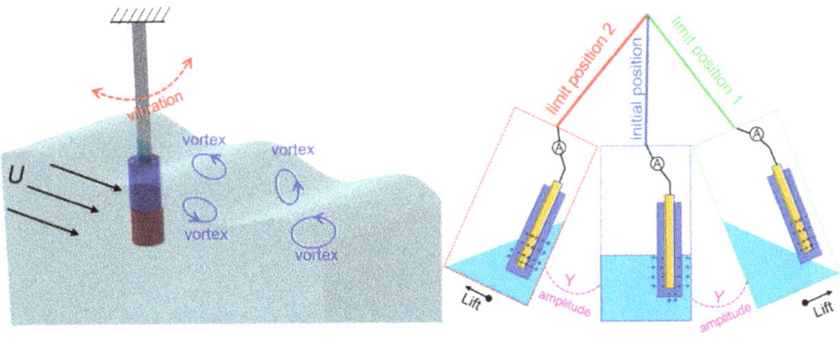

Figure 1. Design concept of VIV-SL-TENG.

2. Materials and Methods

2.1. Experiment Platform and TENG Working Mechanism

We built an experimental platform, as shown in Figure 2a; the VIV-SL-TENG is vertically installed in a flume that provides a continuous and stable water flow. Vortex-induced vibration occurs on the VIV-SL-TENG under the impact of water flow, which converts the kinetic energy of the water flow into vibration energy, making it possible to capture the water flow energy. The VIV-SL-TENG can be installed in situations where water flows, such as rivers and oceans, to continuously power devices. A schematic diagram of a single VIV-SL-TENG is shown in Figure 2b; the VIV-SL-TENG is primarily composed of a cantilever beam, a hollow cylinder, a guiding disc, a carrier sheet, and a solid-liquid triboelectric nanogenerator unit. The SL-TENG is fastened to the carrier sheet, and the carrier sheet is assembled on the guide disc. Sequentially, the guide disc is mounted inside the cylinder. As shown in Figure 2c, a single electrode TENG was prepared using an FEP film as the dielectric film. The TENG consists of a dielectric material and an electrode (copper). A copper film (a thickness of 125 μm) is adhered between two FEP films. The area of the FEP films (30 mm × 60 mm, the thickness of FEP films is 50 μm) is larger than that of the copper film (20 mm × 40 mm) to ensure that the copper film is entirely isolated from water. In addition, we utilized deionized water as the liquid triboelectric material to construct the entire TENG. To achieve resonance, a finite element analysis was conducted for the structure, which yielded a natural frequency of 2 Hz. The Strouhal number (St) was then determined based on the flow velocity, which was limited to 1 m/s. Then the vortex shedding frequency could be calculated according to the flow velocity and the model parameters, as shown in Table 1.

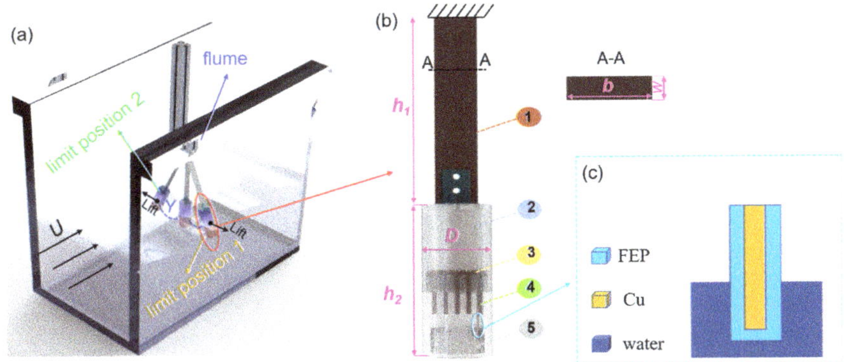

Figure 2. *Cont.*

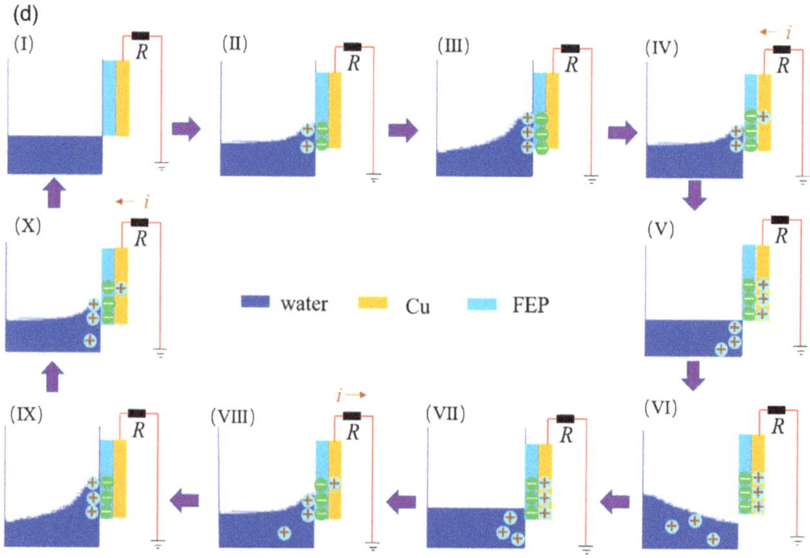

Figure 2. (**a**) TENG system. (**b**) Structural design model of the harvester: (1) cantilever beam, (2) spoiler cylinder, (3) guiding disc, (4) carrier sheet, (5) deionized water. (**c**) Structure of TENG; (**d**) working principle of the SL-TENG.

Table 1. Physical model and structure parameters.

Description	Symbol	Value
Diameter of the cylinder	D (mm)	50
Natural frequency	f (Hz)	2.017
Strouhal number	$St = \frac{f_s \cdot D}{U}$	0.2
Flow speed	U (m/s)	0.5
Width of the vibration beam	b (mm)	30
Length of the spoiler cylinder	h_2 (mm)	130
Length of the vibration beam	h_1 (mm)	290
Thickness of the vibration beam	w (mm)	1

As shown in Figure 2a, the cylinder is entirely immersed in the water of the flume. The entire VIV-SL-TENG reciprocates under the action of VIV. The deionized water inside the cylinder moves along with the VIV-SL-TENG, causing the relative motion between the contact of the deionized water and the FEP. The working mechanism of the solid-liquid TENG is shown in Figure 2d; in the initial state (Figure 2d(I)), we assume that the surface of the dielectric material and water is electrically neutral. When the deionized water shakes along with the VIV-SL-TENG, electrons move from the deionized water to the surface of the FEP due to the different abilities to attract electrons. This results in the surface of the FEP becoming negatively charged and the water interface becoming positively charged, as shown in Figure 2d(II). When the VIV-SL-TENG moves to the limit position, the contact area between the deionized water and the FEP reaches a maximum, and a large number of charges accumulate on the FEP surface (Figure 2d(III)). Subsequently, the deionized water begins to move away from the FEP surface, and the charges on the contact surface are separated. This causes transferred electrons to remain on the FEP, resulting in the potential of the deionized water to be higher than that of the induced electrode. Therefore, a potential difference is generated between the induced electrode and the ground, driving current from the ground to the electrode (Figure 2d(IV)). During the motion of the deionized water, electrons flow continuously until the charge on the FEP reaches saturation (Figure 2d(V,VI)).

When the cylinder and the deionized water reach the other limit position, the deionized water flows towards the FEP again (Figure 2d(VII)). At this point, due to the gradual decrease in the potential difference with the increase in contact area, the current flows back to the ground, as shown in Figure 2d(VIII). Then the electrons continue to flow until the deionized water moves to the limit position again (Figure 2d(IX)), and then the deionized water returns to the initial state and starts the next cycle (Figure 2d(X)).

2.2. Mathematical Model

We consider the cylinder as a one degree-of-freedom elastic-supported rigid cylinder with a diameter of D. The VIV phenomenon is a non-linear process that involves self-excitation, and the VIV can cause the system to exhibit large-amplitude oscillations. A van der Pol oscillator, as a non-linear process, can better capture the dynamics of the system and can be a useful representation of the self-excited behavior seen in VIV. Therefore, a novel van der Pol model was adopted to establish the kinematic coupling equation between the rigid cylinder. The vibration response of the cylinder is obtained by decoupling. As shown in Figure 3, x is the down-flow direction, y is the vibration direction, and U is the flow speed. We assume that the cylinder only moves in one direction, namely, the lateral displacement Y in the y direction. The vibration equation can be written as follows:

$$m\ddot{Y} + r\dot{Y} + hY = S. \tag{1}$$

where $(\cdot)$ is the derivative concerning the dimensional time T, S is the forcing caused only by vorticity in the wake, and the mass m includes the sum of the rigid cylinder mass m_s and the additional mass m_f of the fluid.

$$m = m_s + m_f,\ m_f = C_M \rho^2 \beta \pi / 4,\ \mu = \frac{m_s + m_f}{\rho D^2},\ S = \frac{1}{2} \rho U^2 D C_L \tag{2}$$

where ρ is the density of water, μ is the mass ratio, and C_M is the additional mass coefficient. In Equation (1), r is damping, including structural damping r_n and fluid resistance r_f.

$$r = r_n + r_f,\ r_f = \gamma \Omega \rho D^2,\ \Omega = 2\pi St U / D. \tag{3}$$

where γ is a stall parameter, Ω is the angular frequency, and St is the Strouhal number.

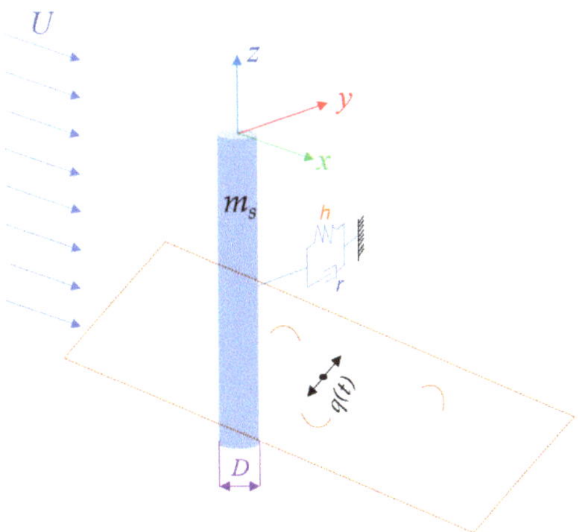

Figure 3. Model of coupled structure and wake oscillators.

In this paper, only the low-order coupling effect of transverse VIV on the wake vibrator was considered, and only the transverse vibration displacement generated was calculated. Therefore, a 1 DOF transverse vibration wake oscillator model was established based on the work of Facchinetti et al. Facchinetti discussed the coupling effects of displacement, velocity, and acceleration on the wake oscillators and concluded that the form of acceleration coupling better reflects the vortex-induced vibration. Therefore, in this article, we also chose the transverse acceleration to represent the forcing term. The vibration characteristics of the vortex street were simulated by a nonlinear oscillator satisfying the van der Pol equation:

$$\ddot{q} + \varepsilon\Omega_f\left(q^2 - 1\right)\dot{q} + \Omega_f^2 q = F. \tag{4}$$

$$\Omega_f = 2\pi\frac{StU}{D}. \tag{5}$$

By dimensionless normalization of the cylinder vibration equation and wake vibrator equation, the dimensionless time $t = T \cdot \Omega_f$, $y = \frac{Y}{D}$ is introduced, respectively, and the two dimensionless quantities are substituted into Equations (1) and (4):

$$\begin{aligned} \ddot{y} + \left(2\xi\delta + \tfrac{\gamma}{\mu}\right)\dot{y} + \delta^2 y = Mq \\ \ddot{q} + \varepsilon(q^2 - 1)q + \dot{q} = A\ddot{y}. \end{aligned} \tag{6}$$

where ξ is structure-reduced damping, M is essentially a mass number and scales the effect of the wake on the structure, A is the coupling force coefficient of the structure on the fluid, and ε is a small parameter in a nonlinear term, and among them,

$$M = \frac{C_{L0}}{16\pi^2 St^2 \mu},\ \delta = \frac{1}{StU_r},\ U_r = \frac{2\pi U}{\Omega_S D}. \tag{7}$$

The amplitude of the linear transfer equation between the displacement of the cylinder and the variable of fluid is

$$\frac{y_0}{q_0} = M[(\delta^2 - \omega^2)^2 + (2\delta\xi + \frac{\gamma}{\mu})^2\omega^2]^{-0.5}. \tag{8}$$

Then, substituting in the wake oscillator equation and only considering the major harmonic contributions of nonlinearity, the equation for the amplitude q_0 and frequency ratio ω can be obtained via elementary algebra:

$$\omega^6 - \left[1 + 2\delta^2 - (2\xi\delta + \gamma/\mu)^2 - AM\right]\omega^4 - \left[-2\delta^2 + (2\xi\delta + \gamma/\mu)^2 - \delta^4 + AM\delta^2\right]\omega^2 - \delta^4 = 0. \tag{9}$$

$$q_0 = 2[1 + \frac{AM}{\varepsilon}\frac{(2\xi\delta + \gamma/\mu)\omega^2}{(\delta^2 - \omega^2)^2 + (2\xi\delta + \gamma/\mu)^2\omega^2}]^{0.5}. \tag{10}$$

The values of ω, y_0, and q_0 can be obtained via Equations (8)–(10), respectively, representing the frequency ratio, dimensionless maximum amplitude, and lift oscillator amplitude, respectively.

As shown in Figure 4a–c, they represent the variation tendencies of the cylinder motion amplitude, lift oscillator, and frequency ratio at different flow velocities when μ = 0.93 and ξ = 0.03. As shown in Figure 4a, when the flow speed increases, the frequency ratio decreases slowly and remains at about 1, indicating that the vibration frequency of the cylinder is close to its natural frequency. As shown in Figure 4b, the lift oscillator first increases and then decreases with the increase of flow speed, reaching the maximum when U_r = 5 because resonance occurs at U_r = 5, according to the design and calculation. When U_r is less than 5, the lift oscillator rises rapidly and then drops slowly. As shown in Figure 4c, the variation trend of the maximum displacement vibration amplitude along with the flow speed is the same as that of the lift oscillator. Before the locked region, when the flow speed increases, the vibration amplitude of the rigid cylinder, namely, y_0, increases significantly, and the lift oscillator also gradually increases, reaching the maximum in the locked region. The locked region is around U_r = 4.5–6, and both before and after this interval are non-locked phases. When U_r is larger than 6, the amplitude slowly decreases with the increase of flow speed. However, the vibration amplitude reaches the maximum when U_r = 6 instead of U_r = 5. As a result, the displacement amplitude caused by locking increases to the maximum point rather than at the frequency ratio of 1, which reflects the phenomenon of mistuning during locking. At the same time, when the vortex shedding frequency is close to the cylinder vibration frequency, the structure is locked.

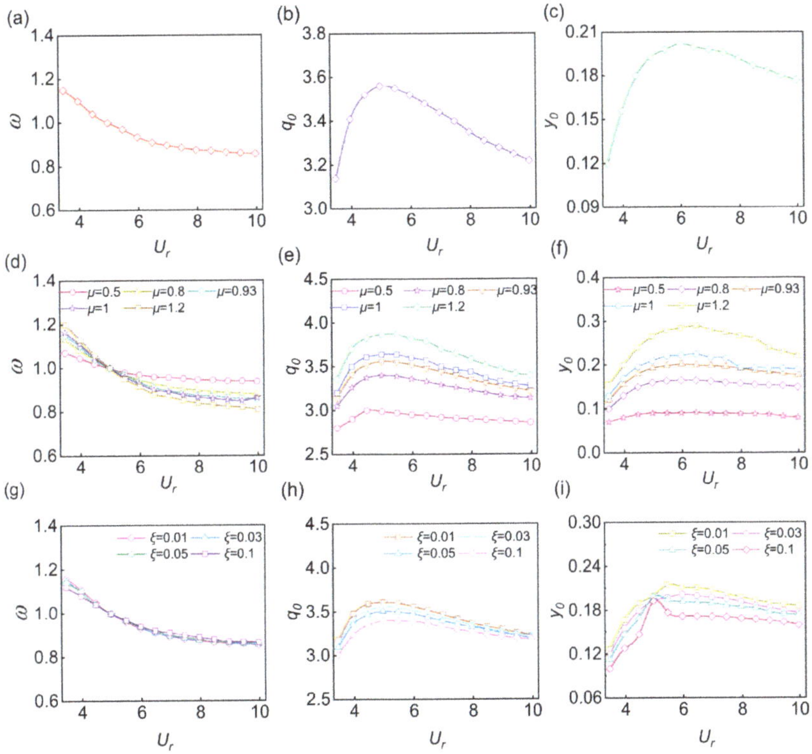

Figure 4. (**a**) Frequency ratio ω, (**b**) lift q_0, and (**c**) amplitude y_0 as a function of reduced speed U_r. Influence of mass ratio on (**d**) frequency ratio, (**e**) lift, and (**f**) amplitude. Influence of structural damping on (**g**) frequency ratio, (**h**) lift force, and (**i**) amplitude.

Next, we studied the VIV response characteristics of a rigid cylinder. The main parameters that affect the VIV of a rigid cylinder are the flow speed, mass ratio, and structure damping ratio. Here, we discussed the influence of the mass ratio μ and the structural damping ratio ξ on the maximum amplitude, lift force, and frequency ratio. Figure 4d–f describe how the mass ratio affects the frequency ratio, maximum amplitude, and lift amplitude. As the mass ratio increases, the lift amplitude and the maximum amplitude decrease, and the range of the locked region also decreases. Then, Figure 4g–i describe how the structural damping ratio affects the maximum amplitude, lift amplitude, and frequency ratio, respectively. They show that the maximum displacement response decreases significantly when the structural damping ratio increases from 0.03 to 0.1, and the locked region is steady, and the influence on lift decreases with the increase of the structural damping ratio. By comparison, the mass ratio presents a significant impact on the VIV characteristics, while the structural damping presents a weaker impact.

2.3. Simulation

To determine the basic characteristics of the cylinder from the water flow, 2D transient numerical simulations were employed to qualitatively analyze the causes for the induced flow-induced force. As an effective approach, researchers have taken advantage of CFD in VIV research in recent years. In this article, we utilized the 2D Reynolds-Averaged Navier–Stokes (RANS) equations accompanied by the k-ω SST turbulence model to simulate the flow field. As shown in Figure 5a, the diameter D of the cylinder is 0.05 m, and the computational domain is 10 $D \times 30$ D. The distance between the cylinder and the speed inlet is 10 D, while the distance between the cylinder and the pressure outlet is 20 D. The vortex contour of the cylinder at 0.5 m/s is presented in Figure 5b, which shows a series of vortices appearing behind the cylinder.

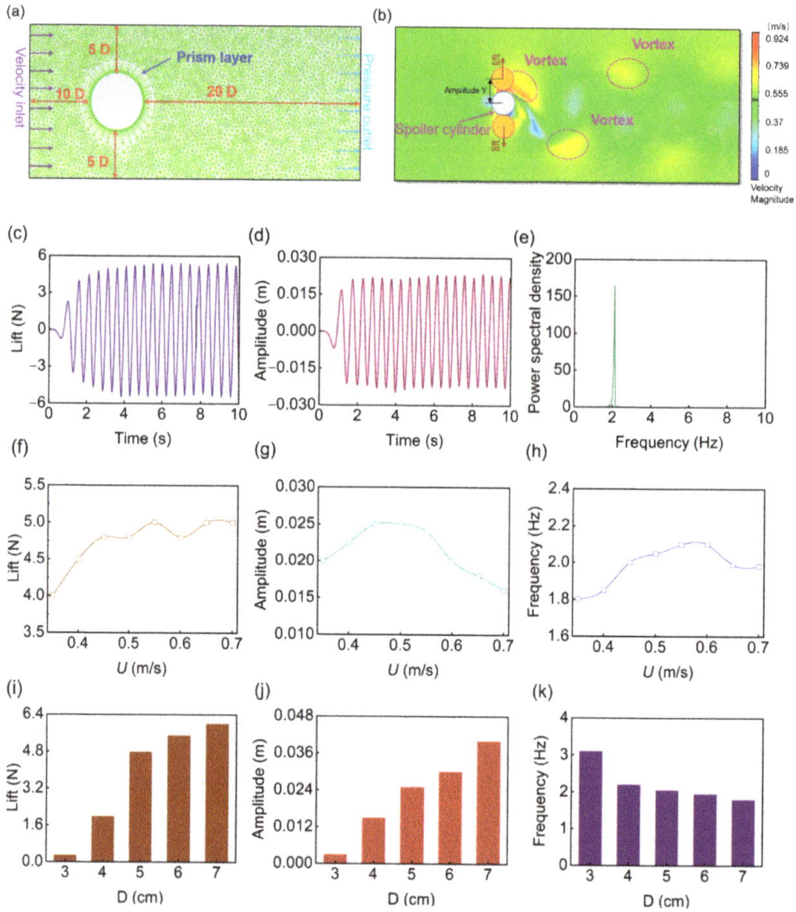

Figure 5. (**a**) Grid of computational domain. (**b**) Scalar diagram of the flow field. (**c**) Lift force, (**d**) displacement, and (**e**) frequency of the cylinder. Effect of flow speed on (**f**) lift force, (**g**) amplitude, and (**h**) frequency. Effect of flow diameter on (**i**) lift force, (**j**) amplitude, and (**k**) frequency.

The lift force and amplitude of the cylinder at 0.5 m/s are shown in Figure 5c,d. The lift force of the cylinder fluctuates with time, the same as amplitude, and both lift and amplitude show periodicity clearly. The frequency of the lift force can be calculated by fluorinated ethylene propylene (FEP), as shown in Figure 5e; the frequency is 2 Hz, which is equal to the vortex shedding frequency calculated by the equation above. In addition, the natural frequency of the harvester is also 2 Hz. Therefore, after design and calculation, the harvester will reach resonance at 0.5 m/s.

To better understand the basic characteristics of VIV in water flow, we investigated the effects of speed and diameter on the lift force, amplitude, and frequency. When the diameter D = 50 mm, the effect of speed is shown in Figure 5f–h, with a flow speed range of 0.35–0.7 m/s. As shown in Figure 5f, the lift force increases slowly with the increase of flow speed. However, as shown in Figure 5g, the amplitude initially increases and then decreases with the flow speed and reaches the maximum value at 0.5 m/s. Figure 5h shows that the frequency is not sensitive to the speed, which is consistent with the above theoretical calculation results. When the flow speed U = 0.5 m/s, the effect of diameter is shown in Figure 5i,j, with a diameter D the range of 3–7 cm. And both the lift and the amplitude increase with the increase of diameter. Figure 5k shows that the frequency is stable at about 2 Hz, which is close to the natural frequency.

To investigate the influential factors of the SL-TENG, we simulated the potential distribution of the TENG by COMSOL Multiphysics software. In this approach, we neglected the influence of fluid speed to facilitate the simulation. Then, we chose ethanol (Figure 6a), deionized water (Figure 6b), and gallium (Figure 6c) as liquid triboelectric materials to investigate the effects of liquids on potential distribution. Furthermore, as shown in Figure 6d–f, we investigated the effects of the contact area between the deionized water and the FEP on potential distribution.

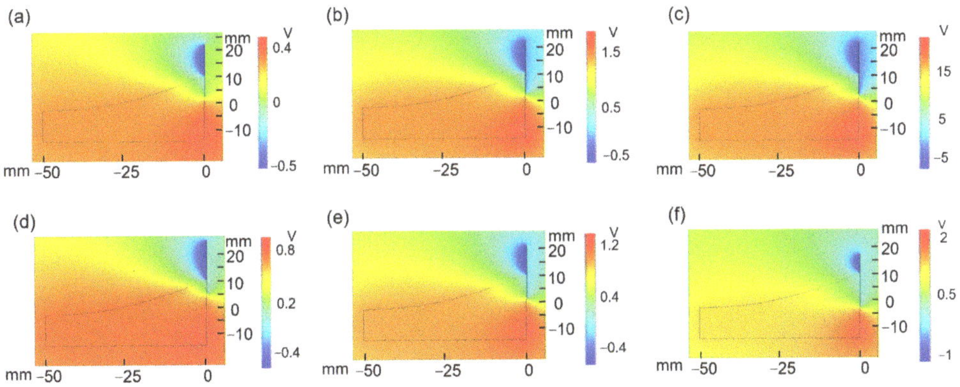

Figure 6. Finite element simulation. Potential distribution when the liquid is (**a**) ethanol, (**b**) deionized water, and (**c**) gallium; and (**d**) potential distribution when the contact area is 5 mm, (**e**) 10 mm, and (**f**) 15 mm.

As shown in Figure 6a–c, the potential distribution of SL-TENGs with different kinds of liquids demonstrated that the output performance of gallium was higher than those of the other two kinds of liquid. Subsequently, Figure 6d–f show that as the contact area between the deionized water and the FEP increased, the potential difference also increased. Thus, we assume that a larger contact area means more transferred charges.

3. Results and Discussion

3.1. Measurement and Characterization

As shown in Figure 7a, an experimental platform was set up on the upper part of the flume. The flume could provide continuous and stable water flow, and the flow speed was measured by a doppler current meter (P 25703), and the current speed monitoring is shown in Figure 7b. For the electrical signal of TENG, we employed an oscilloscope (Tektronix MDO3024) to measure the voltage. The two electrodes of the oscilloscope were connected to the ground and the copper electrode of TENG, respectively.

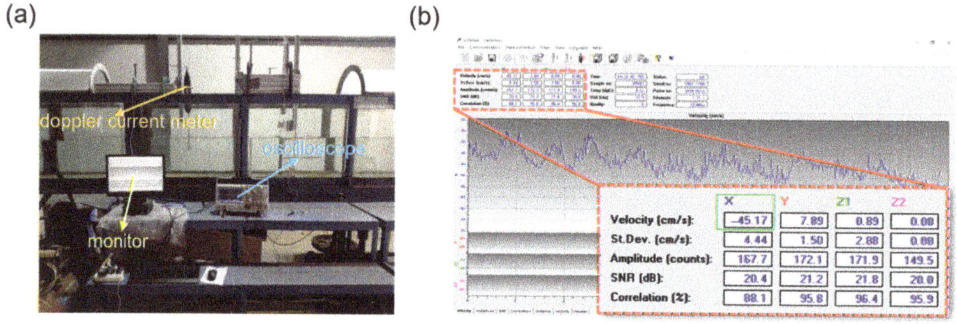

Figure 7. (**a**) Experimental platform. (**b**) Monitor.

3.2. Experimental Results

To illustrate the influence of the flow speed on VIV and the output performance of the SL-TENG, we conducted experiments under the same initial liquid level height and deflection angle, and we only adjusted the valve to control the flow speed of water in the flume. The energy harvester was fixed in the middle of the flume so that the cylinder part was completely submerged in the water, and the flow range was set to 0.35–0.75 m/s. Figure 8a–c show the snapshots of the VIV-SL-TENG at 0.4 m/s, 0.5 m/s, and 0.6 m/s, respectively. Figure 8b shows the cylinder presented a significant amplitude at 0.5 m/s. Furthermore, a series of vortices could also be distinctly seen behind the cylinder (Supplementary Video S1).

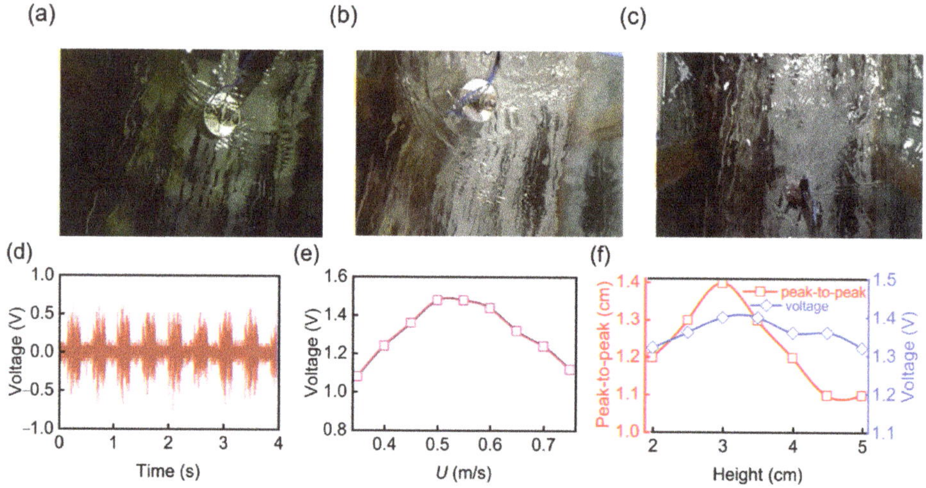

Figure 8. Snapshots of the energy harvester with the cylinder: (**a**) 0.4 m/s, (**b**) 0.5 m/s, (**c**) 0.6 m/s. (**d**) The output voltage of the generator under 0.5 m/s. (**e**) The output voltage of the generator under different flow velocities. (**f**) The output voltage and peak-to-peak of the generator under different heights.

Figure 8d shows the voltage signal of the VIV-SL-TENG when the flow speed was 0.5 m/s. Figure 8d shows that the voltage signal was stable, uniform, and periodic, with two peaks occurring within one second. This periodicity is in good agreement with the simulation of VIV above, indicating that the voltage was generated from the VIV. In addition, the voltage signal was constantly accompanied by noise, which was found to be caused by friction between the outer wall of the energy harvester and the flowing water after the experimental investigation. As shown in Figure 8e, the open-circuit voltage of VIV-SL-TENG increased first and then decreased with the increase of the water flow speed, reaching the maximum value of 1.48 V at about 0.5 m/s, and decreasing to 1.12 V at 0.75 m/s. This trend is consistent with the theoretical calculations presented above, as both lift and amplitude reached their maximum at 0.5 m/s. Furthermore, according to the calculation, the VIV-SL-TENG reached the resonance state at 0.5 m/s, where the amplitude reached the maximum. At this time, the contact area between the deionized water and the friction plate was the maximum, and thus the voltage reaching the maximum value. When the flow speed was faster or slower than 0.5 m/s, the energy harvester could not achieve the maximum amplitude state, and thus the voltage produced was less than that at 0.5 m/s.

To maximize the efficiency of the TENG, it is necessary to increase the contact area and the contact speed between the deionized water and FEP as much as possible. To explore the factors affecting the motion state of the deionized water in the cylinder, we carried out simulation analysis by STAR-CCM+. For different initial liquid level heights, as shown in Figure 8f, with the increase of the initial liquid level height, the peak-to-peak value initially increased and then decreased, and reaching the maximum at 3 cm. In addition, we conducted an experimental study on the influence of the liquid level height. The installation position of the energy harvester was consistent, and the water flow speed was maintained at 0.5 m/s. The influence of the initial liquid level height in the range of 2–5 cm on the voltage output was studied. As shown in Figure 8f, the voltage response also initially increased and then decreased, and reached the maximum output at 3 cm, with a maximum voltage of 1.4 V and a minimum voltage at 5 cm, with a voltage of 1.32 V. The trend of voltage is consistent with the peak value of the liquid level, reaching the maximum value at 3 cm, which once again verifies the influence of contact area on voltage output.

Then, we executed simulations and experimental research on the influence of the deflection direction of the friction plate inside VIV-SL-TENG. First, we simulated the motion state of the liquid inside the cylinder by STAR-CCM+. After preliminary analysis, we varied the deflection angle from 0° to 360°, with an interval of 36°. The influence of the deflection angle on the peak-to-peak value was explored through 3D CFD simulation, with the friction plate positioned at the edge of the cylinder. As shown in Figure 9b, when the vibration direction was perpendicular to the friction plate, the peak-to-peak value reached the maximum, while when the friction plate was parallel to the vibration direction, the peak-to-peak value reached the minimum.

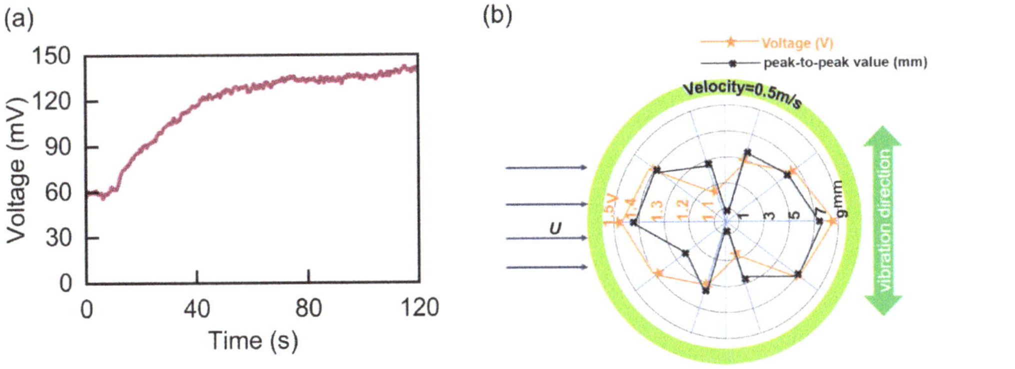

Figure 9. (**a**) Charging performance of the capacitors 0.5 m/s. (**b**) The output voltage and peak-to-peak of the generator under different deflection angles.

In addition, we conducted experiments to explore the effect of deflection angle on the output voltage. From the previous experiments, the VIV-SL-TENG achieved the highest voltage output at a flow speed of 0.5 m/s. Therefore, the deflection angle experiment was set at U = 0.5 m/s, with the initial height of the liquid level was 3 cm, and the friction plate was deflected by 36° each time. As shown in Figure 9b, the vibration direction of the energy harvester was perpendicular to the water flow direction. When the direction of the friction plate was perpendicular to the vibration direction, the voltage reached the maximum, namely, 1.44 V. The voltage gradually decreased with the deflection angle increases until the friction plate was parallel to the vibration direction. Hence, the deflection angle presented a significant impact on the voltage output of the SL-TENG. The peak-to-peak value of the liquid level obtained by the simulation was consistent with the voltage output, indicating that the voltage output of the energy harvester was closely related to the peak-to-peak value of the liquid level. In addition, we carried out a charging experiment under 0.5 m/s, as shown in Figure 9a; we charged a smaller capacitor, which was 22 μF. The corresponding voltage stabilized and reached the maximum at 120 s, with a voltage of 147 mV. In summary, the experimental results indicated that the VIV-SL-TENG performed well in the vortex-induced resonance. And further work will focus on optimizing the system and exploring the influence of the shape of the bluff body on VIV-SL-TENG.

4. Conclusions

In summary, we proposed a novel VIV-SL-TENG. Furthermore, we simulated and analyzed the lift and amplitude of the cylinder under the impact of the water flow via the CFD method. Subsequently, we established a mathematical model to calculate the responses of VIV and discussed the influence factors of VIV through a parametric study. Based on theoretical and simulation analysis, we prototyped the concept and conducted extensive experiments with different flow velocities and deflection angles. The experimental results indicated that the harvester performed well in the resonance state. The voltage output reached a maximum of 1.48 V at a flow velocity of 0.5 m/s. Furthermore, we investigated the influence of initial liquid level height and deflection angle by experiments, and the voltage output reached the peak at approximately 3 cm, and the deflection angle was positively correlated with the peak-to-peak value of the liquid level. The voltage reached the maximum when the direction of the friction plate was perpendicular to the vibration direction. As a whole, the VIV-SL-TENG shows promising potential in developing a fluid kinetic energy harvester.

Supplementary Materials: The following supporting information can be downloaded at: https://www.mdpi.com/article/10.3390/nano13061036/s1, Video S1: The VIV-SL-TENG vibrates at different flow velocities.

Author Contributions: X.L.: Supervision, Resources, Writing—review and editing. D.Z. (Di Zhang): Conceptualization, Methodology, Writing—original draft, Validation, Data curation. D.Z. (Dan Zhang): Supervision, Resources, Writing—review and editing. Z.L.: Supervision, Resources, Writing—review and editing. H.W.: Writing—review and editing. Y.Z.: Writing—review and editing. B.W.: Writing—review and editing. H.G.: Writing—review and editing. Y.P.: Supervision, Resources. All authors have read and agreed to the published version of the manuscript.

Funding: This work was funded by the National Natural Science Foundation of China (No. 62001281), the Shanghai Science and Technology Committee (No. 22dz1204300), and the Shanghai Outstanding Academic Leaders Plan (No. 20XD1421700).

Data Availability Statement: The data that support the findings of this study is available from the corresponding author upon reasonable request.

Acknowledgments: This work was supported by the National Natural Science Foundation of China (No. 62001281), the Shanghai Science and Technology Committee (No. 22dz1204300), and the Shanghai Outstanding Academic Leaders Plan (No. 20XD1421700).

Conflicts of Interest: The authors declare no conflict of interest.

References

1. Ankanahalli Shankaregowda, S.; Sagade Muktar Ahmed, R.F.; Liu, Y.; Bananakere Nanjegowda, C.; Cheng, X.; Shivanna, S.; Ramakrishna, S.; Yu, Z.; Zhang, X.; Sannathammegowda, K. Dry-Coated Graphite onto Sandpaper for Triboelectric Nanogenerator as an Active Power Source for Portable Electronics. *Nanomaterials* **2019**, *9*, 1585. [CrossRef] [PubMed]
2. Chen, T.; Shi, Q.; Li, K.; Yang, Z.; Liu, H.; Sun, L.; Dziuban, J.A.; Lee, C. Investigation of Position Sensing and Energy Harvesting of a Flexible Triboelectric Touch Pad. *Nanomaterials* **2018**, *8*, 613. [CrossRef] [PubMed]
3. Zhang, D.; Wang, D.; Xu, Z.; Zhang, X.; Yang, Y.; Guo, J.; Zhang, B.; Zhao, W. Diversiform sensors and sensing systems driven by triboelectric and piezoelectric nanogenerators. *Coord. Chem. Rev.* **2021**, *427*, 213597. [CrossRef]
4. Chen, H.; Xing, C.; Li, Y.; Wang, J.; Xu, Y. Triboelectric nanogenerators for a macro-scale blue energy harvesting and self-powered marine environmental monitoring system. *Sustain. Energy Fuels* **2020**, *4*, 1063–1077. [CrossRef]
5. Facchinetti, M.L.; de Langre, E.; Biolley, F. Coupling of structure and wake oscillators in vortex-induced vibrations. *J. Fluids Struct.* **2004**, *19*, 123–140. [CrossRef]
6. Bearman, P.W. Vortex shedding from oscillating bluff bodies. *Annu. Rev. Fluid Mech.* **1984**, *16*, 195–222. [CrossRef]
7. Moe, G.; Wu, Z.J. The lift force on a cylinder wibrating in a current. *J. Offshore Mech. Arct. Eng.* **1990**, *112*, 297–303. [CrossRef]
8. Zhou, C.Y.; So, R.M.C.; Lam, K. Vortex-induced vibrations of an elastic circular cylinder. *J. Fluids Struct.* **1999**, *13*, 165–189. [CrossRef]
9. The lift and drag forces on a circular cylinder oscillating in a flowing fluid. *Proc. R. Soc. London. Ser. A Math. Phys. Sci.* **1997**, *277*, 51–75. [CrossRef]
10. Parkinson, G. Phenomena and modelling off low-induced vibrations of bluff bodies. *Prog. Aerosp. Sci.* **1989**, *26*, 169–224. [CrossRef]
11. Wang, J.; Gu, S.; Zhang, C.; Hu, G.; Chen, G.; Yang, K.; Li, H.; Lai, Y.; Litak, G.; Yurchenko, D. Hybrid wind energy scavenging by coupling vortex-induced vibrations and galloping. *Energy Convers. Manag.* **2020**, *213*, 112835. [CrossRef]
12. Mehdipour, I.; Madaro, F.; Rizzi, F.; De Vittorio, M. Comprehensive experimental study on bluff body shapes for vortex-induced vibration piezoelectric energy harvesting mechanisms. *Energy Convers. Manag. X* **2022**, *13*, 100174. [CrossRef]
13. Toma, D.M.; del Rio, J. Underwater energy harvesting system based on plucked-driven piezoelectrics. In *OCEANS 2015-Genova*; IEEE: Piscataway, NJ, USA, 2015.
14. Zhang, B.; Mao, Z.; Wang, L.; Fu, S.; Ding, W. A novel V-shaped layout method for VIV hydrokinetic energy converters inspired by geese flying in a V-Formation. *Energy* **2021**, *230*, 120811. [CrossRef]
15. Gong, Y.; Shan, X.; Luo, X.; Pan, J.; Xie, T.; Yang, Z. Direction-adaptive energy harvesting with a guide wing under flow-induced oscillations. *Energy* **2019**, *187*, 115983. [CrossRef]
16. Zhang, B.; Mao, Z.; Song, B.; Tian, W.; Ding, W. Numerical investigation on VIV energy harvesting of four cylinders in close staggered formation. *Ocean Eng.* **2018**, *165*, 55–68. [CrossRef]
17. Wang, J.; Sun, S.; Tang, L.; Hu, G.; Liang, J. On the use of metasurface for Vortex-Induced vibration suppression or energy harvesting. *Energy Convers. Manag.* **2021**, *235*, 113991. [CrossRef]
18. Zhang, M.; Zhang, C.; Abdelkefi, A.; Yu, H.; Gaidai, O.; Qin, X.; Zhu, H.; Wang, J. Piezoelectric energy harvesting from vortex-induced vibration of a circular cylinder: Effect of Reynolds number. *Ocean Eng.* **2021**, *235*, 109378. [CrossRef]
19. Gu, M.; Song, B.; Zhang, B.; Mao, Z.; Tian, W. The effects of submergence depth on Vortex-Induced Vibration (VIV) and energy harvesting of a circular cylinder. *Renew. Energy* **2020**, *151*, 931–945. [CrossRef]
20. Mei, Y.-F.; Zheng, C.; Aubry, N.; Li, M.-G.; Wu, W.-T.; Liu, X. Active control for enhancing vortex induced vibration of a circular cylinder based on deep reinforcement learning. *Phys. Fluids* **2021**, *33*, 103604. [CrossRef]
21. Han, P.; Huang, Q.; Pan, G.; Wang, W.; Zhang, T.; Qin, D. Energy harvesting from flow-induced vibration of a low-mass square cylinder with different incidence angles. *AIP Adv.* **2021**, *11*, 025126. [CrossRef]
22. Du, T.; Ge, B.; Mtui, A.E.; Zhao, C.; Dong, F.; Zou, Y.; Wang, H.; Sun, P.; Xu, M. A Robust Silicone Rubber Strip-Based Triboelectric Nanogenerator for Vibration Energy Harvesting and Multi-Functional Self-Powered Sensing. *Nanomaterials* **2022**, *12*, 1248. [CrossRef]
23. Ji, S.H.; Cho, Y.S.; Yun, J.S. Wearable Core-Shell Piezoelectric Nanofiber Yarns for Body Movement Energy Harvesting. *Nanomaterials* **2019**, *9*, 555. [CrossRef] [PubMed]
24. Wang, C.; Shi, Q.; Lee, C. Advanced Implantable Biomedical Devices Enabled by Triboelectric Nanogenerators. *Nanomaterials* **2022**, *12*, 1366. [CrossRef] [PubMed]
25. Zhai, L.; Gao, L.; Wang, Z.; Dai, K.; Wu, S.; Mu, X. An Energy Harvester Coupled with a Triboelectric Mechanism and Electrostatic Mechanism for Biomechanical Energy Harvesting. *Nanomaterials* **2022**, *12*, 933. [CrossRef] [PubMed]
26. Sripadmanabhan Indira, S.; Aravind Vaithilingam, C.; Oruganti, K.S.P.; Mohd, F.; Rahman, S. Nanogenerators as a Sustainable Power Source: State of Art, Applications, and Challenges. *Nanomaterials* **2019**, *9*, 773. [CrossRef]
27. Wang, Z.L. From contact electrification to triboelectric nanogenerators. *Rep. Prog. Phys.* **2021**, *84*, 096502. [CrossRef]
28. Fan, F.-R.; Tian, Z.-Q.; Lin Wang, Z. Flexible triboelectric generator. *Nano Energy* **2012**, *1*, 328–334. [CrossRef]
29. Liu, Z.; Zhao, T.; Guan, H.; Zhong, T.; He, H.; Xing, L.; Xue, X. A self-powered temperature-sensitive electronic-skin based on tribotronic effect of PDMS/PANI nanostructures. *J. Mater. Sci. Technol.* **2019**, *35*, 2187–2193. [CrossRef]
30. Liu, J.; Liu, M.; Sun, C.; Lin, Z.; Feng, Z.; Si, S.; Yang, J. Triboelectric hydrophone for underwater detection of low-frequency sounds. *Nano Energy* **2022**, *99*, 107428. [CrossRef]

31. Wu, Y.; Zeng, Q.; Tang, Q.; Liu, W.; Liu, G.; Zhang, Y.; Wu, J.; Hu, C.; Wang, X. A teeterboard-like hybrid nanogenerator for efficient harvesting of low-frequency ocean wave energy. *Nano Energy* **2020**, *67*, 104205. [CrossRef]
32. Xu, L.; Pang, Y.; Zhang, C.; Jiang, T.; Chen, X.; Luo, J.; Tang, W.; Cao, X.; Wang, Z.L. Integrated triboelectric nanogenerator array based on air-driven membrane structures for water wave energy harvesting. *Nano Energy* **2017**, *31*, 351–358. [CrossRef]
33. Wang, H.; Xu, L.; Bai, Y.; Wang, Z.L. Pumping up the charge density of a triboelectric nanogenerator by charge-shuttling. *Nat. Commun.* **2020**, *11*, 4203. [CrossRef] [PubMed]
34. Xiao, T.X.; Liang, X.; Jiang, T.; Xu, L.; Shao, J.J.; Nie, J.H.; Bai, Y.; Zhong, W.; Wang, Z.L. Spherical Triboelectric Nanogenerators Based on Spring-Assisted Multilayered Structure for Efficient Water Wave Energy Harvesting. *Adv. Funct. Mater.* **2018**, *28*, 1802634. [CrossRef]
35. Cheng, B.; Xu, Q.; Ding, Y.; Bai, S.; Jia, X.; Yu, Y.; Wen, J.; Qin, Y. High performance temperature difference triboelectric nanogenerator. *Nat. Commun.* **2021**, *12*, 4782. [CrossRef]
36. Li, X.; Tao, J.; Wang, X.; Zhu, J.; Pan, C.; Wang, Z.L. Networks of High Performance Triboelectric Nanogenerators Based on Liquid-Solid Interface Contact Electrification for Harvesting Low-Frequency Blue Energy. *Adv. Energy Mater.* **2018**, *8*, 1800705. [CrossRef]
37. Pan, L.; Wang, J.; Wang, P.; Gao, R.; Wang, Y.-C.; Zhang, X.; Zou, J.-J.; Wang, Z.L. Liquid-FEP-based U-tube triboelectric nanogenerator for harvesting water-wave energy. *Nano Res.* **2018**, *11*, 4062–4073. [CrossRef]
38. Tang, W.; Jiang, T.; Fan, F.R.; Yu, A.F.; Zhang, C.; Cao, X.; Wang, Z.L. Liquid-Metal Electrode for High-Performance Triboelectric Nanogenerator at an Instantaneous Energy Conversion Efficiency of 70.6%. *Adv. Funct. Mater.* **2015**, *25*, 3718–3725. [CrossRef]
39. Yang, X.; Chan, S.; Wang, L.; Daoud, W.A. Water tank triboelectric nanogenerator for efficient harvesting of water wave energy over a broad frequency range. *Nano Energy* **2018**, *44*, 388–398. [CrossRef]
40. Nie, J.; Wang, Z.; Ren, Z.; Li, S.; Chen, X.; Wang, Z.L. Power generation from the interaction of a liquid droplet and a liquid membrane. *Nat. Commun.* **2019**, *10*, 2264. [CrossRef]
41. Zhu, S.; Yu, G.; Tang, W.; Hu, J.; Luo, E. Thermoacoustically driven liquid-metal-based triboelectric nanogenerator: A thermal power generator without solid moving parts. *Appl. Phys. Lett.* **2021**, *118*, 113902. [CrossRef]
42. You, J.; Shao, J.; He, Y.; Yun, F.F.; See, K.W.; Wang, Z.L.; Wang, X. High-Electrification Performance and Mechanism of a Water-Solid Mode Triboelectric Nanogenerator. *ACS Nano* **2021**, *15*, 8706–8714. [CrossRef] [PubMed]
43. Lu, C.X.; Han, C.B.; Gu, G.Q.; Chen, J.; Yang, Z.W.; Jiang, T.; He, C.; Wang, Z.L. Temperature Effect on Performance of Triboelectric Nanogenerator. *Adv. Eng. Mater.* **2017**, *19*, 1700275. [CrossRef]
44. Zhang, L.; Li, X.; Zhang, Y.; Feng, Y.; Zhou, F.; Wang, D. Regulation and influence factors of triboelectricity at the solid-liquid interface. *Nano Energy* **2020**, *78*, 105370. [CrossRef]
45. Ma, L.; Wu, R.; Patil, A.; Yi, J.; Liu, D.; Fan, X.; Sheng, F.; Zhang, Y.; Liu, S.; Shen, S.; et al. Acid and Alkali-Resistant Textile Triboelectric Nanogenerator as a Smart Protective Suit for Liquid Energy Harvesting and Self-Powered Monitoring in High-Risk Environments. *Adv. Funct. Mater.* **2021**, *31*, 2102963. [CrossRef]
46. Xie, X.; Chen, Y.; Jiang, J.; Li, J.; Yang, Y.; Liu, Y.; Yang, L.; Tu, X.; Sun, X.; Zhao, C.; et al. Self-Powered Gyroscope Angle Sensor Based on Resistive Matching Effect of Triboelectric Nanogenerator. *Adv. Mater. Technol.* **2021**, *6*, 2100797. [CrossRef]
47. Wang, S.; Wang, Y.; Liu, D.; Zhang, Z.; Li, W.; Liu, C.; Du, T.; Xiao, X.; Song, L.; Pang, H.; et al. A robust and self-powered tilt sensor based on annular liquid-solid interfacing triboelectric nanogenerator for ship attitude sensing. *Sens. Actuators A Phys.* **2021**, *317*, 112459. [CrossRef]

nanomaterials

Article

Improving the Performance of Polydimethylsiloxane-Based Triboelectric Nanogenerators by Introducing CdS Particles into the Polydimethylsiloxane Layer

Jianbin Mao and Soonmin Seo *

Department of Bionano Technology, Gachon University, Seongnam 13120, Republic of Korea; maojianbin@gachon.ac.kr
* Correspondence: soonmseo@gachon.ac.kr

Abstract: Energy harvesting and power generation technologies hold significant potential for meeting future energy demands and improving environmental sustainability. A triboelectric nanogenerator (TENG), which harnesses energy from the surrounding environment, has garnered significant attention as a promising and sustainable power source applicable in various fields. In this study, we present a technique to improve the triboelectric performance of a PDMS-based TENG by incorporating nanostructured cadmium sulfide (N-CdS). This study investigates the utilization of CdS nanomaterials in TENG production, where mechanical energy is converted into electrical energy. We conducted a comparative analysis of TENGs utilizing N-CdS/PDMS, commercial CdS/PDMS (C-CdS/PDMS), and pure PDMS substrates. The N-CdS/PDMS substrates demonstrated superior triboelectric performance compared to TENG devices based on pure PDMS and C-CdS/PDMS. The triboelectric open-circuit voltage (V_{oc}) and short-circuit current (I_{sc}) of the N-CdS/PDMS-based TENG device were approximately 236 V and 17.4 μA, respectively, when operated at a 2 Hz frequency. These values were approximately 3 times and 2.5 times higher, respectively, compared to the pure PDMS-based TENGs. They were further studied in detail to understand the effect of different parameters such as contact–separation frequency and contact force on the TENGs' operation. The stability of the TENG devices was studied, and their potential to be integrated into self-powered smart textiles as power sources was demonstrated.

Keywords: metal chalcogenides; CdS; TENG; output performance; nanomaterials

Citation: Mao, J.; Seo, S. Improving the Performance of Polydimethylsiloxane-Based Triboelectric Nanogenerators by Introducing CdS Particles into the Polydimethylsiloxane Layer. *Nanomaterials* **2023**, *13*, 2943. https://doi.org/10.3390/nano13222943

Academic Editor: Ning Wang

Received: 16 October 2023
Revised: 7 November 2023
Accepted: 10 November 2023
Published: 14 November 2023

1. Introduction

The topic of energy sources has long been a subject of significant focus particularly in relation to the utilization and scarcity of fossil fuels [1]. Therefore, energy harvesting is crucial for our daily lives and is receiving significant attention. A TENG is a well-designed method capable of converting mechanical energy from the environment into electrical energy through the coupling of triboelectrification and electrostatic induction [2]. According to the relevant literature, a TENG has the capability to convert mechanical energy from various external forces, such as wind, water, and even body motion, into electrical energy [3–5]. Since their introduction and invention by Wang's group in 2012, TENGs have been recognized as a promising method for energy harvesting and a sustainable renewable resource [6]. TENGs offer numerous advantages when compared to other methods of energy acquisition. These benefits include high-efficiency power conversion, a simple structure, ease of fabrication, low cost, durability, scalability, and the ability to harvest energy from low-frequency, irregular input sources [7,8]. Overall, due to these advantages, TENGs are becoming increasingly integrated into people's daily lives.

However, increasing the power output of TENGs is essential to meet their various practical applications [9], as the current output power of TENGs remains significantly below the ideal value. Moreover, there are various factors that can compromise the efficiency of

TENGs [10,11]. Hence, conducting comprehensive research to enhance performance and address various challenges faced by TENGs is crucial to advance their practical applications. Selecting and modifying materials, as well as employing surface treatment techniques, are crucial approaches for effectively enhancing a TENG's power output. For example, our group deposited Galinstan onto a polydimethylsiloxane (PDMS) substrate and designed a Galinstan-PDMS-based TENG, resulting in a significant increase in output power [12–14]. Kavarthapu et al. discovered that the output performance of TENG could be enhanced through the modulation of the loading concentration of $ZnSnO_3$ nanoparticles (NPs) in polyvinylidene fluoride hexafluoropropylene (PVDF-HFP) fibrous films [15]. Li et al. suggest the utilization of ordered mesoporous SiO_2 (OMS) nanoparticles, characterized by a substantial specific surface area, as efficient sites for body charge storage within OMS-PDMS to augment the output performance of TENGs [16]. The above studies highlight the significance of material modification and selection.

Metal chalcogenides are intriguing materials that have attracted significant attention [17]. In recent years, metal chalcogenides have found extensive applications, including batteries, photovoltaics, nanoelectronics, nanotribology, catalysis, and more [18–22]. Furthermore, metal chalcogenides have also entered the field of energy harvesting. For example, because ZnS possesses piezoelectric and triboelectric properties, Mishra and colleagues introduced ZnS nanosheets onto a surface-modified aluminum substrate [23]. Their report demonstrated that ZnS can significantly enhance the charge transfer rate between two tribolayers. CdS, as a representative member of metal chalcogenides and one of the most widely used II-VI semiconductors, has been extensively studied owing to its captivating physical and chemical properties [24]. CdS exists in two crystalline forms: hexagonal (wurtzite) and cubic (zinc blende), with the ability to produce CdS films in both phases [25]. CdS is categorized as an n-type semiconductor with a direct band gap of 2.42 eV in its cubic structure and 2.57 eV in its hexagonal form [26]. CdS offers numerous advantages, including a high absorption coefficient ($>10^4$ cm^{-1}), excellent photoelectric properties, strong reflectance in the infrared range, high transmittance in the visible spectrum, low work function, superior electron affinity, high carrier concentration (10^{16}–10^{18} cm^{-3}), high carrier mobility (0.1–10 cm^2 V^{-1} s^{-1}), high electron mobility (300 cm^2 V^{-1} s^{-1}), low resistivity, and exceptional thermochemical and electrochemical stability [27–30]. CdS is widely used in photosensors, photovoltaic devices, and solar cells [31]. Additionally, owing to the semiconductor and piezoelectric characteristics of CdS, it is employed in energy conversion and harvesting. Furthermore, there have been reports of piezoelectric nanogenerators (PENGs) incorporating CdS [32]. However, the use of CdS as a material for TENGs remains relatively underreported.

The polymer film PDMS is commonly used as a negative triboelectric layer in the fabrication of TENGs. In this study, we introduced a novel approach by utilizing the triboelectrification of a dielectric composite film composed of CdS and PDMS, resulting in the creation of an easily manufacturable and cost-effective TENG for the first time. The CdS/PDMS film was combined with a copper film as the triboelectric contact material, and the resulting CdS/PDMS-based TENG operated in a contact–separation mode. We systematically investigated the impact of introducing CdS on the output performance of the TENG by measuring the voltage and current when CdS was incorporated into the PDMS film. Initially, through a comparison with pure PDMS, we observed a significant improvement in the output performance of the TENG following the introduction of CdS. This enhancement primarily stemmed from the augmented contact area, accelerated charge transfer, and improved conductivity facilitated by the introduction of CdS. Our preparation involved the transformation of commercially available micron-level CdS into nanostructured CdS. The test results revealed that the performance of the nanostructured N-CdS was further amplified upon integration with PDMS, demonstrating a maximum output voltage of 236 V and a current of 17.4 μA. Furthermore, simulation under real circuit conditions, utilizing an external load resistor, indicated an output energy density of 4 $\mu W/m^2$ based on N-CdS/TENG. Finally, our TENG successfully illuminated 17 LED lights concurrently, underscoring its potential for practical applications.

2. Materials and Methods

2.1. Preparation of Nanostructured CdS

CdS nanomaterials were synthesized using commercial CdS (99.999% purity, Alfa Aesar, Tewksbury, MA, USA) through an ultrasonication method. Specifically, 4 g of commercial CdS was dispersed in deionized water. Subsequently, the solution was ultrasonicated for 16 h (ultrasound equipment: SONICS-VCX500, SONICS, Newtown, CT, USA). Additionally, the set power is 200 W. After ultrasonication, the CdS solution was allowed to settle for 2 h. The upper layer was then centrifuged at 3000 rpm for 40 min (centrifuge: UNIVERSAL 320R, Hettich, Beverly, MA, USA). Subsequently, the CdS nanomaterials were collected and dried in an oven.

2.2. Fabrication of TENGs

The CdS-based TENG fabrication process is illustrated in Figure 1. First, the PDMS (Sylgard 184, Dow Corning, Midland, MI, USA) was prepared by mixing base and curing agents at the weight percentage of 10/1 (*w*/*w*), and then CdS (commercial CdS, and nanostructured CdS) was dispersed into the PDMS elastomer at the concentration of 0.6 wt%. After mixing thoroughly, the mixture was placed in a container and subjected to a vacuum for 0.5 to 1 h. This step aimed to remove any dissolved air from the mixture. The solution was then used to fabricate a 200 μm thick layer on a polyethylene terephthalate (PET) substrate via a bar coating using a 4-side applicator. The PET tape was washed several times with water and ethanol. A 200 μm wire-wound rod was used to deposit the CdS/PDMS-based thin films. After depositing the CdS/PDMS on the PET, the composite thin membrane was dried in an oven at constant temperature of 60 °C for approximately 2 h. After the curing step, the cured composite thin film with CdS was sliced to 2.5 × 2.5 cm^2. The TENG device was fabricated using a simple two-electrode system. For the bottom electrode, copper foil was utilized as the electrode with the thin composite film (CdS/PDMS based PET tape) directly attached to it. The upper electrode, made of copper, functioned as both the charge-collecting electrode (CCE) and the frictional electrode (FE). Electrical contact was established through Cu wire connecting the bottom and upper electrodes. Subsequently, the TENG was constructed.

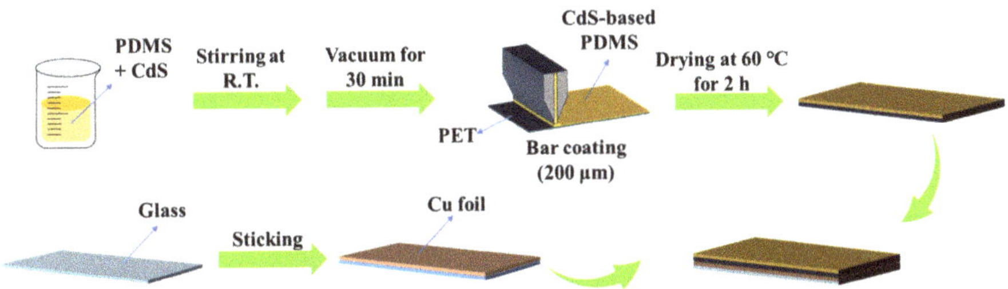

Figure 1. Schematic illustration of the CdS-based TENG fabrication process.

2.3. Characterization

Field emission scanning electron microscope (FESEM, JSM-7500F, JEOL, Tokyo, Japan) analysis was conducted. UV-visible absorption spectra were obtained using a Shimadzu UV-2550 spectrometer. FT-IR spectra were obtained using a Thermo Fisher Scientific (iS50) FT-IR spectrometer.

2.4. Measurement of TENGs

To investigate the enhanced triboelectricity generation due to the nanostructured CdS particle layer, we measured the voltage and current of the device. The open circuit voltage was measured using an oscilloscope (MDO34, Tektronix, Beaverton, OR, USA), while an electrometer (Keithley-6514, Keithley, Cleveland, OH, USA) quantified the short-circuit current. A linear motor served as a controlled vibration source during electrical performance tests, with the motor integrated with a force sensor to measure applied force on the TENG.

3. Results and Discussion

To investigate the surface morphology and size of CdS nanomaterials prepared using ultrasonication, we conducted a study using FE-SEM and compared them to C-CdS (referred to as original CdS), as depicted in Figure 2. Figure 2a,b display the morphology and size of the C-CdS. From Figure 2a, it is evident that C-CdS predominantly exhibits a blocky morphology reminiscent of variously shaped and sized gravel. We illustrate the size distribution of C-CdS through size measurements. In Figure 2b, the size distribution of C-CdS ranges from 1 to 10 μm. Based on our measurements, C-CdS with a size between 2 and 5 μm constitutes the majority, indicating that the C-CdS is at the micron level in size. Figure 2c–f depict the morphology and size distribution of N-CdS obtained through ultrasonic treatment of the C-CdS. A comparison of Figure 2c with Figure 2a reveals a substantial reduction in the size of CdS. This result suggests that with a specific ultrasonic duration, the CdS size can be diminished, which is consistent with prior research findings [33]. Further size analysis indicates that cadmium sulfide nanomaterials primarily range from a few hundred nanometers to 2.5 μm with their maximum size considerably smaller than the original CdS. Furthermore, the result confirms that the ultrasonic method can produce sub-micrometer-sized CdS. While the size distribution diagram reveals some CdS exceeding 1 μm, this could be attributed to uneven ultrasonication or timing issues. We anticipate that extending the ultrasonication duration will yield smaller CdS. In Figure 2e,f, we discern a clearer view. Here, C-CdS treated by ultrasonication predominantly exhibits particle agglomerates of varying sizes primarily within the range of several hundred nanometers. The variations in size are likely due to differences in the original CdS, and the particle agglomerates morphology may be linked to the layered structure of CdS and the bulk morphology of C-CdS, as the ultrasonic process may resemble a stripping process. Furthermore, however, in comparison to the C-CdS, N-CdS exhibits a tendency to stack, primarily attributed to the size reduction, which promotes material agglomeration. In summary, the findings above demonstrate the successful transformation of micron-level CdS into nanoscale CdS through a straightforward ultrasonication process.

Next, we examined the optical properties of CdS nanomaterials. Figure 3a exhibits the UV-visible spectrum of CdS nanomaterials and commercial CdS sample obtained through the recording of absorption spectra using DI water as a reference. Consistent with prior findings, a distinct peak at 290 nm was observed, confirming its association with CdS nanomaterials [34]. This result further attests to the successful synthesis of nanostructured N-CdS via a straightforward ultrasonication method and highlights the strong electron confinement effect in N-CdS. Moreover, we employed FT-IR analysis to investigate CdS, PDMS, and CdS/PDMS. Figure 3b does not display the anticipated peak corresponding to CdS. Upon comparing the primary absorption peaks of PDMS before and after the introduction of CdS, no noticeable disparity was observed between the two. In the infrared spectrum (Figure 3b), the absorption peak near 2964 cm^{-1} is attributed to the C-H stretching vibration of CH_3, and the peak near 1259 cm^{-1} arises from the symmetrical bending of CH_3 in Si-CH_3. The appearance of peaks near 1067 cm^{-1} and 1010 cm^{-1} can be linked to the formation of Si-O-Si bonds, while the 787 cm^{-1} absorption peak emerges due to the swinging motion of CH_3 in Si-CH_3. These results indicate that the introduction of CdS did not significantly alter the primary chemical bonds of PDMS, which was possibly due to the minimal quantity of CdS introduced.

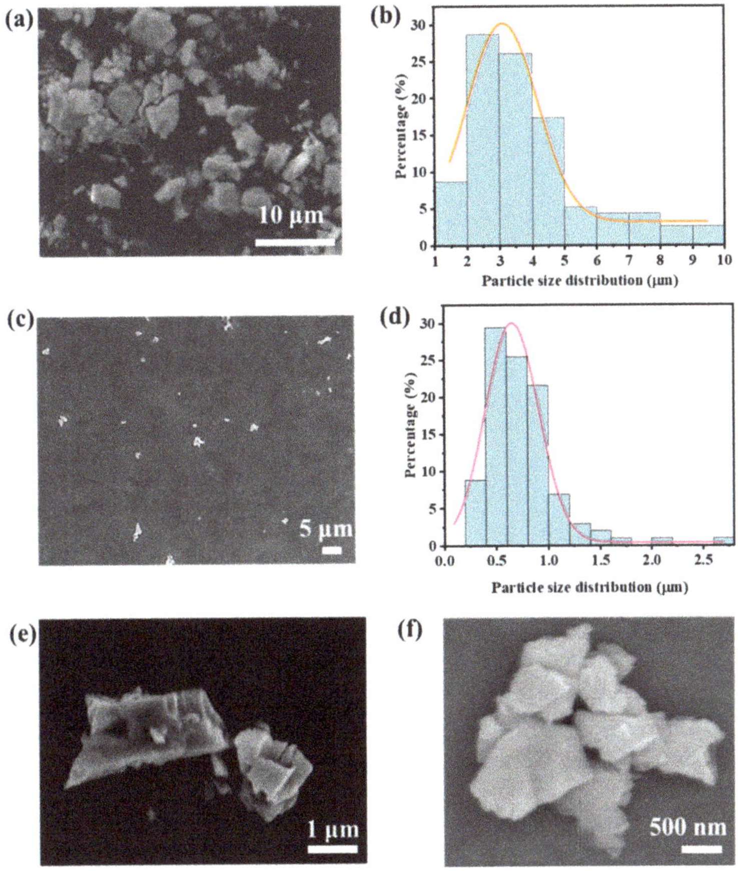

Figure 2. Morphology and size distribution of CdS powder. (**a**) SEM image of C-CdS powder (bar = 10 μm). (**b**) Particle size distribution of C-CdS powder. (**c**) SEM image of N-CdS powder (bar = 5 μm). (**d**) Particle size distribution of N-CdS powder. (**e**,**f**) Morphology of N-CdS power (bar = 1 μm and 500 nm).

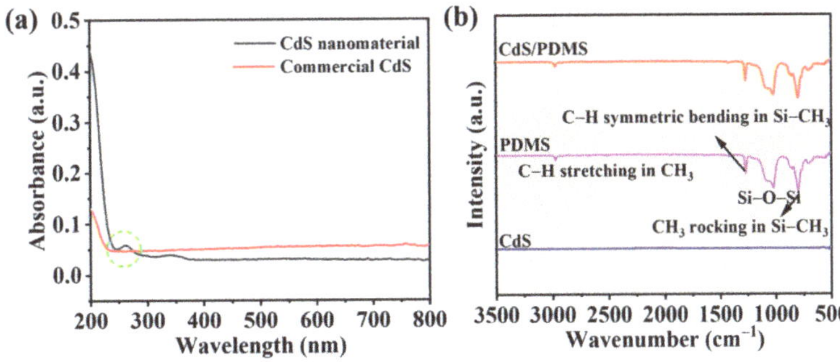

Figure 3. (**a**) UV-visible absorption spectrum for N-CdS and C-CdS dispersion in water. (**b**) FTIR spectra of CdS, PDMS, and CdS/PDMS.

In this work, CdS was utilized as a co-triboelectric layer material in the TENG. Figure 4 illustrates a schematic diagram depicting the operating mechanism of the CdS-PDMS-based TENG. When no external force is applied, the TENG is in its initial state (Figure 4-i), where the top and bottom sides are separated. When a force is applied to the TENG (Figure 4-ii), the top and bottom sides come into contact, inducing negative and positive electrostatic charges on the surfaces the upper electrode and the lower electrode, respectively. The separation of the Cu layer and the CdS/PDMS composites results in the generation of negative and positive electrostatic charges on the surfaces of both the upper electrode layer and the lower electrode layer, respectively, through electrostatic induction. Consequently, current flows from the top to the bottom side of the TENG (Figure 4-iii). Once the electrostatic charges reach equilibrium, no current flows in the external circuit (Figure 4-iv). Moreover, when a force is applied to the TENG, reverse current flows from the bottom to the top side (Figure 4-v). Thus, current continuously flows between the bottom and top sides of the TENG as a force is continuously applied and removed.

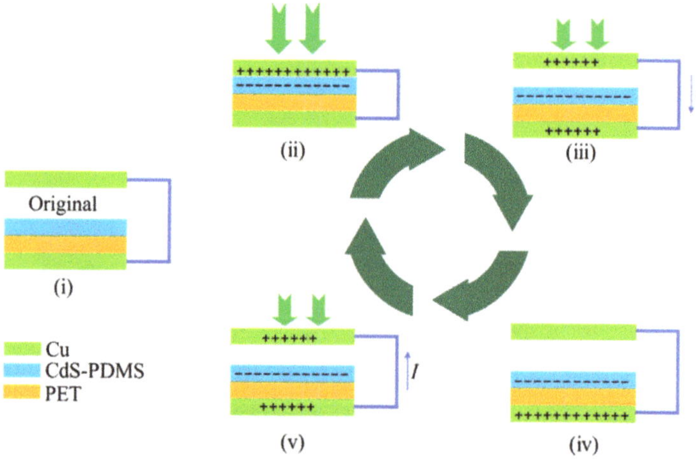

Figure 4. Scheme of working mechanisms and charge generation processes. (i) In initial position with no charge on their faces. (ii) The pressed state. (iii) The releasing processes. (iv) The released state. (v) The pressing processes.

A TENG with an area of 2.5 cm × 2.5 cm was used to measure triboelectric energy. The spacing between the friction layers was 45 mm, and it operated at a frequency of 2 Hz with an applied force of 5 N, as shown in Figure 5a. Initially, we altered the arrangement of the interconnection between the friction layers to assess any impact on the output voltage and current. The results are depicted in Figure 5b,c. It was observed that changing the interconnection direction solely affected the polarity of the voltage and current while leaving their magnitudes unaffected. This finding underscores the independence of the output voltage and current from the interconnection direction, which aligns with earlier research findings [35]. The performance of the TENG primarily depends on the spacing between the friction layers, the dielectric constant, the design structure, the friction frequency, and other relevant factors. Different friction layers, including PDMS, C-CdS/PDMS, and N-CdS/PDMS, were used to investigate and compare their output properties within the TENG system. As shown in Figure 5d, the voltage output when using PDMS as the friction layer measured 80 V, while C-CdS/PDMS and N-CdS/PDMS produced significantly higher output voltages of 204 V and 236 V, respectively, corresponding to approximately 2.6 times and 3 times that of pure PDMS. Likewise, Figure 5e illustrates that the current output with PDMS as the friction layer was 7.2 μA, while using C-CdS/PDMS and N-CdS/PDMS as the friction layer resulted in output currents of 16 μA and 17.2 μA, respectively, representing

2.2 times and 2.4 times the original PDMS output. Evidently, incorporating CdS into the PDMS as the friction layer markedly amplifies both the output voltage and current of the TENG. These results substantiate the significant performance enhancement achieved by the introduction of CdS into the PDMS-based TENG system. The superior output performance of the CdS/PDMS-based TENG can be attributed to the following reasons: (i) the introduction of CdS significantly enhances the charge transfer rate and conductivity; and (ii) CdS provides a rough surface, offering a larger contact area compared to pure PDMS. CdS is a piezoelectric material, as highlighted in the Introduction. Combining the piezoelectric and triboelectric effects can significantly enhance the performance of TENG [36]. When a mechanical force is applied, the triboelectric material PDMS, containing piezoelectric properties of CdS, undergoes deformation, leading to an immediate manifestation of piezoelectricity. The supplementary surface charges resulting from the piezoelectric effect of CdS reinforce the triboelectric charge during simultaneous contact electrification, ultimately boosting the device's electrical outputs. For TENGs utilizing N-CdS/PDMS fabricated from nanostructured CdS as the friction layer, the observed output performance surpasses that of C-CdS/PDMS derived from micron-sized CdS. This enhancement can be primarily attributed to the smaller size of the nanoscale N-CdS, as evident from the SEM image in Figure 2. Additionally, the evaluation of their transferred charge densities reveals that N-CdS/PDMS exhibits the highest transferred charge density, measuring 55 μC/m2, as shown in Figure 5f. This outcome is a result of the accelerated charge transfer by the nanostructure. Nevertheless, the extent of performance improvement in comparison to C-CdS/PDMS as the friction layer is not substantial, potentially owing to the limited dispersion effect of N-CdS within the PDMS matrix. In this work, N-CdS demonstrates inferior dispersibility in PDMS compared to C-CdS. This inadequate dispersion is primarily attributed to the tendency of nanomaterials to agglomerate. Finally, as illustrated in Figure 5g, the superior output performance of N-CdS/PDMS verifies the effectiveness of our approach in enhancing TENG performance by inserting N-CdS into the PDMS matrix.

For a comprehensive evaluation of the electrical output performance, we tested the TENG composed of Cu foil and an N-CdS/PDMS membrane at various frequencies and tapping pressures. Figure 6a–c illustrate the variations in output voltage, short-circuit current, and charge transfer density of the TENG across different frequencies. All these parameters exhibited an increasing trend as the frequencies were raised. Specifically, within the frequency range of 0.5 to 2 Hz, the peak-to-peak voltage increased from 57.6 to 236 V, the peak-to-peak short-circuit current increased from 6.4 to 17.4 μA, and the charge transfer density increased from 41 to 55 $\mu m/m^2$, as shown in Figure 6d. Based on the data presented in Figure 6a–d, it is evident that increasing the frequency substantially enhances the output performance, aligning with findings from prior research. The experimental results reaffirm that augmenting the frequency is a viable method for enhancing the performance of TENGs utilizing N-CdS/PDMS as the friction layer. Furthermore, we examined the impact of an applied force on the TENG's output performance. The results revealed that with an increase in applied pressure from 1 to 5 N, the output voltage increased from 104 to 234 V, as depicted in Figure 6e. This observation indicates that increasing the external pressure effectively boosts the TENG's output voltage. Similarly, in our investigation of the effect of an external force on the TENG's output current, as illustrated in Figure 6f, we observed that as the applied pressure rose from 1 to 5 N, the output current increased from 10.2 to 17.4 μA. Thus, it can be inferred that increasing the applied pressure results in a corresponding increase in the output current. Finally, we conducted a study on the impact of an applied force on the charge transfer of the TENG. In contrast to the output voltage and current, our findings indicated that augmenting the applied force has no discernible effect on the charge transfer of the TENG. In TENG studies, it is a common observation that the output power rises with both frequency and applied pressure. Increasing the frequency enhances energy transfer efficiency, consequently raising the maximum output power. Additionally, the increase in pressure leads to a rise in friction force ($f = \mu N$, where f represents the friction force, μ is the coefficient of friction, and N is the normal force), thus generating more frictional energy and augmenting the triboelectric output. We demonstrated that the output performance increased tremendously with frequency and applied pressure. Figure 6i illustrates the relationship between the force and output voltage. Increasing the applied force results in the rise of the output voltage of the TENG. In essence, with adjustments to certain parameters and calibration of the instrument, it holds significant promise for its application in mechanical sensing. The TENG's response to varying forces aligns with trends reported in prior research [37]. Hence, the fabricated TENG has the potential to serve as a force sensor when calibrated against a known force reference.

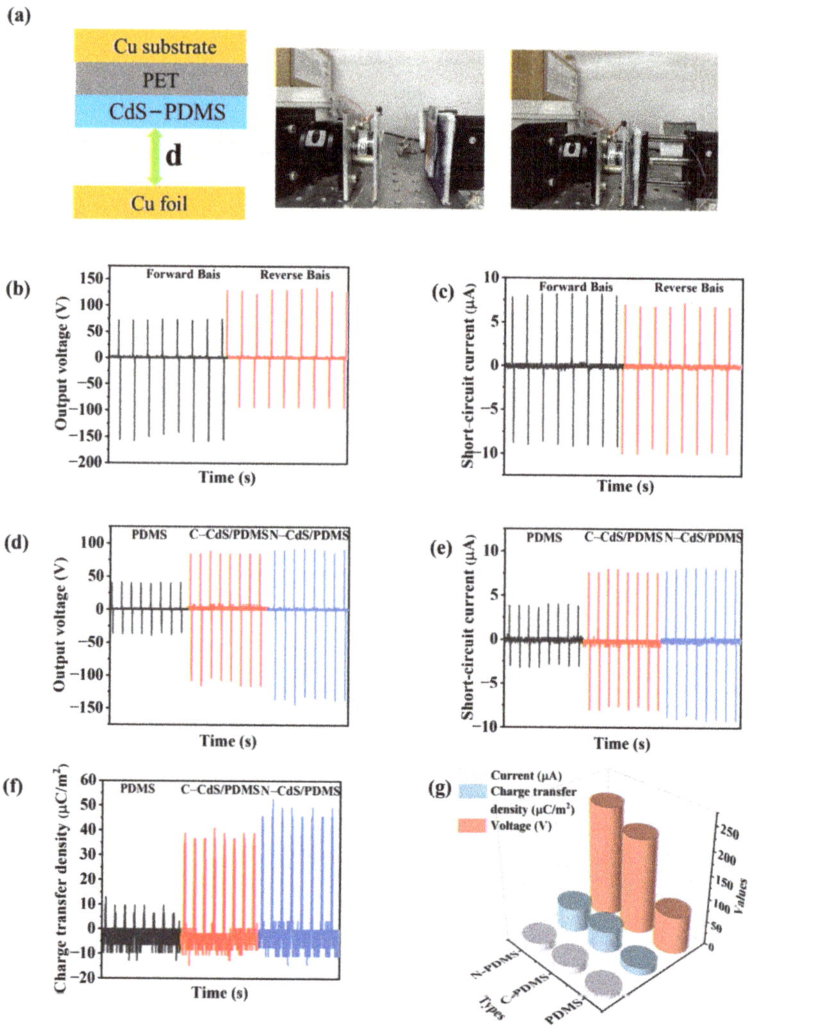

Figure 5. (**a**) Schematic illustration of the CdS-PDMS-based TENG and the image of test setup. (**b**) Forward–reverse open-circuit output voltage characteristics of TENG. (**c**) Forward–reverse short-circuit current of TENG. The output voltage (**d**), current (**e**), and charge transfer density (**f**) signals were harvested from the TENG device with pure PDMS, C-CdS/PDMS, and N-CdS/PDMS. (**g**) Comparison of the output performance values of TENGs based on PDMS, C-CdS/PDMS, and N-CdS/PDMS, including output voltage, current, and charge transfer density.

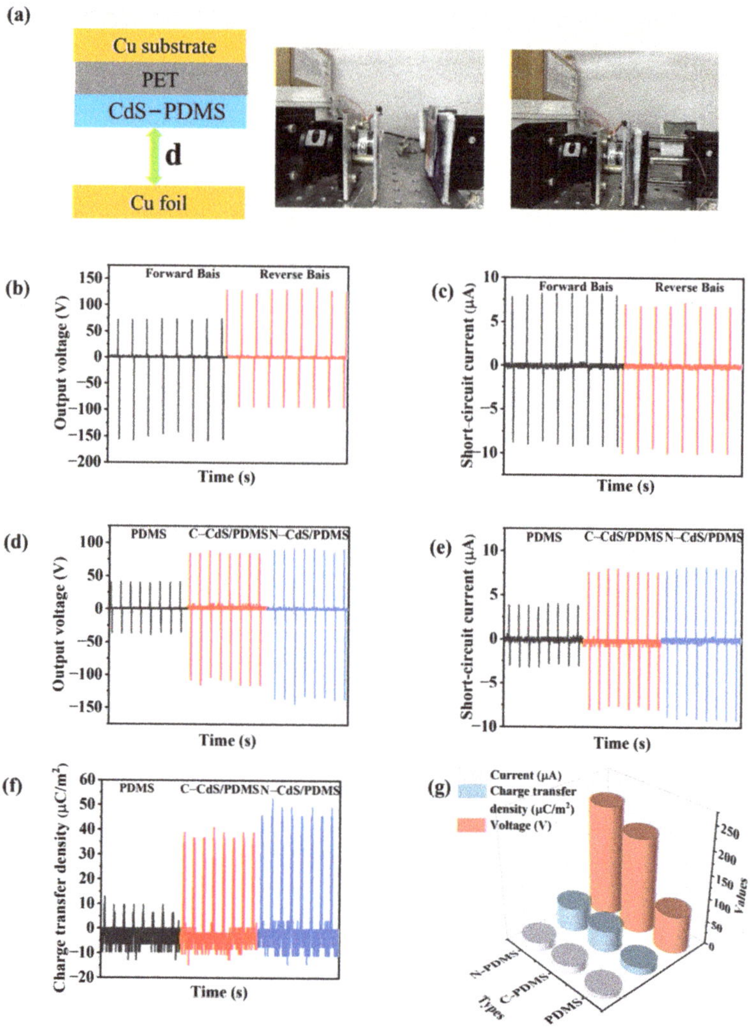

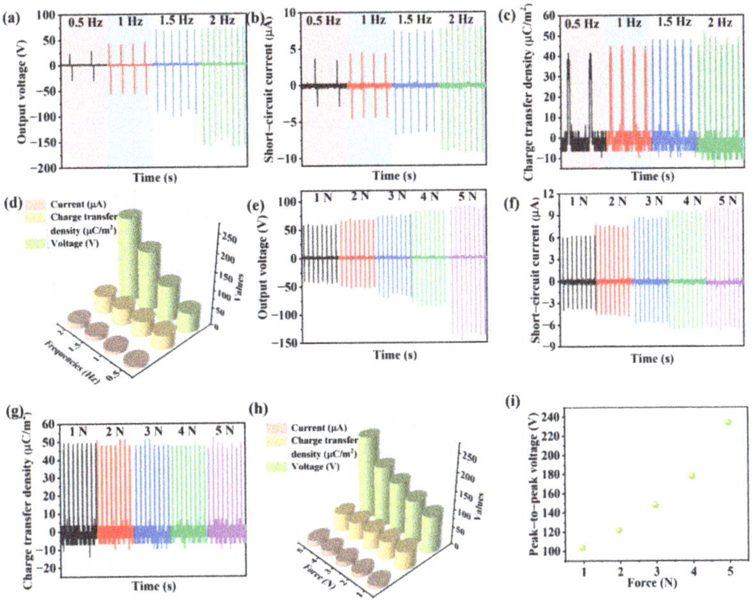

Figure 6. (**a**) Output voltage, (**b**) output current, and (**c**) charge transfer density under loading frequencies of 0.5, 1, 1.5, and 2 Hz. (**d**) Comparison of the output performance values of TENGs based on different loading frequencies, including output voltage, current, and charge transfer density. The output voltage (**e**), output current (**f**), and charge transfer density (**g**) under different loading forces. (**h**) Comparison of the output performance values of TENGs based on different loading forces, including output voltage, current, and charge transfer density. (**i**) Average output voltage as a function of different applied forces.

To explore the maximum output performance of N-CdS/PDMS-based TENGs in practical circuits and facilitate a comparison with C-CdS-PDMS-based TENGs, we conducted tests on their output energy density with an external load resistor connected. We applied load resistance to the circuit using a custom-made resistor box capable of providing resistances ranging from 10 KΩ to 900 MΩ. We then measured the resulting output voltage and current under a force of 5 N at a frequency of 2 Hz. Figure 7a,b display the output voltage and current of the N-CdS/PDMS-based TENG under various external load resistances, respectively. For a clearer comparison, we plotted the peak values of the output voltage and current against different external load resistors in Figure 7c. As shown in Figure 7c, the response of the output voltage and current varies significantly with increasing external load resistance. The output voltage steadily increased with higher external load resistance, reaching 226 V in the end. Notably, the voltage exhibits rapid growth when the load resistance was below 100 MΩ, which was followed by a considerable slowdown beyond this threshold, nearly reaching stagnation. On the contrary, the behavior of the output current is distinctly different. As the external load resistance rises, the output current notably declines, particularly before the 100 MΩ mark, which was followed by a slower rate of decrease beyond this point. The highest output current achieved at 200 KΩ was 23.2 μA. These divergent trends in voltage and current changes are primarily attributed to the circuit voltage and inherent factors of the TENG. The power density exhibited a rapid increase to a maximum value and subsequently decreased (Figure 7d). The peak power of a TENG can be calculated using the following equation:

$$P_{peak} = UI = U^2/R \quad (1)$$

where U is the output voltage, I is the short-circuit current, and R is the load resistance. The calculated P_{peak} values initially rose, which was followed by a decline as R increased. The TENG achieved a maximum power density of approximately 4 W/m^2 at a load resistance of 9 MΩ. Subsequently, to gain a comprehensive understanding of the impact of small-sized CdS on TENG output performance, we investigated TENGs based on C-CdS/PDMS. As illustrated in Figure 7e, upon the addition of a load resistor, the output voltage of the C-CdS/PDMS-based TENG initially demonstrates rapid growth, eventually stabilizing around 184 V. Similarly, the current exhibits a similar pattern, peaking at 15 μA before rapidly declining and then gradually stabilizing. Although both C-CdS/PDMS and N-CdS/PDMS-based TENGs exhibit analogous changes in output current and voltage following the introduction of a load resistor, the peak voltage and current of the N-CdS/PDMS-based TENG surpass those of the C-CdS/PDMS-based counterpart. This observation further underscores the effective enhancement of TENG performance through modifying the sized of CdS. Furthermore, the maximum output performance of the TENG based on C-CdS/PDMS was calculated, and it is presented in Figure 7f. The results indicate a maximum output energy density of approximately 2.5 W/m^2 for the TENG based on C-CdS/PDMS, which is less than that of the TENG based on N-CdS/PDMS. This outcome serves as additional evidence supporting the concept that reducing the size of CdS effectively amplifies the output power of the TENG.

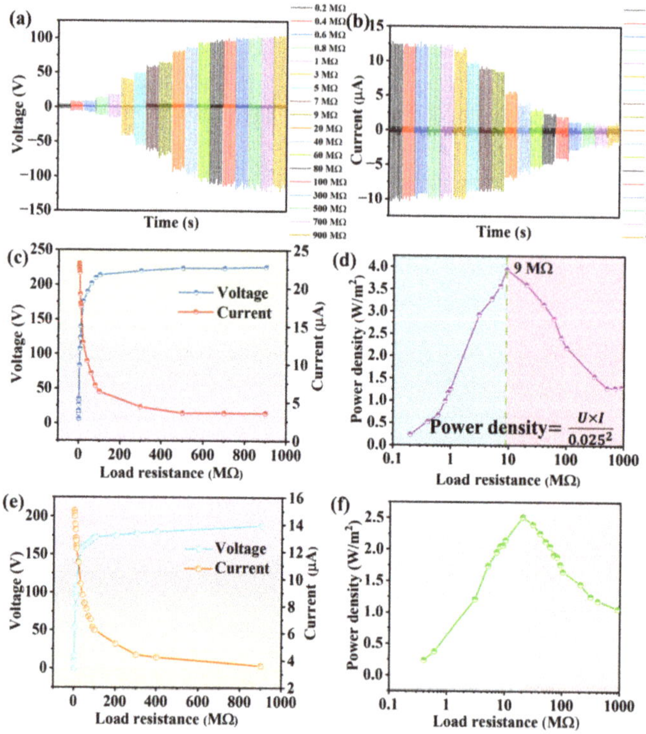

Figure 7. (**a**) Output current and (**b**) output voltage of the N-CdS/PDMS-based TENG as a function of load resistance. (**c**) Under different load resistance, the output current and voltage of the N-CdS/PDMS based TENG. (**d**) Under different load resistance, the output power of the N-CdS/PDMS-based TENG. (**e**) Under different load resistance, the output current and voltage of the C-CdS/PDMS-based TENG. (**f**) Under different load resistance, the output power of the C-CdS/PDMS-based TENG.

We rectified the output voltage of the N-CdS/PDMS-based TENG using a full bridge rectifier, as depicted in Figure 8a. The conversion of the non-rectified output voltage into rectified voltage resulted in a negligible difference, as illustrated in Figure 8b. This suggests that the TENG's output signals can be efficiently converted into rectified signals without any decrease in performance. To harness the stable power generated by the TENG, we connected LEDs in series, as shown in Figure 8c. Furthermore, the TENG was capable of directly powering 17 LEDs continuously. We also conducted the tests on our device at different time intervals, including 3 days and 5 days, to assess its repeatability, as illustrated in Figure 8d. This demonstrates that TENG based on N-CdS/PDMS has relatively good stability. Our results suggest that CdS holds promise as a potential candidate for cost-effective and high-output energy-harvesting devices.

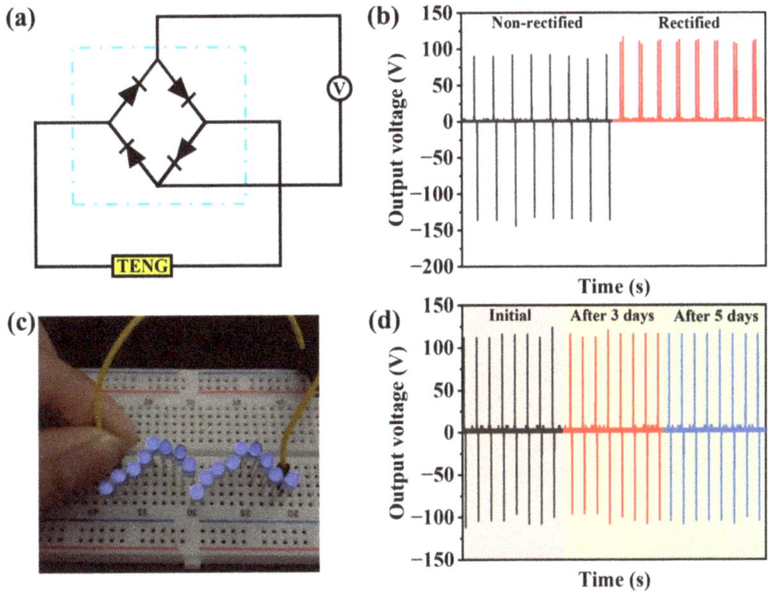

Figure 8. (**a**) Diagram of rectified circuit. (**b**) Rectified output voltage with a full-bridge rectifier. (**c**) Powering blue LEDs with TENG. (**d**) The stability test of TENG over different time intervals.

4. Conclusions

In summary, we present a high-performance TENG that integrates nanostructure CdS and PDMS. This cost-effective TENG can efficiently generate stable electrical outputs during periodic mechanical compression and release. Our research demonstrates that the introduction of N-CdS to PDMS significantly enhances the output performance of the TENG in comparison to a TENG based solely on pure PDMS. More precisely, upon the introduction of 0.6 wt% CdS to the PDMS, the CdS/PDMS-based TENG achieved output voltages of 236 V and currents of 17.4 μA, respectively. These values represent a threefold improvement in output voltage and a 2.5-fold improvement in current compared to the original PDMS-based TENG. We also compared the performance to that of the original CdS and found that reducing the size of CdS to the sub-micrometer level effectively enhances the TENG's output performance. Although the performance enhancement is not substantial, this might be attributed to the poor dispersion characteristics in PDMS, which is likely stemming from the tendency of nanostructured CdS to aggregate. Addressing this issue will be a significant focus for our future research efforts. Furthermore, we investigated the effects of pressure force and frequency on the TENG's performance, including voltage, current, and charge transfer. We achieved a maximum power density of 4 W/m^2 under the external load resistance of 9 MΩ, demonstrating TENG's ability to power LEDs. The results indicate that CdS is a promising candidate for TENGs. However, due to the tendency of CdS nanoparticles to easily agglomerate within the polymer matrix and potentially impact the output performance of the TENG, addressing this issue will be one of our future research priorities.

Author Contributions: J.M.: Conceptualization, methodology, formal analysis, investigation, data curation, visualization, and writing—original draft preparation; S.S.: Conceptualization, formal analysis, visualization, writing—review and editing, supervision, and funding acquisition. All authors have read and agreed to the published version of the manuscript.

Funding: This work was supported by the Korea government (MSIT) (IRIS RS-2023-00240826) and by the Gachon University research fund of 2022 (GCU-202208900001).

Data Availability Statement: Data are contained within the article.

Conflicts of Interest: The authors declare no conflict of interest.

References

1. Garemark, J.; Ram, F.; Liu, L.; Sapouna, I.; Cortes Ruiz, M.F.; Larsson, P.T.; Li, Y. Advancing Hydrovoltaic Energy Harvesting from Wood through Cell Wall Nanoengineering. *Adv. Funct. Mater.* **2023**, *33*, 2208933. [CrossRef]
2. Cheng, T.; Shao, J.; Wang, Z.L. Triboelectric nanogenerators. *Nat. Rev. Method. Prim.* **2023**, *3*, 39. [CrossRef]
3. Shi, B.; Wang, Q.; Su, H.; Li, J.; Xie, B.; Wang, P.; Qiu, J.; Wu, C.; Zhang, Y.; Zhou, X.; et al. Progress in recent research on the design and use of triboelectric nanogenerators for harvesting wind energy. *Nano Energy* **2023**, *116*, 108789. [CrossRef]
4. Qiu, H.; Wang, H.; Xu, L.; Zheng, M.; Wang, Z.L. Brownian motor inspired monodirectional continuous spinning triboelectric nanogenerators for extracting energy from irregular gentle water waves. *Energy Environ. Sci.* **2023**, *16*, 473–483. [CrossRef]
5. Chen, Q.; Li, W.; Yan, F.; Maniar, D.; van Dijken, J.; Rudolf, P.; Pei, Y.; Loos, K. Lightweight Triboelectric Nanogenerators Based on Hollow Stellate Cellulose Films Derived from *Juncus effusus* L. Aerenchyma. *Adv. Funct. Mater.* **2023**, 2304801. [CrossRef]
6. Fan, F.-R.; Tian, Z.-Q.; Lin Wang, Z. Flexible triboelectric generator. *Nano Energy* **2012**, *1*, 328–334. [CrossRef]
7. Feng, L.; Wang, Z.L.; Cao, X.; Zhang, L. Accordion-inspired parallelly assembled triboelectric nanogenerator: For efficient biomechanical energy harvesting and music responding. *Nano Today* **2023**, *49*, 101760. [CrossRef]
8. Wang, D.; Wang, X.-X.; Jin, M.L.; He, P.; Zhang, S. Molecular level manipulation of charge density for solid-liquid TENG system by proton irradiation. *Nano Energy* **2022**, *103*, 107819. [CrossRef]
9. Peng, Z.; Xiao, X.; Song, J.; Libanori, A.; Lee, C.; Chen, K.; Gao, Y.; Fang, Y.; Wang, J.; Wang, Z.; et al. Improving Relative Permittivity and Suppressing Dielectric Loss of Triboelectric Layers for High-Performance Wearable Electricity Generation. *ACS Nano* **2022**, *16*, 20251–20262. [CrossRef]
10. Li, Y.; Yu, J.; Wei, Y.; Wang, Y.; Feng, Z.; Cheng, L.; Huo, Z.; Lei, Y.; Sun, Q. Recent Progress in Self-Powered Wireless Sensors and Systems Based on TENG. *Sensors* **2023**, *23*, 1329. [CrossRef]
11. Duan, Q.; Peng, W.; He, J.; Zhang, Z.; Wu, Z.; Zhang, Y.; Wang, S.; Nie, S. Rational Design of Advanced Triboelectric Materials for Energy Harvesting and Emerging Applications. *Small Methods* **2023**, *7*, e2201251. [CrossRef]
12. Kim, J.H.; Seo, S. Fabrication of an imperceptible liquid metal electrode for triboelectric nanogenerator based on gallium alloys by contact printing. *Appl. Surf. Sci.* **2020**, *509*, 145353. [CrossRef]
13. Kim, J.H.; Kim, J.-H.; Seo, S. Amplifying the Output of a Triboelectric Nanogenerator Using an Intermediary Layer of Gallium-Based Liquid Metal Particles. *Nanomaterials* **2023**, *13*, 1290. [CrossRef] [PubMed]
14. Park, T.H.; Kim, J.; Seo, S. Facile and Rapid Method for Fabricating Liquid Metal Electrodes with Highly Precise Patterns via One-Step Coating. *Adv. Funct. Mater.* **2020**, *30*, 2003694. [CrossRef]
15. Kavarthapu, V.S.; Graham, S.A.; Manchi, P.; Paranjape, M.V.; Yu, J.S. Electrospun $ZnSnO_3$/PVDF-HFP Nanofibrous Triboelectric Films for Efficient Mechanical Energy Harvesting. *Adv. Fiber Mater.* **2023**, *5*, 1685–1698. [CrossRef]
16. Li, W.; Xiang, Y.; Zhang, W.; Loos, K.; Pei, Y. Ordered mesoporous SiO_2 nanoparticles as charge storage sites for enhanced triboelectric nanogenerators. *Nano Energy* **2023**, *113*, 108539. [CrossRef]
17. Zhou, J.; Zhu, C.; Zhou, Y.; Dong, J.; Li, P.; Zhang, Z.; Wang, Z.; Lin, Y.-C.; Shi, J.; Zhang, R.; et al. Composition and phase engineering of metal chalcogenides and phosphorous chalcogenides. *Nat. Mater.* **2023**, *22*, 450–458. [CrossRef] [PubMed]
18. Colombo, R.; Versaci, D.; Amici, J.; Bella, F.; Para, M.L.; Garino, N.; Laurenti, M.; Bodoardo, S.; Francia, C. Reduced Graphene Oxide Embedded with ZnS Nanoparticles as Catalytic Cathodic Material for Li-S Batteries. *Nanomaterials* **2023**, *13*, 2149. [CrossRef]
19. Monika, S.; Mahalakshmi, M.; Pandian, M.S. TiO_2/CdS/CdSe quantum dots co-sensitized solar cell with the staggered-gap (type-II) heterojunctions for the enhanced photovoltaic performance. *Ceram. Int.* **2023**, *49*, 8820–8826. [CrossRef]
20. Conceição, P.; Barata, J.F.B.; Hernández-Rodríguez, M.A.; Lacerda, P.S.S.; Neves, M.G.P.M.S.; Carlos, L.D.; Trindade, T. Ratiometric nanothermometry via porphyrin inner filter effect applied to colloidal ZnS quantum dots. *Sens. Actuator A-Phys.* **2023**, *357*, 114382. [CrossRef]

21. Hajra, S.; Panda, S.; Song, S.; Panigrahi, B.K.; Pakawanit, P.; Jeong, S.M.; Kim, H.J. Flexible composite material for self-powered applications via triboelectricity and mechanoluminescence: PDMS/ZnS:Cu composites. *Nano Energy* **2023**, *114*, 108668. [CrossRef]
22. Kanakaraju, D.; Chandrasekaran, A. Recent advances in TiO_2/ZnS-based binary and ternary photocatalysts for the degradation of organic pollutants. *Sci. Total Environ.* **2023**, *868*, 161525. [CrossRef] [PubMed]
23. Mishra, S.; Supraja, P.; Haranath, D.; Kumar, R.R.; Pola, S. Effect of surface and contact points modification on the output performance of triboelectric nanogenerator. *Nano Energy* **2022**, *104*, 107964. [CrossRef]
24. Tao, J.; Wang, M.; Zhang, X.; Lu, L.; Tang, H.; Liu, Q.; Lei, S.; Qiao, G.; Liu, G. A novel CoP@AAH cocatalyst leads to excellent stability and enhanced photocatalytic H2 evolution of CdS by structurally separating the photogenerated carriers. *Appl. Catal. B-Environ.* **2023**, *320*, 122004. [CrossRef]
25. Enríquez, J.P.; Mathew, X. Influence of the thickness on structural, optical and electrical properties of chemical bath deposited CdS thin films. *Sol. Energy Mater. Sol. Cells* **2003**, *76*, 313–322. [CrossRef]
26. Xiong, S.; Xi, B.; Qian, Y. CdS Hierarchical Nanostructures with Tunable Morphologies: Preparation and Photocatalytic Properties. *J. Phys. Chem. C* **2010**, *114*, 14029–14035. [CrossRef]
27. Gou, G.; Dai, G.; Qian, C.; Liu, Y.; Fu, Y.; Tian, Z.; He, Y.; Kong, L.; Yang, J.; Sun, J.; et al. High-performance ultraviolet photodetectors based on CdS/CdS:SnS_2 superlattice nanowires. *Nanoscale* **2016**, *8*, 14580–14586. [CrossRef] [PubMed]
28. Isik, M.; Gullu, H.H.; Delice, S.; Parlak, M.; Gasanly, N.M. Structural and temperature-dependent optical properties of thermally evaporated CdS thin films. *Mater. Sci. Semicond. Process* **2019**, *93*, 148–152. [CrossRef]
29. Yılmaz, S.; Ünverdi, A.; Tomakin, M.; Polat, İ.; Bacaksız, E. Surface modification of CBD-grown CdS thin films for hybrid solar cell applications. *Optik* **2019**, *185*, 256–263. [CrossRef]
30. Beggas, A.; Benhaoua, B.; Attaf, A.; Aida, M.S. Growth study of CdS thin films deposited by chemical bath. *Optik* **2016**, *127*, 8423–8430. [CrossRef]
31. Zhai, T.; Fang, X.; Li, L.; Bando, Y.; Golberg, D. One-dimensional CdS nanostructures: Synthesis, properties, and applications. *Nanoscale* **2010**, *2*, 168–187. [CrossRef] [PubMed]
32. Zhang, W.; Yang, H.; Li, L.; Lin, S.; Ji, P.; Hu, C.; Zhang, D.; Xi, Y. Flexible piezoelectric nanogenerators based on a CdS nanowall for self-powered sensors. *Nanotechnology* **2020**, *31*, 385401. [CrossRef] [PubMed]
33. Zhao, Q.; Guo, Y.; Zhou, Y.; Yan, X.; Xu, X. Solution-processable exfoliation and photoelectric properties of two-dimensional layered MoS_2 photoelectrodes. *J. Colloid Interface Sci.* **2017**, *490*, 287–293. [CrossRef] [PubMed]
34. Murugan, R.; Kumar, M.R.; Chander, D.S.; Kishore, S.C.; Lei, X.; Sophia, P.J. Facile and large scale aqueous synthesis of CdS nanoparticles at room temperature towards optoelectronic applications. *Mater. Res. Express* **2018**, *5*, 105003. [CrossRef]
35. Mishra, S.; Rakshita, M.; Divi, H.; Potu, S.; Rajaboina, R.K. Unique Contact Point Modification Technique for Boosting the Performance of a Triboelectric Nanogenerator and Its Application in Road Safety Sensing and Detection. *ACS Appl. Mater. Interfaces* **2023**, *15*, 33095–33108. [CrossRef] [PubMed]
36. Sriphan, S.; Vittayakorn, N. Hybrid piezoelectric-triboelectric nanogenerators for flexible electronics: Recent advances and perspectives. *J. Sci. Adv. Mater. Devices.* **2022**, *7*, 100461. [CrossRef]
37. Thakur, D.; Seo, S.; Hyun, J. Three-dimensional triboelectric nanogenerator with carboxymethylated cellulose nanofiber and perfluoroalkoxy films. *J. Ind. Eng. Chem.* **2023**, *123*, 220–229. [CrossRef]

nanomaterials

MDPI

Article

Low-Cost Manufacturing of Monolithic Resonant Piezoelectric Devices for Energy Harvesting Using 3D Printing

Marcos Duque and **Gonzalo Murillo** *

Department of Nano and Microsystems, Instituto de Microelectrónica de Barcelona—Centro Nacional de Microelectrónica (Consejo Superior de Investigaciones Cientificas) (IMB-CNM, CSIC), 08193 Bellaterra, Spain; marcos.duque@imb-cnm.csic.es

* Correspondence: gonzalo.murillo@csic.es

Abstract: The rapid increase of the Internet of Things (IoT) has led to significant growth in the development of low-power sensors. However; the biggest challenge in the expansion of the IoT is the energy dependency of the sensors. A promising solution that provides power autonomy to the IoT sensor nodes is energy harvesting (EH) from ambient sources and its conversion into electricity. Through 3D printing, it is possible to create monolithic harvesters. This reduces costs as it eliminates the need for subsequent assembly tools. Thanks to computer-aided design (CAD), the harvester can be specifically adapted to the environmental conditions of the application. In this work, a piezoelectric resonant energy harvester has been designed, fabricated, and electrically characterized. Physical characterization of the piezoelectric material and the final resonator was also performed. In addition, a study and optimization of the device was carried out using finite element modeling. In terms of electrical characterization, it was determined that the device can achieve a maximum output power of 1.46 mW when operated with an optimal load impedance of 4 MΩ and subjected to an acceleration of 1 G. Finally, a proof-of-concept device was designed and fabricated with the goal of measuring the current passing through a wire.

Keywords: energy harvesting; IoT; 3D printing; piezoelectric; monolithic; magnetic field

check for updates

Citation: Duque, M.; Murillo, G. Low-Cost Manufacturing of Monolithic Resonant Piezoelectric Devices for Energy Harvesting Using 3D Printing. *Nanomaterials* **2023**, *13*, 2334. https://doi.org/10.3390/nano13162334

Academic Editor: Jyh-Ming Wu

Received: 31 March 2023
Revised: 21 June 2023
Accepted: 25 June 2023
Published: 14 August 2023

1. Introduction

The large increase in the Internet of Things (IoT) has led to a huge growth in the development of very low-power sensors that communicate with each other and form a large wireless network [1–4]. The greatest difficulty in the expansion of the IoT is the energy dependence of the sensors, limited by the use of their batteries. For this reason, we are facing a great barrier due to the high cost of replacing batteries and the environmental impact that it entails. A promising solution that provides autonomic energy to the IoT sensor nodes is the harvesting of residual environmental energy, such as movement, vibrations, magnetic fields, light or heat, and converting them into electrical energy (Energy Harvesting (EH)) [1,3,5–9]. Mechanical energy can be converted into electrical energy using piezoelectric transducers [10–14].

In vibrational energy harvesting, piezoelectric materials are one of the most widely used transducers. For the last 20 years, piezoceramics, such as lead zirconate titanate (PZT), have been the dominating materials. After the classification of lead as a toxic material, restrictions on its use have increased, opening the opportunity to investigate new lead-free materials such as polyvinylidene fluoride (PVDF) and its copolymers.

Three-dimensional printing or additive manufacturing (AM) is a process of making three-dimensional solid objects from a digital file [15–18]. It is increasingly used for the production of any type of custom design in the fields of agriculture, healthcare, automotive and aerospace.

AM technologies are based on three types: Synthesis, in which the temperature of the material is raised without liquefying it to create high-resolution prototypes, fusion, in which powders are fused by means of electron beams, and stereolithography, which uses a method called photopolymerization, which uses an associated ultraviolet laser [19].

According to the American Society for Testing and Materials (ASTM), AM has been divided into seven processes that include:

Blinder jetting (BJ), an additive manufacturing process in which a liquid bonding agent is selectively deposited to join powder materials; directed energy deposition, an additive manufacturing process in which focused thermal energy is used to fuse materials by melting them as they are being deposited; material extrusion, an additive manufacturing process in which material is selectively dispensed through a nozzle or orifice; material jetting, an additive manufacturing process in which droplets of build material are selectively deposited; powder bed fusion (PBF), additive manufacturing process in which thermal energy selectively fuses regions of a powder bed; sheet lamination, an additive manufacturing process in which sheets of material are bonded to form an object; vat photopolymerization, an additive manufacturing process in which liquid photopolymer in a vat is selectively cured by light-activated polymerization.

Fused deposition modeling (FDM) is a material extrusion process used to make thermoplastic parts through heated extrusion and deposition of materials layer by layer. FDM is the most popular technology for 3D printing.

There are many 3D printing materials such as; PLA (polylactic acid), ABS (acrylonitrile butadiene styrene), PVA (polyvinyl alcohol), PP (polypropylene), PLA Tough, CPE (copolyester), PET-G (glycol-modified polyethylene terephthalate), Nylon (polyamide) and PC (polycarbonate), but the most common material is PLA. PLA is a biodegradable and recyclable polymer that is a reliable and easy-to-print material that can be printed at low temperatures. It has a low shrinkage factor and does not require the use of a heated build plate. A special type of PLA, called Tough PLA, combines the ease of printing with increased mechanical performance. Specifically, tough PLA has increased impact resistance, avoiding the typical brittle failure of normal PLA. Tough PLA is used for functional prototyping tooling and manufacturing aids.

Through 3D printing, it is possible to create monolithic functional structures (i.e., resonators, sensors, energy harvesters, etc.). This reduces costs as it eliminates the need for subsequent assembly tools. It is a technology that does not require additional molds, allowing for quick design variations and the creation of more complex structures. Furthermore, thanks to computer-aided design (CAD), the devices can be specifically tailored to the environmental conditions of the application. Figure 1 illustrates the advantage of integrating piezoelectric materials by 3D printing to create monolithic 3D-printed piezoelectric devices. In this way, it is possible to adapt the harvester to our needs, making it fully customizable, and the cost of the final device is reduced because subsequent assembly is not required.

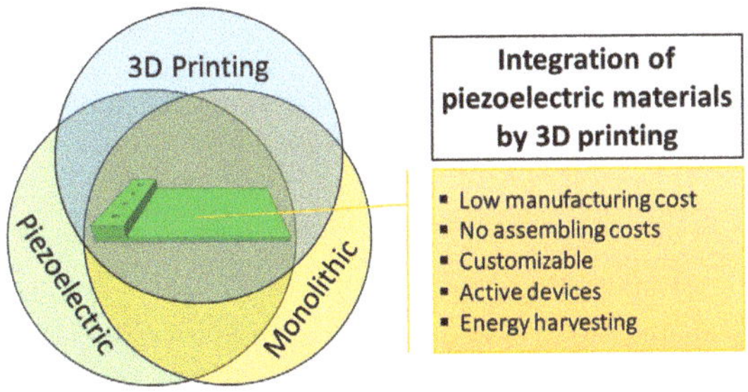

Figure 1. Advantages of using a hybrid solution of 3D printing and energy harvesting.

In the literature, there are several studies on the use of 3D printing to create energy harvesters. In most of these works, the printed piece is used as a support and the harvester is assembled later [20–30]. In this work, a resonant piezoelectric energy harvester has been designed and manufactured, using 3D printing by hybridizing this technology with a piezoelectric material (PVDF) to achieve a monolithic energy harvester. In addition, a study and optimization of the device have been carried out by means of finite element modeling (COMSOL Multiphysics®). Finally, a use-case resonator was designed and fabricated with the goal of measuring the current carried by a wire.

2. Material and Methods

2.1. Structural and Functional Materials

The piezoelectric material used in the harvester is a commercially available PVDF film with two parallel electrodes, from TE Connectivity. According to the manufacturer's specification, the PVDF material has a thickness of 110 μm and 6 μm for each silver ink electrode. The piezoelectric coefficient of the commercial PVDF is -33×10^{-12} m/V, according to the datasheet. The selected material for the structural harvester is Ultimaker Tough PLA green from Ultimaker, with a Young's modulus of 2.8 ± 0.15 GPa and a density of 1.22 g/cm^3. A neodymium (Nd) mass with a density of 7 g/cm^3, from Supermagnete, was used.

2.2. Finite Element Modelling

Finite element modeling (FEM) of the energy harvester was conducted using COMSOL Multiphysics® 6.0. The AC/DC (Electrostatics, Electrical Circuits and Piezoelectricity) and Structural Mechanics (Solid Mechanics and Piezoelectricity) modules were utilized for the simulation. The model was created using a fine triangular mesh, consisting of over 25,000 elements.

2.3. Manufacturing Process and Characterizations

The Ultimaker S3 printer with AA 0.25 mm nozzle was employed for the manufacturing process, allowing printing layers from 60 μm to 150 μm thick.

The initial characterization involved a stepper motor from Zaber Technologies LSQ075B-T3-MC03 (Vancouver, BC, Canada), capable of performing movements from 1 mm to 6 mm on the Z-axis. Concurrently, the voltage generated by the harvester was measured using a Sourcemeter Keithley 2470 (Cleveland, OH, USA). A force sensor Mark-10 (Copiage, NY, USA) MR03-20 fixed onto the stepper motor recorded the data with the assistance of a dynamometer Mark-10 M5I. An ad-hoc LabVIEW program controlled the setup comprising the devices.

The second electrical characterization was conducted using an electrodynamic shaker VR9500 from Vibration Testing Controller (Jenison, MI, USA)to emulate environmental vibrations at different input acceleration magnitudes. This process was controlled by a customized MATLAB program, enabling automation, acquisition, and processing of voltage measurements. All data were measured using an acquisition system from National Instruments (Austin, TX, USA) PCI-6132 and a BNC-2110 terminal block.

The piezoelectric coefficient of the material was measured using a Wide-Range d33 piezometer from APC International, Ltd. (Mackeyville, PA, USA).

3. Results

3.1. Design of the Harvester

3.1.1. Definition Fixed Parameters

There are different sources for energy harvesting, such as vibrations, human motion, residual magnetic field, etc. In this case, we have chosen the magnetic fields generated by a current-carrying conductor. For this reason, the resonance frequency is fixed at 50 Hz (European electric grid frequency).

In order to study structure optimization, an effective area and resonance frequency of the device must be set. In this case, the dimensions are 50 mm × 40 mm (length × width).

3.1.2. Device Structure

There are different types of structures for vibrational harvesters, such as cantilevers, doubly supported beams or diaphragms. The cantilever results in the highest power output when used as a piezoelectric energy harvester. Furthermore, the cantilever structure leads to the lowest resonance frequency for a given size. For this reason, the cantilever structure has been chosen.

The resonance frequency of a spring-mass system can be expressed as [31–33]:

$$f_r = \frac{1}{2\pi}\sqrt{\frac{k_{eff}}{m_{eff}}}, \tag{1}$$

Here, k_{eff} and m_{eff} represent the effective suspension stiffness and the effective mass values, respectively.

The neutral axis of a beam is the layer in the cross-section of a beam that experiences no longitudinal strain under bending. For composite beam structures, the effect of the different materials of each layer must be summed up to find the neutral axis:

$$Z = \frac{\sum_{i=1}^{n} z_i Y_i h_i}{\sum_{i=1}^{n} Y_i h_i}, \tag{2}$$

where Z is the neutral axis position of the composite beam, Z_i is the height of the centroid of layer i, h_i is the thickness of layer i and Y_i is Young's modulus of layer i.

3.2. Shape Optimizations and Finite Elements Modeling

3.2.1. Obtaining Resonance Frequency

In addition to the reduction in the manufacturing cost, by having a monolithic device, anchor losses are avoided due to the improved resonator anchor structure. To achieve the desired resonance frequency, a mass with a very large thickness was required due to the low density of the PLA. Furthermore, to achieve actuation with magnetic fields by means of Lorentz's force, a ferromagnetic material is needed as a cantilever tip mass. For this reason, Nd magnets have been used as an inertial mass.

As shown in Figure 2a, our harvester consists of a PLA cantilever with a length composed of the length of the beam (L_b), the mass (L_m) and the length of the cantilever (L_c). Moreover, the thickness of the beam has been defined as (T_b), the thickness of the mass as (T_m) and the depth of the piezoelectric material with respect to the surface as (P_d).

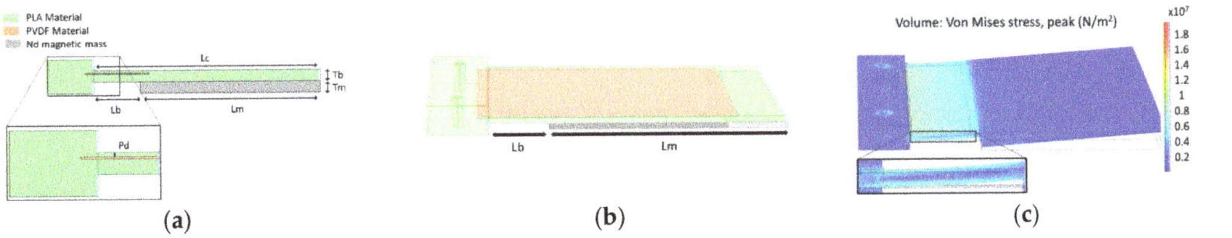

Figure 2. (**a**) Schematic of the cross-section of the energy harvester with the dimensions; (**b**) 3D design used and (**c**) stress distribution when the harvester resonates at its resonance frequency and detail of the cross-section showing the neutral axis in COMSOL Multiphysics®.

The 3D model shown in Figure 2b consists of the energy harvester with the piezoelectric layer inside the structure and an external circuit with a load resistor to calculate the power.

A piezoelectric material produces an electrical charge from mechanical stress. The d33 coefficient represents the piezoelectric charge coefficient along a specific direction, reflecting the material's response to applied stress. It can be expressed as:

$$d33 = \frac{\Delta Q}{\Delta S}, \quad (3)$$

where ΔQ is the induced charge density and ΔS is the applied stress along a particular axis.

From Equation (2), it can be deduced that the highest stress resulting from a beam deformation in a cantilever is located on the surface. In Figure 2c, the stress distribution is shown when the device resonates at its resonance frequency. It can be observed that the stress on the surface of the material is at its maximum, while it is zero at the center of the beam. Consequently, during the manufacturing process, it is crucial to position the encapsulated piezoelectric material as close to the surface as feasible. However, there is a trade-off, if the piezoelectric layer is too close to the surface, it will not be properly embedded and can get delaminated.

A parametric simulation of L_b and L_m is performed, to find the different combinations to obtain a resonance frequency of 50 Hz. The dimensions of the magnasets are more restrictive because they are defined by the manufacturer. A parametric study from 10 mm to 40 mm of L_m is performed. Table 1 shows the different dimension combinations obtained to achieve a resonance frequency of 50 Hz.

Table 1. Comparative with the different combinations of the parameters to obtain the fixed resonance frequency at 50 Hz.

L_b (mm)	L_m (mm)	L_c (mm)	Ratio L_b/L_c (%)	Frequency (Hz)
41.20	10.00	51.20	80	50.04
33.10	15.00	48.10	69	49.95
26.80	20.00	46.80	57	50.02
21.60	25.00	46.60	46	50.00
17.15	30.00	47.15	36	50.01
13.35	35.00	45.35	29	50.05
10.20	40.00	50.20	20	49.99

3.2.2. Study of Generated Power

The voltage and power of the different dimensions are studied using an external circuit that includes a resistor. Performing a sweep of the external resistor, the optimal load resistor can be determined to achieve maximum power. As shown in Figure 3a, the lower ratio between L_b and L_c, the higher the power output. Furthermore, a smaller piezoelectric area is needed, leading to a decrease in device cost since PVDF is the most expensive material in the harvester.

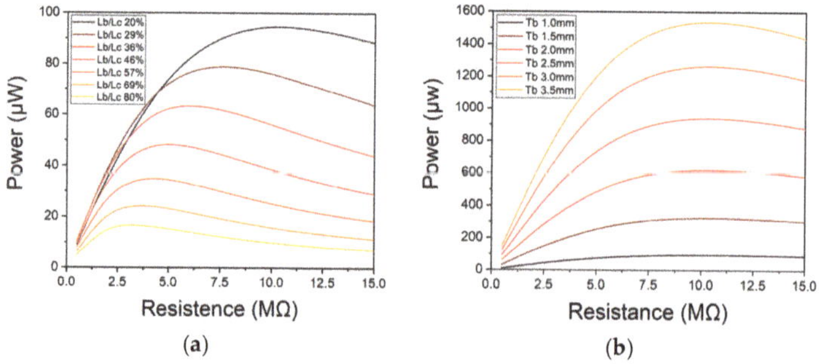

Figure 3. (**a**) Maximum power generated by the device with different L_c/L_b ratios; (**b**) Maximum power generated modifying the neutral axis of the beam increasing T_b and the density of the mass to maintain the same resonance frequency.

A parametric thickness of the beam under the piezoelectric sheet was simulated to improve the power generated by the piezoelectric generator. Due to the increase in L_b, the natural frequency of the device increases. To compensate for this augment, the density or T_m must be raised. In this occasion, the density of the mass has been augmented to compensate for the frequency. In this way, the dimension of the mass is always the same and simplifies the meshing process during the simulation. As can be seen in Figure 3b, the greater T_b, the greater the power generated.

In the next Table 2, different T_m can be seen to compensate for the increased T_b. In addition, the data of the maximum generated power, volume and power density of each harvester obtained from the simulation are also shown.

Table 2. Comparative with the different thicknesses of mass to compensate for the increased thickness of the beam and data of the maximum power generated, volume and power density of each harvester.

T_b (mm)	T_m (mm)	Power (μW)	Volume (cm^3)	Power Density ($\mu W/cm^3$)
1.00	1.00	95	4.10	23.08
1.50	3.48	324	10.20	31.77
2.00	8.13	620	20.80	29.85
2.50	15.46	938	36.80	25.50
3.00	26.12	1261	59.60	21.15
3.50	40.40	1534	89.90	17.06

With the data obtained in the simulation shown in Table 2, the volume and cost of each harvester are calculated. As seen in Figure 4a, the cost and the volume of the harvester are closely related because most of the volume is mass. Furthermore, it can be seen how the highest power density is for a T_b of 1.5 mm.

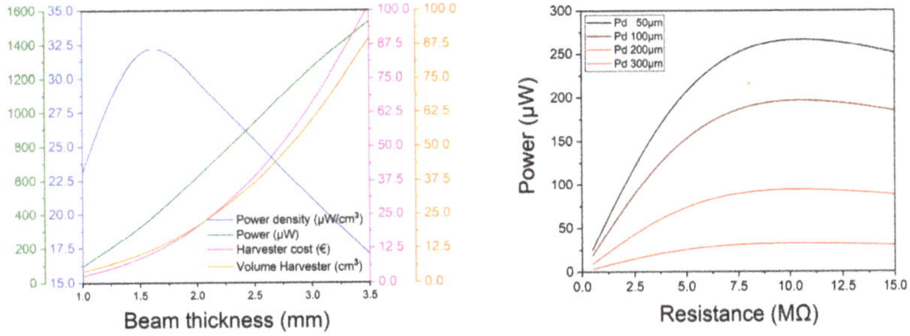

Figure 4. (**a**) Graphic of the maximum power generated, power density, harvester cost and volume of the harvester as a function of beam thickness; (**b**) Power curve generated by the harvester as a function of the depth of the piezoelectric material with respect to the surface.

Finally, a parametric simulation of the depth of the piezoelectric material regarding the surface has been performed. As can be seen in Figure 4b and as mentioned above in the design section. The closer the material is to the surface, the more power it generates since it is farther from the neutral line of the beam.

With the results obtained from the simulations, it can be concluded that the best specification for the harvester is the design with a dimension of 10.2 mm × 40 mm (L_c and L_m). In addition, to obtain a higher power density, the beam should be 1.5 mm thick and the piezoelectric material should be as close to the surface as possible.

3.3. Manufacturing Process

In 3D printing, as can be seen in Figure 5 the process begins with designing the desired piece using 3D design software (Tinkercard, https://www.tinkercad.com), typically

CAD software. Once the design is completed, the next step is exporting the file in STL (Stereolithography) format. The STL file contains all the necessary information about the 3D object. Subsequently, the CAD model is sliced into layers using slicing software Ultimaker Cura 4.13.1. Finally, the object is printed by transferring the file, positioning and printing the design layer-by-layer.

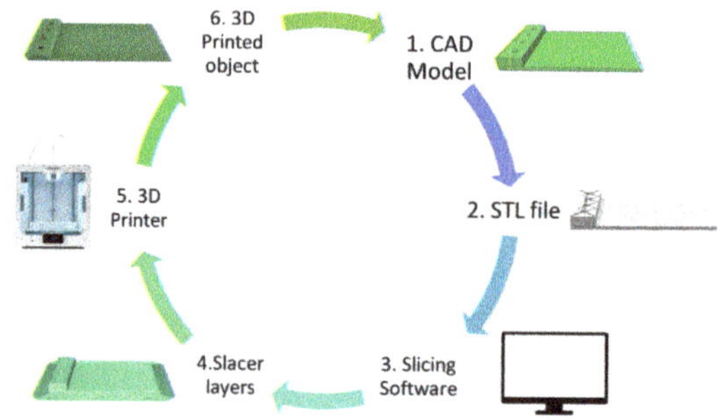

Figure 5. Basic process of 3D printers to create 3D objects.

Our manufacturing process is very similar to a typical 3D object; however, two additional steps are required to introduce the bottom electrode and the PVDF material. The first step, shown in Figure 6a, is the printing of the brim around the object to hold the edges of your piece and subsequently printing the layer that encapsulates the piezoelectric material within the structure. As previously mentioned in the generated power simulations, a thinner layer leads to greater power output. Since the minimum layer the nozzle can print is 60 µm, and to ensure good quality and avoid pushing the limits of the printer head, it was decided to print two or three layers of 100 µm. It is crucial to maintain high quality in this layer because it contributes to the stress received by the piezoelectric material. Additionally, a hole is left on the material's surface to place the lower copper electrode.

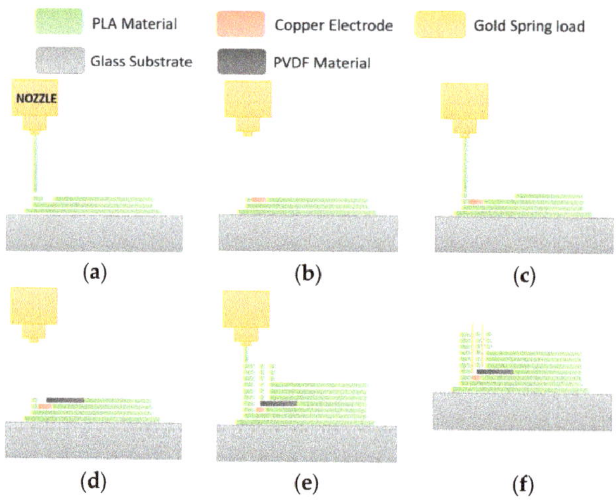

Figure 6. (**a**) Schematic of our 3D printing process and the sequences of steps to create the device (**a**) PLA base layer printing for encapsulation of the piezoelectric material; (**b**) placement of copper material to make the connection of the bottom electrode; (**c**) PLA layer printing for later assembly of the piezoelectric material; (**d**) placement of the piezoelectric material (PVDF); (**e**) printing of the rest of the PLA structure leaving two holes for subsequent connections and (**f**) insertion of gold connectors for the connection of the two electrodes of the piezoelectric material.

Subsequently, a copper piece is inserted to establish the connection with the bottom electrode, and the printing process continues depositing the piezoelectric material layer. At this point, the printing process is paused again to add the piezoelectric material, followed by the printing of the remaining structure. During this final stage of the printing, the rest of the beam and support with two holes to secure the harvester are created. Upon completing the print, gold connectors are inserted to facilitate the connection between the two electrodes of the piezoelectric material. In Figure 7, it can be seen the images of the steps of our 3D printing process to create the harvester. Three harvesters were manufactured with the piezoelectric material at 200 μm from the surface and other three devices with 300 μm.

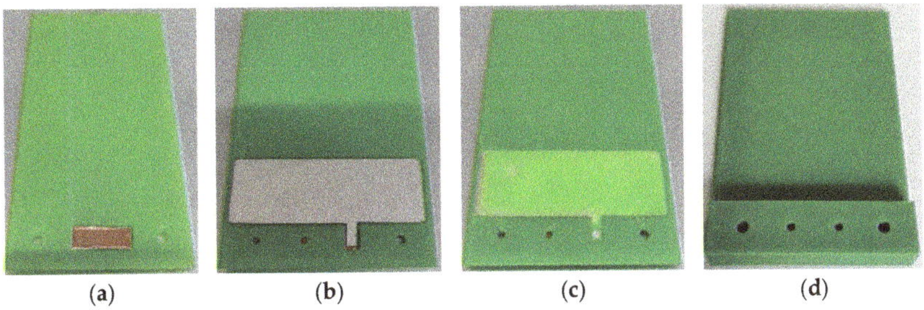

Figure 7. Images of steps of our 3D printing process to create the harvester (**a**) PLA base layer printing for encapsulation of the piezoelectric material and placement of copper material to make the connection of the bottom electrode; (**b**) placement of the piezoelectric material; (**c**) next layer printed after the placement of the PVDF to see how it is encapsulated and (**d**) printing of the rest of the PLA structure leaving two holes for subsequent connections.

3.4. *Characterization*

3.4.1. Electrical Characterization

Once the harvesters are manufactured, we proceed to the electrical characterization of our final devices. Two electrical characterizations have been conducted. The first characterization, shown in Figure 8a, involves a stepper motor that performs movements from 1 mm to 6 mm of distance on the Z-axis. Simultaneously, the voltage generated by the harvester is measured using a Sourcemeter. A force sensor, attached to the stepper motor, records these data through a dynamometer. An ad-hoc LabVIEW program controls all the setup comprising these devices. As can be seen in Figure 8b, a displacement of 6 mm along the Z-axis results in an average voltage of 71.7 V for the harvester with the piezoelectric material positioned 300 μm far from the surface. Reducing the distance of the piezoelectric material from 300 μm to 200 μm increases the average voltage to 85.2 V, representing a 19% increment in the maximum voltage generated. According to the force data represented in Figure 8c, the depth of the piezoelectric material does not affect the force needed to achieve 6 mm of travel at the tip of the cantilever.

For the second electrical characterization, as can be shown in Figure 9a, a magnetic mass is attached to adjust the resonance frequency to 50 Hz. An electrodynamic shaker is used to emulate environmental vibrations at different input acceleration levels. This is controlled through an ad-hoc MATLAB program that allows automatizing, acquiring and processing the voltage measurements. All these data are captured by an acquisition system. Subsequently, fixing a certain acceleration magnitude, a sweep of load resistors values is performed. This allows us to determine the optimal load resistor value for the highest generated power.

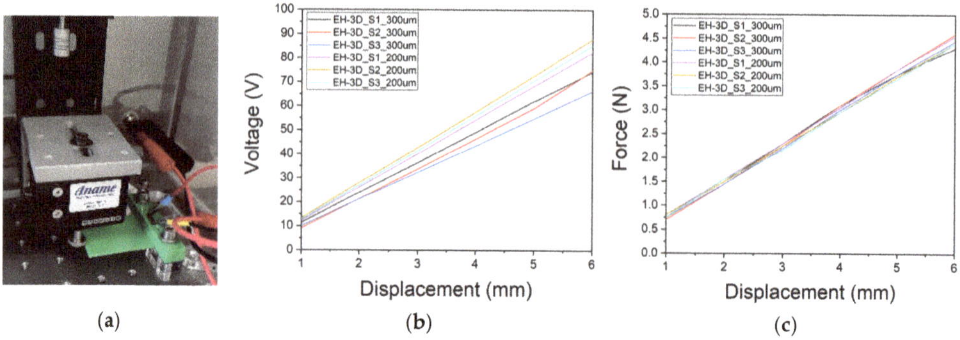

Figure 8. (**a**) Setup for electrical characterization using a stepper motor, sourcemeter and dynamometer; (**b**) Maximum voltage achieved with a displacement of the stepper motor from 1 mm to 6 mm of distance on the Z-axis and (**c**) Force measured with a displacement of the stepper motor from 1 mm to 6 mm of distance on the Z-axis.

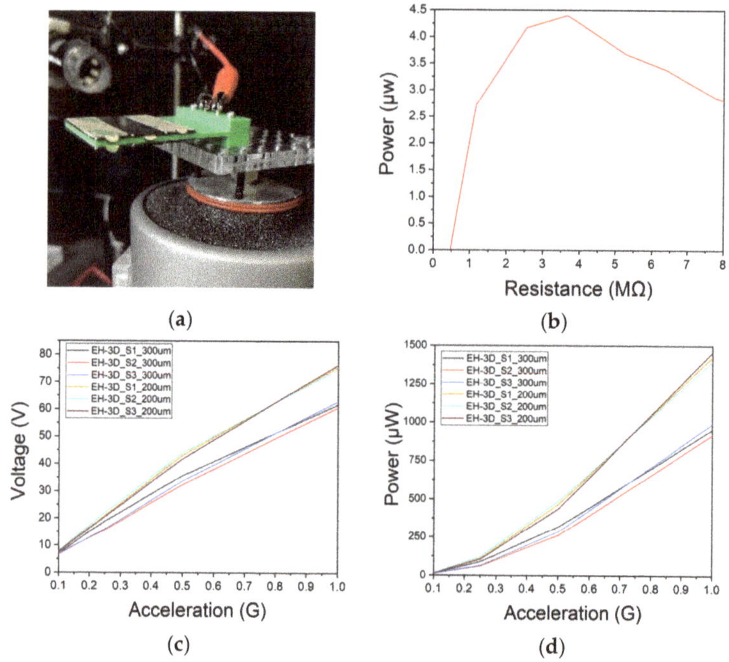

Figure 9. (**a**) Setup for electrical characterization using an electrodynamic shaker; (**b**) Optimal load resistance for maximum power generation with 0.1 G of acceleration; (**c**,**d**) Voltage and maximum power for an acceleration sweep from 0.1 G to 1 G and using a frequency of 50 Hz and an optimal load resistance of 4 MΩ.

Due to the voltage limitations of the acquisition system, which cannot measure voltages exceeding 10 V, the load resistance sweep is initially carried out with a small acceleration of 0.1 G, ensuring that it remains below 10 V. Figure 9b shows that the optimal load resistance for maximum power generation is 4 MΩ, resulting a maximum power of 4.4 µW.

Once the optimal load resistance is known, use a voltage divider, composed of a set of 500 KΩ and 3.5 MΩ resistors. It is possible to measure the voltages generated by harvesters with accelerations greater than 0.1 G.

The voltage generated by the piezoelectric generator was measured while performing an acceleration sweep ranging from 0.1 G to 1 G, as shown in Figure 9c. For an acceleration of 1 G, the harvester with the piezoelectric material positioned 300 µm from the surface achieved an average voltage of 62.3 V. By reducing the distance of the piezoelectric material to 200 µm, an average voltage of 75.6 V is achieved, resulting in a 21.4% increase in the maximum voltage generated. Using the aforementioned voltage data, the maximum power for the different accelerations was calculated and plotted in Figure 9d. It can be observed that reducing the difference between the two depths of the piezoelectric material resulted in an increase of up to 47.3% in the maximum power generated. This led to a maximum output power of 1.46 mW.

3.4.2. Physical Characterization of PVDF Material

An inspection using a scanning electron microscope (SEM) was performed to validate the thickness of materials provided by the manufacturer and, as can be seen in Figure 10a, the measured thickness closely matches the specifications.

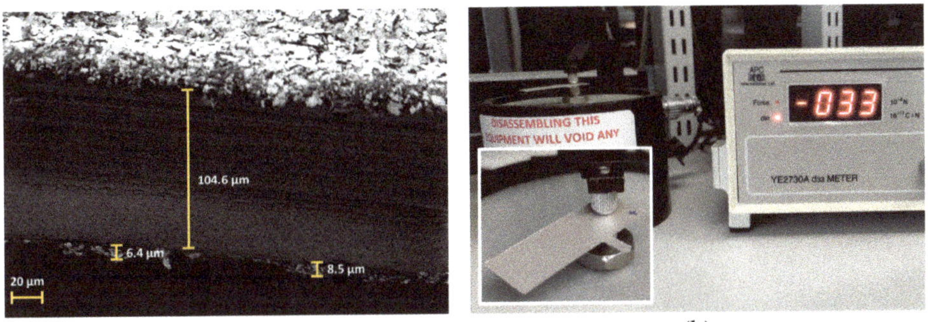

(**a**) (**b**)

Figure 10. (**a**) SEM image of the thicknesses of the layers of the piezoelectric material; (**b**) Image of the measurement of the piezoelectric coefficient.

Using a piezometer, we can measure the exact coefficient of our PVDF piece. As shown in Figure 10b, the coefficient is the same as the manufacturer's specifications.

3.4.3. Material Measurement of Young's Modulus of Our Harvester

The Young's modulus of the used PLA is 2.8 ± 0.15 GPa, as stated in the technical datasheet. When introducing a material within the PLA, the overall effective Young's modulus can be changed. Therefore, the Young's modulus of the harvesters will be calculated based on the force data measured with a stepper motor displacement from the previous section.

A displacement of a cantilever beam δ is related to the applied load, P and the Young modulus, E by the next Equation (4):

$$\delta = \frac{1}{3}\frac{PL^3}{EI}, \tag{4}$$

where L is the length of the cantilever, and I the second moment of area (moment of inertia).

For a prismatic beam with a rectangular section (depth h and with w) the value of the second moment inertia of cantilever beam I is given by:

$$I = \frac{wh^3}{12} = \frac{0.04\ m \times 0.0015\ m^3}{12} = 11.25 \times 10^{-12}\ m^4, \tag{5}$$

From Equation (4) and using the force data measured with stepper motor displacement, the Young's modulus of harvesters is calculated:

$$E = \frac{1}{3}\frac{PL^3}{\delta I} = \frac{4.5\ N \times 0.05\ m^3}{3 \times 0.006\ m \times 11.25 \times 10^{-12}\ m^4} = 2.78\ GPa, \quad (6)$$

From Young's modulus obtained from the above Equation (6), it can be seen that the modulus is within the manufacturer's specifications and that the piezoelectric material has little effect on the final Young's modulus of the harvesters.

3.5. Finite Elements Simulation

The energy harvesters were simulated again with COMSOL Multiphysics® to compare the results of the electrical characterization with a theoretical model. Physical parameters such as Young's modulus, thickness, coefficient of the piezoelectric material, and harvester dimensions were included in the simulation to achieve a realistic representation.

A parametric sweep ranging from 48 Hz to 54 Hz was performed to determine the resonance frequency of the resonators with an optimal load resistance of 4 MΩ. The acceleration used for this simulation was 0.5 G. It can be seen, in Figure 11a, the resonance frequency for the COMSOL Multiphysics® harvesters was approximately 50.5 Hz, while for the electrical characterization of the manufactured harvesters yielded a resonance frequency of 49.5 Hz. Initially, the harvesters were tuned to resonate at 50 Hz with an acceleration of 0.1 G. However, due to spring softening with increasing acceleration, the resonance frequency is shifted from 0.5 Hz to 49.5 Hz.

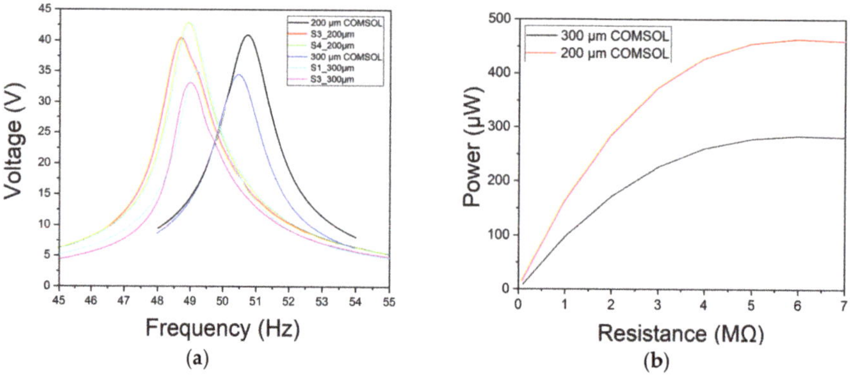

Figure 11. Simulated results of the piezoelectric device by using COMSOL Multiphysics®: (**a**) Comparative of simulation and electrical characterization of the maximum open-circuit voltage that can be generated by the piezoelectric generator submitted to a frequency sweep with an acceleration of 0.5 G; (**b**) A comparison of the maximum power generated in relation to the depth of the piezoelectric material and the load impedance.

As in the electrical characterization with the shaker, a sweep of the load resistance is performed to calculate the maximum power. As shown in the Figure 11b, the maximum power generation is around 5 MΩ, with a maximum power of 283 µW for the harvester with a 300 µm distance of the piezoelectric material and 464 µW for the 200 µm distance. These values are very similar to the maximum powers measured in Figure 9c, with 288 µW and 462 µW, respectively.

3.6. Example of Application

As mentioned earlier, the energy harvester can be customized using CAD software (e.g (Tinkercard, https://www.tinkercad.com) to adapt its specifications for various application scenarios. An example application for our device would be its use as a sensor rather than an energy harvester. In this particular case, it can be employed to measure the current flowing through a wire in a high-voltage tower. When an electric current passes through the wires, it generates a magnetic field that is directly proportional to the current flowing through the wire. By using the magnetic mass, our device can be made to resonate, and depending on the current it passes, we can measure an instantaneous proportional voltage in our sensor.

In this example application, the maximum dimensions for the sensor are 30 mm × 20 mm with a height of 10 mm. The resonance frequency for this resonator is the same as the frequency of the current flowing through the wire being measured, which is 50 Hz. The magnets for the magnetic mass have dimensions of 10 mm × 10 mm and a thickness of 2.7 mm. The support will be fixed at dimensions of 20 mm × 10 mm and a height of 10 mm to ensure good anchoring and easy fastening. A simulation is performed to obtain the resonance frequency at 50 Hz. As shown in Figure 12a, measurements of 12 mm × 20 mm and a thickness of 700 μm for the cantilever beam are obtained. Two magnets will be required for the magnetic mass. Figure 12b shows the stress distribution in the beam when the cantilever is resonating at an acceleration of 0.1 G.

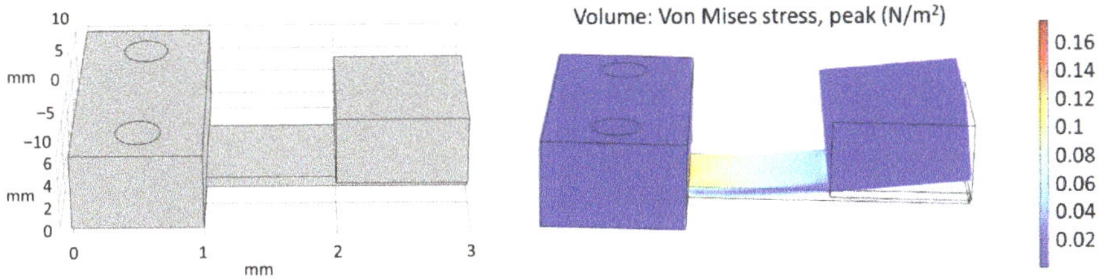

(a) (b)

Figure 12. Simulated results of the piezoelectric current sensor by using COMSOL Multiphysics®: (**a**) Structure and dimensions used in the simulation; (**b**) 3D FEM simulation of the stress distribution in the cantilever beam.

As can be shown in Figure 13a,b, the simulated sensor is mounted inside a sensor node provided by the startup company Energiot Devices SL (www.energiot.com). To test the sensor in the laboratory, two heaters with two power selections have been connected. This setup allows for the generation of four different currents depending on the selected power: 1500 W (6.5 A), 3000 W (13 A), 4500 W (19.5 A) and 6000 W (26 A). As can be seen in Figure 13c, the different current cases passing through the cable can be measured using the corresponding voltage readings, demonstrating a linear response.

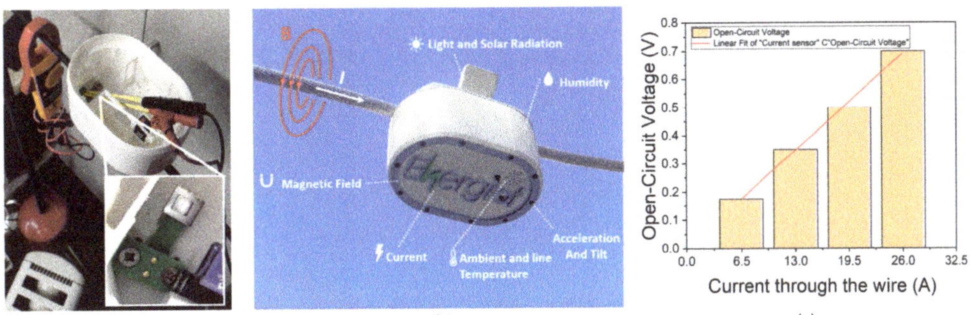

(a) (b) (c)

Figure 13. (**a**,**b**) Piezoelectric current sensor manufactured using 3D printing technology mounted inside a sensor node from the start-up Energiot; (**c**) Open circuit voltage generated by the piezoelectric sensor, depending on the selected power of two heaters, which can obtain up to four different currents (1500 W, 3000 W, 4500 W and 6000 W).

Furthermore, the magnetic field can also be used to power wireless sensors that measure various parameters such as magnetic field, ambient and line temperature, acceleration, tilt, humidity and light intensity and transmit the gathered information to the cloud.

4. Conclusions

In this article, we designed, manufactured and electrically characterized a resonant piezoelectric energy harvester by exploiting 3D printing technology and encapsulating PVDF within the structure, resulting in a monolithic harvester. With this, an economically viable energy harvester was achieved that mechanically improves the anchoring of the resonator thanks to its structure attached to the fixing support. The device was optimized through the finite element simulation software COMSOL Multiphysics® 6.0.

Furthermore, we demonstrated an example of the practical application of our resonator by designing and fabricating a specific use case for measuring the current passing through a wire using magnetic fields. Through 3D printing, not only did it enable the manufacturing of a compact and monolithic harvester but also facilitated the customization and optimization of the device for specific requirements.

Overall, this work contributes to the advancement of piezoelectric energy harvesting technology by providing a practical approach to designing and manufacturing resonant harvesters using 3D printing. The demonstrated use case application shows the potential to use such devices in a myriad of real-world scenarios.

Finally, this article represents a small contribution to the ongoing trends and future sustainability of self-powered sensors. By harvesting ambient energy, self-powered sensors have the potential to reduce the environmental impact associated with batteries. Furthermore, current advancements are paving the way for more efficient and reliable self-powered sensors. However, further research and development are needed to overcome challenges such as energy conversion efficiency and long-term reliability improvement.

Author Contributions: Conceptualization, M.D. and G.M.; methodology, M.D. and G.M.; investigation, M.D. and G.M.; writing—original draft preparation, M.D.; writing—review and editing, G.M.; supervision, G.M. All authors have read and agreed to the published version of the manuscript.

Funding: This research was funded by La Caixa Foundation under the Junior Leader Retaining program (LCF/BQ/PR19/11700010), and EUR2020-112082 under the Dynamization Actions "Europa Excelencia 2020" funded by the NextGenerationEU Instrument.

Institutional Review Board Statement: Not applicable.

Informed Consent Statement: Not applicable.

Data Availability Statement: Data are contained within the article.

Acknowledgments: G. Murillo thanks the financial support from the La Caixa Foundation under the Junior Leader Retaining Fellowship (LCF/BQ/PR19/11700010), the Agencia Estatal de Investigación from Spanish Government (PID2020-119350RA-I00, MICROBIO project) and project EUR2020-112082 under the Dynamization Actions "Europa Excelencia 2020" funded by the NextGenerationEU Instrument. In addition, this work also received support by Generalitat de Catalunya-AGAUR (2021 SGR 00497). This work has been carried out in collaboration with ENERGIOT DEVICES SL (www.energiot.com) and in the framework of the doctoral program in Electronic and Telecommunication Engineering of the Autonomous University of Barcelona.

Conflicts of Interest: The authors declare no conflict of interest.

References

1. Yau, C.-W.; Kwok, T.T.-O.; Lei, C.-U.; Kwok, Y.-K. Energy Harvesting in Internet of Things. In *Internet of Everything: Algorithms, Methodologies, Technologies and Perspectives*; Springer: Berlin/Heidelberg, Germany, 2017; pp. 35–79.
2. Rose, K.; Eldridge, S.; Chapin, L. *The Internet of Things: An Overview Understanding the Issues and Challenges of a More Connected World*; Internet Society: Reston, VA, USA, 2015; Volume 80.

3. Probst, L.; Pedersen, B.; Dakkak-Arnoux, L. Energy harvesting to power the rise of the Internet of Things. *Digit. Transform. Monit.* **2017**, 1–8.
4. Paliwoda, B.; Krzysztof, W.; Biega, M. Internet of Things in Industry: Research Profiling, Application, Chal-lenges and Opportunities—A Review. *Energies* **2022**, *15*, 1806.
5. DeFeo, C. *Energy Harvesting and the Internet of Things*; Elsevier: Amsterdam, The Netherlands, 2015.
6. Sanislav, T.; Mois, G.D.; Zeadally, S.; Folea, S.C. Energy Harvesting Techniques for Internet of Things (IoT). *IEEE Access* **2021**, *9*, 39530–39549. [CrossRef]
7. Ratha, S.; Sahoo, S.; Mane, P.; Polai, B.; Sathpathy, B.; Chakraborty, B.; Nayak, S.K. Experimental and computational investigation on the charge storage performance of a novel Al_2O_3-reduced graphene oxide hybrid electrode. *Sci. Rep.* **2023**, *13*, 5283. [CrossRef]
8. Sahoo, S.; Ratha, S.; Rout, C.S.; Nayak, S.K. Self-charging supercapacitors for smart electronic devices: A con-cise review on the recent trends and future sustainability. *J. Mater. Sci.* **2022**, *57*, 4399–4440. [CrossRef]
9. Askari, H.; Hashemi, E.; Khajepour, A.; Khamesee, M.B.; Wang, Z.L. Towards self-powered sensing using nan-ogenerators for automotive systems. *Nano Energy* **2018**, *53*, 1003–1019. [CrossRef]
10. Duque, M.; Leon-Salguero, E.; Sacristán, J.; Esteve, J.; Murillo, G. Optimization of a piezoelectric Energy harvester and design of a charge pump converter for CMOS-MEMS monolithic integration. *Sensors* **2019**, *19*, 1985. [CrossRef]
11. Covaci, C.; Gontean, A. Piezoelectric Energy Harvesting Solutions: A Review. *Sensors* **2020**, *20*, 3512. [CrossRef]
12. Sezer, N.; Koç, M. A comprehensive review on the state-of-the-art of piezoelectric energy harvesting. *Nano Energy* **2020**, *80*, 105567. [CrossRef]
13. Sharma, S.; Kiran, R.; Azad, P.; Vaish, R. A review of piezoelectric energy harvesting tiles: Available designs and future perspective. *Energy Convers. Manag.* **2022**, *254*, 115272. [CrossRef]
14. Murillo, G.; Lozano, H.; Cases-Utrera, J.; Lee, M.; Esteve, J. Improving Morphological Quality and Uniformity of Hydrothermally Grown ZnO Nanowires by Surface Activation of Catalyst Layer. *Nanoscale Res. Lett.* **2017**, *12*, 4–11. [CrossRef]
15. Azlin, M.N.M.; Ilyas, R.A.; Zuhri, M.Y.M.; Sapuan, S.M.; Harussani, M.M.; Sharma, S.; Nordin, A.H.; Nurazzi, N.M.; Afiqah, A.N. 3D Printing and Shaping Polymers, Composites, and Nanocomposites: A Review. *Polymers* **2022**, *14*, 180. [CrossRef]
16. Yun, B.K.; Kim, H.S.; Ko, Y.J.; Murillo, G.; Jung, J.H. Interdigital electrode based triboelectric nano-generator for effective energy harvesting from wáter. *Nano Energy* **2017**, *36*, 233–240.
17. Wu, H.; Fahy, W.; Kim, S.; Kim, H.; Zhao, N.; Pilato, L.; Kafi, A.; Bateman, S.; Koo, J. Recent developments in polymers/polymer nanocomposites for additive manufacturing. *Prog. Mater. Sci.* **2020**, *111*, 100638. [CrossRef]
18. Shahrubudin, N.; Lee, T.; Ramlan, R. An Overview on 3D Printing Technology: Technological, Materials, and Applications. *Procedia Manuf.* **2019**, *35*, 1286–1296. [CrossRef]
19. Jandyal, A.; Chaturvedi, I.; Wazir, I.; Raina, A.; Haq, M.I.U. 3D printing–A review of processes, materials and applications in industry 4.0. *Sustain. Oper. Comput.* **2021**, *3*, 33–42. [CrossRef]
20. Wu, Z.; Cheng, T.; Wang, Z.L. Self-Powered Sensors and Systems Based on Nanogenerators. *Sensors* **2020**, *20*, 2925. [CrossRef]
21. Adamski, K.; Adamski, J.; Dziuban, J.A.; Walczak, R. Inkjet 3D Printed Miniature Water Turbine Energy Harvester-Flow Meter for Distributed Measurement Systems. *Proceedings* **2017**, *1*, 578. [CrossRef]
22. Paranjape, M.V.; Graham, S.A.; Patnam, H.; Manchi, P.; Yu, J.S. 3D printed bidirectional rotatory hybrid nan-ogenerator for mechanical energy harvesting. *Nano Energy* **2021**, *88*, 106250. [CrossRef]
23. Han, N.; Zhao, D.; Schluter, J.U.; Goh, E.S.; Zhao, H.; Jin, X. Performance evaluation of 3D printed miniature electromagnetic energy harvesters driven by air flow. *Appl. Energy* **2016**, *178*, 672–680. [CrossRef]
24. Ahmed, A.; Azam, A.; Wang, Y.; Zhang, Z.; Li, N.; Jia, C.; Mushtaq, R.T.; Rehman, M.; Gueye, T.; Shahid, M.B.; et al. Additively manufactured nano-mechanical energy harvesting systems: Advancements, potential applications, challenges and future perspectives. *Nano Converg.* **2021**, *8*, 1–42. [CrossRef] [PubMed]
25. Bianconi, J.; Hallett, J.; Pealo, J.; Rashidi, R. A Hybrid Piezoelectric and Inductive Rotational Energy Harvester. *IOP Conf. Ser. Mater. Sci. Eng.* **2019**, *626*, 012004. [CrossRef]
26. Kawa, B.; Śliwa, K.; Lee, V.C.; Shi, Q.; Walczak, R. Inkjet 3D Printed MEMS Vibrational Electromagnetic Energy Harvester. *Energies* **2020**, *13*, 2800. [CrossRef]
27. Mahmud, M.A.P.; Zolfagharian, A.; Gharaie, S.; Kaynak, A.; Farjana, S.H.; Ellis, A.V.; Chen, J.; Kouzani, A.Z. 3D-Printed Triboelectric Nanogenerators: State of the Art, Applications, and Challenges. *Adv. Energy Sustain. Res.* **2021**, *2*, 2000045. [CrossRef]
28. Seol, M.-L.; Ivaškevičiūtė, R.; Ciappesoni, M.A.; Thompson, F.V.; Moon, D.-I.; Kim, S.J.; Han, J.-W.; Meyyappan, M. All 3D printed energy harvester for autonomous and sustainable resource utilization. *Nano Energy* **2018**, *52*, 271–278. [CrossRef]
29. Gao, S.; Zhu, Y.; Chen, Y.; Tian, M.; Yang, Y.; Jiang, T.; Wang, Z.L. Self-power electroreduction of N_2 into NH_3 by 3D printed triboelectric nanogenerators. *Mater. Today* **2019**, *28*, 17–24. [CrossRef]
30. Gawron, P.; Wendt, T.M.; Stiglmeier, L.; Hangst, N.; Himmelsbach, U.B. A Review on Kinetic Energy Harvesting with Focus on 3D Printed Electromagnetic Vibration Harvesters. *Energies* **2021**, *14*, 6961. [CrossRef]
31. Al-Ashtari, W.; Hunstig, M.; Hemsel, T.; Sextro, W. Analytical determination of characteristic frequencies and equivalent circuit parameters of a piezoelectric bimorph. *J. Intell. Mater. Syst. Struct.* **2011**, *23*, 15–23. [CrossRef]

32. Eggborn, T. Analytical Models to Predict Power Harvesting with Piezoelectric Materials. Master's Thesis, Virginia Polytechnic Institute and State University, Blacksburg, VA, USA, 2003; Volume 5. p. 11.
33. Kamel, T.M.; Elfrink, R.; Renaud, M.; Hohlfeld, D.; Goedbloed, M.; de Nooijer, C.; Jambunathan, M.; van Schaijk, R. Modeling and characterization of MEMS-based piezoelectric harvesting devices. *J. Micromech. Microeng.* **2010**, *20*, 105023. [CrossRef]

nanomaterials

Article

Research on an Optimized Quarter-Wavelength Resonator-Based Triboelectric Nanogenerator for Efficient Low-Frequency Acoustic Energy Harvesting

Xiu Xiao *, Ling Liu, Ziyue Xi, Hongyong Yu, Wenxiang Li, Qunyi Wang, Cong Zhao, Yue Huang and Minyi Xu *

Dalian Key Lab of Marine Micro/Nano Energy and Self-Powered System, Marine Engineering College, Dalian Maritime University, Dalian 116026, China
* Correspondence: xiaoxiu@dlmu.edu.cn (X.X.); xuminyi@dlmu.edu.cn (M.X.)

Abstract: Sound wave is an extensively existing mechanical wave, especially in marine and industrial plants where low-frequency acoustic waves are ubiquitous. The effective collection and utilization of sound waves provide a fresh new approach to supply power for the distributed nodes of the rapidly developing Internet of Things technology. In this paper, a novel acoustic triboelectric nanogenerator (QWR-TENG) was proposed for efficient low-frequency acoustic energy harvesting. QWR-TENG consisted of a quarter-wavelength resonant tube, a uniformly perforated aluminum film, an FEP membrane, and a conductive carbon nanotube coating. Simulation and experimental studies showed that QWR-TENG has two resonance peaks in the low-frequency range, which effectively extends the response bandwidth of acoustic–electrical conversion. The structural optimized QWR-TENG has excellent electrical output performance, and the maximum output voltage, short-circuit current and transferred charge are 255 V, 67 μA, and 153 nC, respectively, under the acoustic frequency of 90 Hz and sound pressure level of 100 dB. On this basis, a conical energy concentrator was introduced to the entrance of the acoustic tube, and a composite quarter-wavelength resonator-based triboelectric nanogenerator (CQWR-TENG) was designed to further enhance the electrical output. Results showed that the maximum output power and the power density per unit pressure of CQWR-TENG reached 13.47 mW and 2.27 $WPa^{-1}m^{-2}$, respectively. Application demonstrations indicated that QWR/CQWR-TENG has good capacitor charging performance and is expected to realize power supply for distributed sensor nodes and other small electrical devices.

Keywords: acoustic energy harvesting; low frequency; quarter-wavelength resonator; broadband

Citation: Xiao, X.; Liu, L.; Xi, Z.; Yu, H.; Li, W.; Wang, Q.; Zhao, C.; Huang, Y.; Xu, M. Research on an Optimized Quarter-Wavelength Resonator-Based Triboelectric Nanogenerator for Efficient Low-Frequency Acoustic Energy Harvesting. *Nanomaterials* **2023**, *13*, 1676. https://doi.org/10.3390/nano13101676

Academic Editor: Earl H. Dowell

Received: 3 April 2023
Revised: 11 May 2023
Accepted: 16 May 2023
Published: 19 May 2023

1. Introduction

With the rapid development of artificial intelligence (AI) and Internet of Things (IoT) technologies, the demand for distributed energy resources has increased dramatically [1–5]. As a clean, widespread, and sustainable energy source, sound waves are almost ubiquitous in the environment. In general, sound waves with frequencies ranging from 0 to 500 Hz are defined as low-frequency acoustic waves [6–8]. Such kinds of sound waves are characterized by low energy density and long wavelengths, and are able to penetrate obstacles and propagate over long distance. Notably, the acoustic waves in marine and industrial plants are mainly characterized by low-frequency and high sound pressure levels [9,10]. The application of acoustic energy collection technology is expected to provide a novel way to achieve acoustic–electrical conversion and supply power for the widely distributed sensors in the above-mentioned scenarios [11–13].

The basic principle of sound energy collection technology is to amplify sound waves using acoustic amplification devices such as Helmholtz resonator, tube resonator and acoustic metamaterials, and then convert the mechanical vibration energy into electrical output based on electromagnetic induction, piezoelectric effect, triboelectric effect or hybrid mechanism. In specific, acoustic energy harvesting devices based on electromagnetic induction use sound waves to drive conductors to cut magnetic lines in the magnetic field and generate induced current. Khan et al. [14] combined the Helmholtz resonator cavity with an electromagnetic generator for sound energy collection. The device generated a root-mean-square load voltage of 319.8 mV and a maximum power output of 1966.77 μW. On this basis, Izhar et al. [15] used a conical Helmholtz resonator to improve the electrical output of the sound energy collector. Results showed that the harvester had two resonance frequencies of 330.3 and 1332 Hz. At the first resonance frequency, the device generated a peak power of 177.2 mW under the incident sound pressure of 100 dB, which effectively improved the electrical output. However, due to the large scale of electromagnetic generator structures and the low conversion efficiency between sound waves and electromagnetic waves, external magnetic fields are usually required, resulting in low sound energy collection efficiency. Piezoelectric materials have good vibration sensitivity and can undergo mechanical deformation under the action of sound waves to generate electric fields. Yuan et al. [16] designed a sound energy collector consisting of an adjustable Helmholtz resonator, a fixed piezoelectric disk and a correction mass body. The device generated an output power of 3.49 μW and an energy conversion efficiency of 38.4% under an incident sound pressure of 100 dB. In addition to the traditional resonators, Qi et al. [17] proposed a new concept of using planar acoustic metamaterial to absorb and utilize high-frequency sound waves. Generally, the piezoelectric materials are susceptible to external interference and the output efficiency is relatively low. The triboelectric nanogenerator (TENG), driven by Maxwell displacement currents, can effectively convert distributed and disordered mechanical energy into electrical energy, and has shown great potential in energy collection [18–22] and self-powered systems [23–29]. In particular, the high vibration sensitivity of TENG makes it a possible high-efficient acoustic collection technology [20,30,31].

In recent years, researchers have conducted some studies on acoustic energy harvesting using TENG technology. In 2016, Yang et al. [32] designed a sound energy harvester by combining an adjustable Helmholtz resonant cavity with a contact-separation triboelectric nanogenerator. The design of a flexible film-based acoustic triboelectric nanogenerator was achieved for the first time, and good acoustic–electrical conversion efficiency was obtained at a resonant frequency of 240 Hz. Recently, Zhao et al. [33] proposed a dual-tube Helmholtz resonator-based triboelectric nanogenerator (HR-TENG). Compared to the previous acoustic TENGs based on traditional Helmholtz resonant cavities, the HR-TENG has a better output performance, with the maximum output voltage increased by 83%. The open-circuit voltage and short-circuit current reached 132 V and 32 μA at the optimal acoustic frequency. On this basis, Yuan et al. [34] presented an acoustic triboelectric nanogenerator using a conical Helmholtz resonator, which further improved the output performance of the sound energy harvester. Furthermore, Fan et al. [35] designed a paper-based triboelectric nanogenerator of 125 μm thickness to collect sound wave energy. The device generated a power density of 121 mW/m^2 at the frequency of 320 Hz and a sound pressure level of 117 dB. Xu et al. [36] developed a self-powered laminated electrospun nanofiber triboelectric nanogenerator for sound energy harvesting. The open-circuit voltage was 170 V when two TENGs were stacked and operated at a frequency of 200 Hz. The above research shows that triboelectric nanogenerator technology has provided an effective way for efficient sound energy collection. However, the current researches mainly focus on acoustic waves with relatively high frequency, and can only achieve acoustic–electrical conversion in a narrow bandwidth, which greatly constrains the application of sound energy harvesters.

Quarter-wavelength resonator (QWR) is a typical acoustic amplifier with superior characteristics of simple structure, excellent sound pressure amplification effect, and wide resonance bandwidth, which has great potential for low-cost large-scale production [37–39]. Therefore, this paper presented a quarter-wavelength resonator-based triboelectric nanogenerator QWR-TENG for low-frequency acoustic energy harvesting. Based on the bimodal resonance characteristic of the acoustic–electrical coupling system in the low-frequency range, QWR-TENG has an excellent electrical output performance and effectively broadens the response bandwidth of acoustic TENG. Furthermore, a composite quarter-wavelength resonator-based triboelectric nanogenerator CQWR-TENG was proposed by introducing a conical energy concentrator to further improve the sound-electrical conversion efficiency and the electrical output. The as-designed sound energy harvester QWR/CQWR-TENG is expected to provide a low-power and cost-effective power solution for IoT technology.

2. Results and Discussion

2.1. Structure Design and Working Principle of QWR-TENG

Figure 1a depicts the application scenario of QWR/CQWR-TENG in various low-frequency sound sources such as ships and industrial plants. The QWR/CQWR-TENG can provide electrical energy for distributed wireless sensor nodes through acoustic energy harvesting. QWR-TENG consists of a quarter-wavelength resonant tube and a contact-separation TENG fixed on the closed side of the tube. As shown in Figure 1b, the TENG is composed of a uniformly perforated aluminum film, a flexible FEP film, and a carbon nanotube conductive ink layer. To improve the electrical output of QWR-TENG, the FEP film was sanded with 10,000 grit sandpaper. Figure 1b also shows the SEM images of the surface morphology of the FEP film before and after polishing. It can be seen that the surface roughness and the micro/nanostructures of the FEP film are significantly increased after sanding.

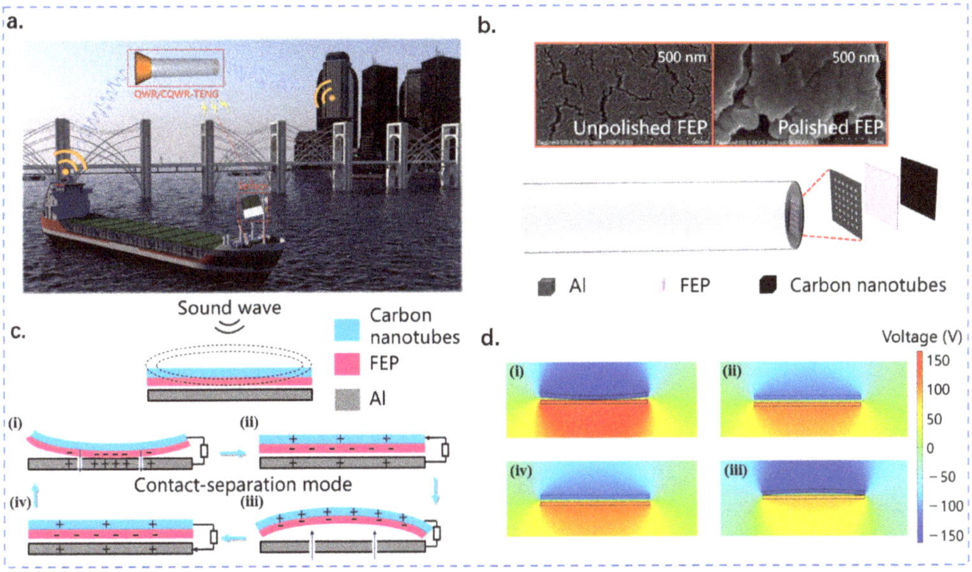

Figure 1. Structure and working principle of QWR-TENG. (**a**) Application scenario of QWR/CQWR-TENG in acoustic energy harvesting; (**b**) structure design of QWR-TENG; (**c**) working principle of QWR-TENG; (**d**) electrostatic field simulation by COMSOL Multiphysics.

The sound waves produced by the sound source propagate in the form of vibration. In QWR-TENG, the quarter-wavelength resonant tube is used to amplify the acoustic waves, and then TENG technology is applied to convert the sound waves from vibration energy into electricity. The specific working principle is shown in Figure 1c. The FEP membrane is initially separated from the aluminum electrode, and the electrons in the aluminum film are free electrons. Under the excitation of sound waves, periodic pressure changes occur between the aluminum electrode and the FEP film, which cause the FEP membrane to vibrate and generate contact-separation with the aluminum film. When the FEP film comes into contact with the aluminum electrode, the FEP film becomes negatively charged due to its high electronegativity, and an equal amount of positive charge is generated on the aluminum electrode (Figure 1(ci)). Under the action of the changing sound pressure, the FEP membrane is separated from the aluminum electrode, and the positive and negative charges no longer overlap in the same plane, resulting in a dipole moment and potential difference between the surfaces. Therefore, free electrons are driven to flow between the top carbon nanotube electrode and the bottom aluminum electrode through an external circuit to balance the local electric field (Figure 1(cii)). The flow of electrons ceases when the separation between the two contact surfaces reaches its maximum (Figure 1(ciii)). Thereafter, the FEP film starts to move in the opposite direction and approach the bottom aluminum electrode. At this stage, the potential difference between the two electrodes weakens, and free electrons flow back to the top carbon nanotube electrode, thus generating a reverse current (Figure 1(civ)). Finally, the FEP film and the aluminum electrode come into contact again to complete a full power generation cycle (Figure 1(ci)). Figure 1d displays the potential changes on different electrodes obtained by COMSOL Multiphysics simulation. Apparently, the simulation results were consistent with the above analysis. Therefore, continuous alternative current (AC) pulses were generated in the external circuit of QWR-TENG and the conversion of mechanical energy into electrical energy was realized.

2.2. Bimodal Resonance Characteristic of QWR-TENG

Figure 2a shows the schematic diagram of the experimental system for acoustic energy harvesting. The signal generator generates a sinusoidal electrical signal and drives the loudspeaker to produce sound waves. The frequency and sound pressure of the sound waves are controlled by the frequency and voltage of the electrical signal. During the experiment, the acoustic energy harvester is placed in an acrylic cover with soundproofing cotton on all sides to form a good sound insulation and shock absorption environment and ensure the accuracy of the experimental values. Under the excitation of sound waves, the triboelectric nanogenerator generates electrical signals through contact and separation between the dielectric material and the metal electrode. The electrical output is detected by an electrostatic high-impedance meter and collected by a data acquisition card.

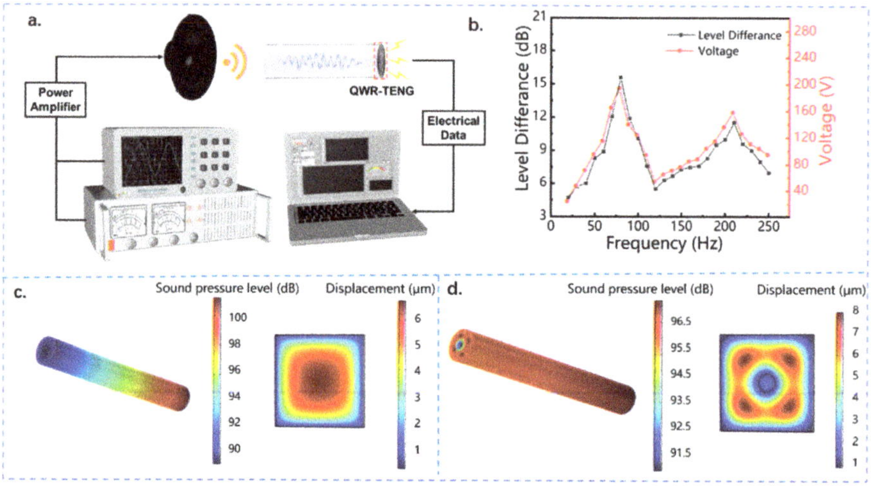

Figure 2. The bimodal resonance characteristic of QWR-TENG. (**a**) The diagram of the experimental system; (**b**) the sound pressure difference and open circuit voltage of QWR-TENG as a function of acoustic frequency; the acoustic field distribution in the quarter-wavelength resonant tube and the vibration mode of the FEP film (**c**) at the first-order resonant frequency and (**d**) at the second-order resonant frequency.

The quarter-wavelength resonator is an important component of the acoustic energy harvester QWR-TENG, and its acoustic performance directly affects the electrical output of the device. QWR is a common acoustic amplification device, and its basic structure is a straight pipe with one end open and the other end closed. In order to systematically study the acoustic field of the quarter-wavelength resonator and its influence on the electrical output of QWR-TENG, the sound pressure difference of the quarter-wavelength tube and the output voltage of the triboelectric nanogenerator were measured at the inlet acoustic frequency of 30 to 250 Hz. As shown in Figure 2b, two peaks of sound pressure difference appeared in the resonator of QWR-TENG within the frequency band, which corresponded to the two natural resonance frequencies of 80 and 210 Hz, respectively. It is worth noting that the sound pressure difference caused by the first-order resonance in the quarter-wavelength resonant tube is significantly higher than that formed by the second-order resonance, that is, the sound pressure amplification factor attenuates at higher resonant mode. Specifically, the sound pressure difference reached 15.6 dB at the resonant frequency of f = 80 Hz, while the value was only 11.5 dB at the resonance frequency of f = 210 Hz. Consistent with the variation trend of the sound pressure difference, there were two output peaks in the open-circuit voltage of QWR-TENG as the frequency increased, which were 194 and 158 V, respectively. Since a larger acoustic pressure difference can enhance the contact-separation between the dielectric material and the aluminum electrode of TENG, a higher electrical output was generated at the first resonance frequency. According to the acoustic theory of pipelines, each resonance frequency corresponds to a specific vibration mode, and the resonant of the system is named the first-order characteristic mode and the second-order characteristic mode in order of increasing resonant frequency [40,41]. Due to the bimodal resonance characteristics of QWR, the QWR-TENG can be used to collect sound wave energy with a broad bandwidth in the low-frequency region.

Furthermore, the acoustic field distribution in the quarter-wavelength resonant tube and the vibration characteristic of the FEP film in the acoustic field were studied by using the acoustic-solid coupling numerical simulation method based on COMSOL 6.0 software. Figure 2c,d show the variation of sound pressure level (SPL) in the quarter-wavelength resonator and the vibration mode of the FEP membrane at the first-order resonance and the second-order resonance, respectively, in which the FEP film is subjected to a tensile stress of 10 $N{\cdot}m^{-1}$. It can be seen that the quarter-wavelength resonator has a significant sound pressure amplification effect regardless of the resonant mode. Especially in the first-order resonance mode, the sound pressure level difference between the two ends of the resonant tube can reach 10 dB. In this resonant mode, the vibration displacement of the FEP membrane varies between 1 to 6 μm, and the displacement peak appears in the middle of the membrane and gradually decreases along the circumference. This indicates that both the highest positive pressure level and the lowest negative pressure level appear in the middle of the resonant cavity, which is beneficial for the contact-separation between the FEP membrane and the metal film and the electrical output of the acoustic TENG. When the inlet acoustic frequency exceeds the first resonant frequency, the acoustic field in the quarter-wavelength resonant tube gradually transforms to the second resonant mode. As demonstrated in Figure 2d, the sound pressure amplification effect decreases at the second-order resonance frequency, and the maximum sound pressure difference between the two ends of the resonator is about 5 dB. More importantly, the sound pressure in the resonator no longer exhibits a circular distribution, but presents multiple symmetrical distributions of sound pressure peaks. Correspondingly, the FEP membrane reveals multiple displacement peaks under the second resonant mode. For the proposed acoustic–electrical coupling system, the stress-strain behavior of the FEP film directly affects its contact-separation with the aluminum electrode, which in turn determines the electrical output of the device. According to the working principle of TENG [8], its output voltage follows:

$$V_{oc} = \frac{\sigma x(t)}{\varepsilon_0} \quad (1)$$

where V_{oc} is the open-circuit voltage, σ is the charge density, $x(t)$ is the film displacement and ε_0 is the dielectric constant. It can be seen that when the material is determined, the dielectric constant is fixed, and the electrical output performance of TENG is related to the film displacement and the charge density of the surface. In the second-order resonance mode, although the peak displacement of the FEP film is enlarged, the specific vibration mode leads to a decrease in the effective contact area between the FEP membrane and the aluminum film, which in turn results in a reduction in the surface charge density. Therefore, the output voltage of QWR-TENG under the second-order resonance mode is reduced compared with that under the first-order resonance mode.

2.3. Output Performance of QWR-TENG

In order to optimize the electrical output performance of QWR-TENG, sensitivity studies were conducted on the structural parameters of the acoustic–electrical coupling system, including the power generation unit area of TENG, the thickness of the FEP film, the length, and diameter of the quarter-wavelength resonant tube. The side length and thickness of the square FEP film are defined as d_1 and d_2, respectively, as shown in Figure 3a. Figure 3b displays the influence of the TENG power generation area on the electrical output performance of QWR-TENG. As the side length d_1 of the FEP film increases from 35 to 65 mm, the first-order resonance frequency decreases from 120 to 70 Hz, but the change in second-order resonance frequency is small. This indicates that the response bandwidth of QWR-TENG increases as the effective power generation unit area is enlarged. In addition, the output voltage of QWR-TENG is positively correlated with the power generation unit area. As the power generation area increases, the deformation of the FEP film and its effective contact area with the aluminum electrode is increased. Accordingly, the peak voltage under the first-order and second-order resonance modes increases from 76 to 165 V and from 59 to 125 V, respectively. Similar variations in the output short-circuit current and transferred charge can also be found in Figure S1 of the Supplementary Materials. Therefore, for the acoustic triboelectric nanogenerator with a fixed quarter-wavelength resonant tube, the adoption of a larger TENG dielectric layer can not only increase the electrical output of the sound energy collector, but also extend the response bandwidth and operating range of the device.

The effect of the FEP film thickness d_2 on the electrical output of QWR-TENG is shown in Figure 3c. Three types of FEP film with thicknesses of 30, 50, and 100 μm were tested and the effective area was set at 55 × 55 mm. The input acoustic wave frequency varied from 30 to 220 Hz with a step of 10 Hz during the experiment. On the one hand, as the thickness of the dielectric layer increased from 30 to 100 μm, the first-order resonance frequency decreased from 100 to 50 Hz, while the second-order resonance frequency remained basically unchanged. Therefore, increasing the thickness of the FEP film can expand the working bandwidth of QWR-TENG to a certain extent. Figure S2 in the supplementary reveals the short-circuit current and transferred charge. On the other hand, the electrical output of QWR-TENG did not change monotonically with the increase of film thickness, but showed a trend of first increasing and then decreasing. Specifically, under the first-order resonance mode, the output voltage of QWR-TENG reached 82 V when the FEP film thickness was set to 50 μm, which is 73% and 93% higher than that of the QWR-TENGs with a dielectric layer thickness of 30 and 100 μm, respectively. Furthermore, the peak output voltage was increased by 54% and 82%, respectively, at the second-order resonance frequency. The reason for this phenomenon is that when the thickness of the dielectric layer is increased (30 to 50 μm), its load impedance on the resonator enlarges gradually and is more suitable for the output impedance of the resonator, so the output voltage of QWR-TENG is increased. However, as the thickness of the FEP film is further enlarged (50 to 100 μm), the stiffness of the FEP film enhances significantly, which leads to a decrease in both the elastic deformation and effective contact area. As a result, the electrical output of QWR-TENG is decreased. Thus, in order to achieve the best output performance of QWR-TENG, it is necessary to select a dielectric layer material with a large area and an appropriate thickness for membrane-acoustic cavity coupling.

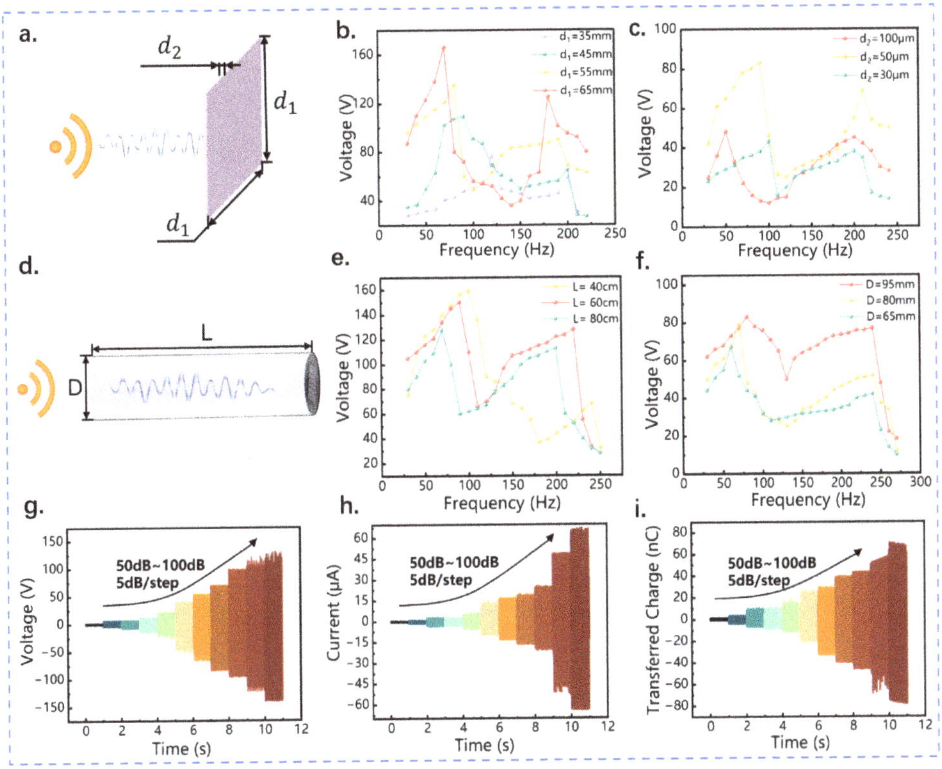

Figure 3. Effects of structural parameters and acoustic conditions on the electrical output of QWR-TENG. (**a**) Schematic diagram of the structural parameters of FEP membrane in the TENG power generation unit; effects of (**b**) power generation unit area and (**c**) FEP film thickness on the output voltage of QWR-TENG; (**d**) schematic diagram of the structural parameters of the quarter-wavelength resonator; effects of (**e**) the length and (**f**) diameter of the quarter-wavelength resonator on the output voltage of QWR-TENG; effects of incident pressure on the (**g**) open-circuit voltage, (**h**) short-circuit current; and (**i**) transferred charge of QWR-TENG.

Transmission loss (TL) is the key indicator for acoustic performance evaluation [42]. Generally, it can be expressed as:

$$TL = 20lg|P_i/P_t| \quad (2)$$

where P_i and P_t represent the incident sound pressure and the transmitted sound pressure at the exit of the resonator, respectively. Therefore, a higher acoustic TL indicates a better resonance effect inside the resonant cavity. Acoustic studies show that the transmission loss TL of the quarter-wavelength resonator is related to quality factors, which mainly include the length, cross-sectional area, and impedance of the resonator. Therefore, the effects of the tube length and diameter of the quarter-wavelength resonator on the output performance of the QWR-TENG were studied.

The length and diameter of the quarter-wavelength tube are defined as L and D, respectively, as shown in Figure 3d. Figure 3e shows the influence of quarter-wavelength tube length on the resonance frequency and output voltage of the QWR-TENG, in which the length is set as 40, 60, and 80 cm. It can be seen that the QWR-TENG has two output peaks regardless of the length, which corresponds to the two resonance frequencies of QWR. As the length of QWR decreased from 80 to 40 cm, the first-order resonance frequency increased from 70 to 100 Hz, but the output performance was enhanced. As shown in Figure S3 of the supplementary, similar variations can also be observed in short-circuit current and transferred charge. This indicates that the length of QWR-TENG can be adjusted to adapt to various acoustic frequencies. Since the present QWR-TENG is designed to harvest sound energy with low frequency, the QWR with a length of 60 cm was adopted for subsequent research by comprehensively considering the response frequency and electrical output performance.

The cross-sectional area of the quarter-wavelength tube is determined by its diameter, so the electrical output of three QWR-TENGs with different tube diameters was tested in the frequency range of 30 to 220 Hz. Figure 3f shows the output open-circuit voltage of the QWR-TENG, in which the diameter of the quarter-wavelength tube is 95, 80, and 65 mm, respectively. With the increase of pipe diameter and cross-sectional area, the first-order resonance frequency of QWR-TENG increases, but the second-order resonance frequency remains unchanged. More importantly, the electric output of QWR-TENG is effectively enhanced as the cross-sectional area increases. This is because a larger cross-sectional diameter indicates a higher impedance and a greater transmission loss of the resonator, thereby resulting in an increased electrical output of QWR-TENG. When the diameter of the quarter-wavelength resonant tube reached 95 mm, the peak open-circuit voltage of QWR-TENG reached 83 V, as shown in Figure 3f. Figure S4 in the supplementary reveals the short-circuit current and transferred charge, in which similar variation profiles can be observed.

In addition, the impact of the inlet sound pressure level on the electric output of QWR-TENG is described in Figure 3g–i. The increase in inlet sound pressure amplifies the sound pressure difference at the outlet of the resonant tube, resulting in an enlarged deformation and radial displacement of the FEP membrane. Therefore, the electrical output of QWR-TENG increases with the incident sound pressure. More specifically, as the inlet sound pressure level raised from 50 to 100 dB, the open-circuit voltage, short-circuit current, and transferred charge of QWR-TENG increased from 4 to 255 V, from 1.2 to 67 μA, and from 2 to 153 nC, respectively. Figure S5 in the Supplementary Materials further demonstrates the electrical output of QWR-TENG under the incident sound pressure of 100 dB.

2.4. Structure Design and Optimization of CQWR-TENG

The conical cavity structure can effectively guide the propagation of sound waves, and improve the sensitivity and effectiveness of sound radiation [43]. To further improve the output performance of the proposed acoustic–electrical coupling system, a conical energy concentrator was introduced to the open end of the quarter-wavelength tube to form a composite quarter-wavelength tube resonator-based triboelectric nanogenerator CQWR-TENG.

The physical structure of CQWR-TENG and the simplified two-dimensional model of the conical cavity are shown in Figure 4a. Along the direction of sound wave propagation, the variation of the conical cavity diameter satisfies [44]:

$$r(x) = r(l) - \frac{x}{l}[r(l) - r(0)] \tag{3}$$

where $r(0)$ and $r(l)$ are the inlet and outlet diameter of the conical cavity, respectively, l is the total length of the conical tube, and x is the length from the inlet end to the certain position inside the conical tube. According to the acoustic amplification theory of the conical cavity, the acoustic amplitude at the outlet of the conical cavity satisfies:

$$A(x) = \frac{1}{r(x)} = \frac{l}{[lr(l) - x(r(l) - r(0))]} \tag{4}$$

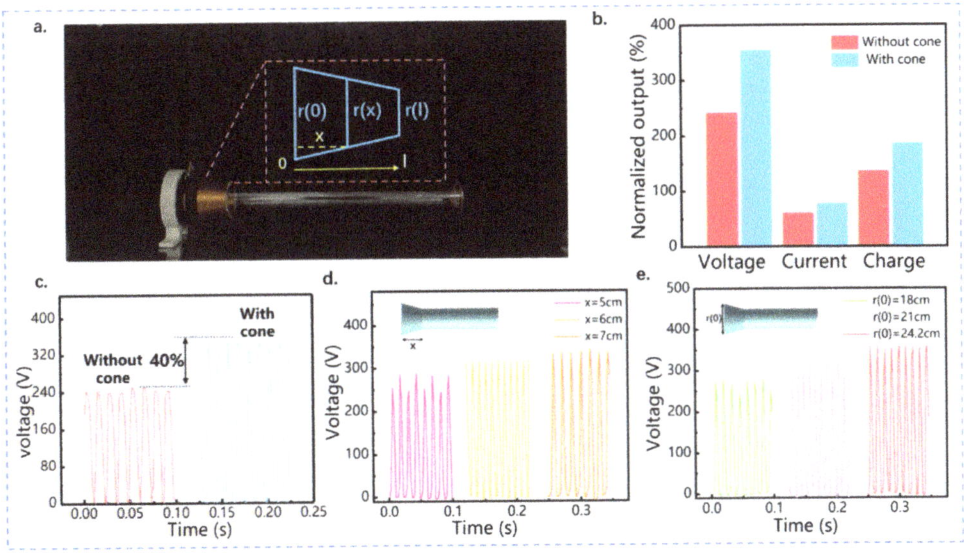

Figure 4. Structure design and output performance of CQWR-TENG. (**a**) Structure scheme of CQWR-TENG; (**b**) effect of conical cavity on the electrical output of the acoustic triboelectric nanogenerator; (**c**) comparison diagram of the output open-circuit voltage of QWR-TENG and CQWR-TENG; effects of (**d**) the length and (**e**) the open-end diameter of the conical cavity on the electrical output of CQWR-TENG.

For a tapered conical cavity, $r(x)$ decreases with the increase of x, so the amplitude of the sound wave $A(x)$ is magnified along the outlet direction of the conical tube. The sound pressure of acoustic waves is proportional to sound frequency and the square of the amplitude. Therefore, the increase in amplitude directly causes the amplification of sound pressure and sound intensity in the resonance cavity, thereby improving the electrical output of the acoustic energy collector. Figure 4b,c validate the enhancement effect of the conical concentrator on the output of the acoustic triboelectric nanogenerator. Under the same acoustic condition, the peak open-circuit voltage of QWR-TENG and CQWR-TENG are 242 and 348 V, respectively. Figure S6 in the supplementary displays the detailed output of QWR-TENG and CQWR-TENG under the acoustic frequency of 100 Hz and sound pressure level of 95.8 dB. Compared to QWR-TENG, the application of the conical concentrator increases the open-circuit voltage, short-circuit current, and transferred charge of the TENG by 43%, 33%, and 38%, respectively.

In addition, the structural parameters of the conical concentrator have a significant impact on the electrical output of TENG. The effect of the length of the conical cavity on the output performance of CQWR-TENG is displayed in Figure 4d, in which the lengths x are 5, 6, and 7 cm, respectively. Obviously, with the increase in the length of the conical concentrator, the output of CQWR-TENG is effectively enhanced, with the peak open-circuit voltage increasing from 285.2 to 339.1 V. The fact is that a larger length of the conical tube can make the reflection and interference phenomena more significant and enhance the accumulation of acoustic energy in the tube. This causes the enlargement of sound pressure and the increase of electrical output of CQWR-TENG. Figure 4e exhibits the effect of the opening diameter $r(0)$ of the conical concentrator on the output voltage. The open-circuit voltage of CQWR-TENG increased from 275.3 to 358 V as the opening diameter raised from 18 to 24.2 cm. This is because a larger opening diameter can diminish the reflection of sound waves at the entrance. In addition, the energy dissipation of sound waves along propagation can be reduced with a larger entrance diameter, thus maintaining a high amplification effect of sound pressure. In summary, increasing the opening diameter and the length of the conical energy concentrator is beneficial to the enhancement of sound pressure in CQWR-TENG. Therefore, the adoption of a conical energy concentrator is recommended within the available environmental space to increase the electrical output of acoustic TENG and improve the efficiency of acoustic–electrical energy conversion.

2.5. Demonstration of QWR/CQWR-TENG

Demonstration experiments were carried out to verify the sound energy harvesting and power generation performance of QWR-TENG and CQWR-TENG, as shown in Figure 5. Figure 5a shows the circuit diagram and the results of QWR-TENG charging different capacitors. It takes about 9, 42, and 131 s for QWR-TENG to charge the capacitors of 47, 330, and 1000 μF to 3 V. This indicates that QWR-TENG has a good charging ability for capacitors and can supply electricity for low-power electronics. In addition, as exhibited in Figure 5b, after charging the 1000 μF capacitor to about 2.7 V, the humidity and temperature sensor was successfully lit by QWR-TENG. The Video S1 in the Supplementary Material shows that the sensor can operate continuously with the power supply of QWR-TENG. Figure 5c demonstrates the experiment of QWR-TENG lighting up 196 LEDs simultaneously under a sound pressure level of 90.1 dB and a frequency of 100 Hz, and the corresponding video can be found in Video S2 in the Supplementary Material. A demonstration experiment was also conducted in an onshore machinery compartment to test the sound energy harvesting ability of QWR-TENG. The experimental diagram can be found in Figure S6 in the supporting document, in which the sound waves were generated by the operation of a diesel engine. It can be seen that the QWR-TENG can output a voltage signal of 5~8 V at a certain distance from the sound source. The significant difference from the resonance frequency of QWR-TENG results in the limited electrical output of the device. In the future, the design of QWR-TENG structural parameters can be based on the actual industrial application scenarios to make its resonance frequency consistent with the frequency of the equipment, which is conducive for the application of QWR-TENG in sound energy harvesting.

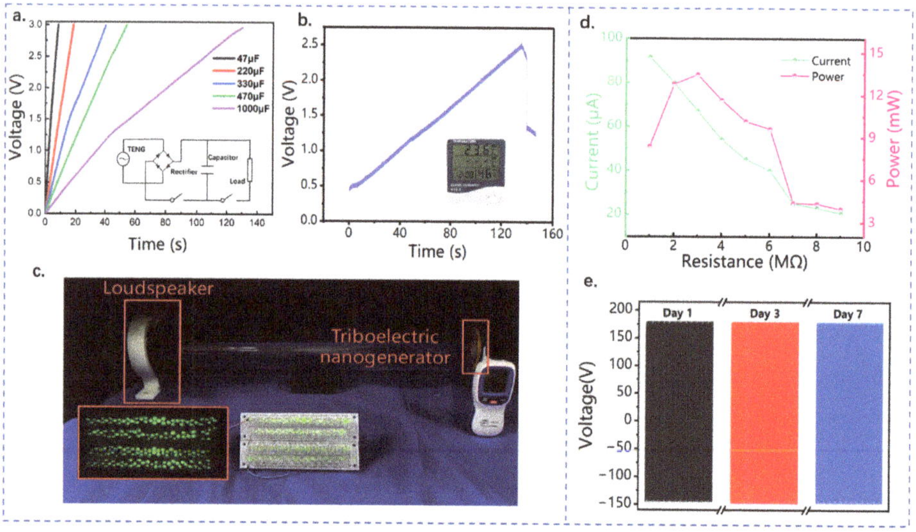

Figure 5. Power generation performance demonstration and output power test of QWR-TENG and CQWR-TENG. (**a**) Demonstration experiment of QWR-TENG charging capacitors with different capacities; (**b**) power supply experiment for temperature and humidity sensor; (**c**) schematic diagram of QWR-TENG simultaneously lighting 196 LEDs; (**d**) the variation profiles of the output current and power of CQWR-TENG with external load resistance; (**e**) durability demonstration of CQWR-TENG.

The introduction of the conical accumulator can further enhance the sound energy collection and electrical output performance of the acoustic triboelectric nanogenerator. Figure 5d depicts the variation of CQWR-TENG output with external resistance under the acoustic condition of 90 Hz and 100 dB. As the external load resistance increased from 1 to 10 MΩ, the output current of CQWR-TENG gradually decreased from 91.5 to 20.7 μA. However, the output power of CQWR-TENG increased first and then decreased. Specifically, the maximum output power and the power density per unit pressure of CQWR-TENG reached 13.47 mW and 2.27 $WPa^{-1}m^{-2}$, respectively. The comparison with the earlier electromagnetic, piezoelectric, and triboelectric nanogenerator-based sound energy harvesters is presented in Table S1 of the supporting document [8,14,17,32,33,38,39,45–48], and the results indicate that the as-presented QWR-TENG has a superior performance in low-frequency acoustic–electrical conversion. Furthermore, the durability test result of CQWR-TENG is shown in Figure 5e. After seven days of operation, the electrical output decreased by about 1%, which confirmed the robustness of CQWR-TENG. Therefore, the as-designed QWR/CQWR-TENG has great application potential in the field of acoustic power generation, and is expected to be applied to specific acoustic scenarios for environmental low-frequency sound energy harvesting.

3. Experimental Section

3.1. Fabrication of QWR-TENG

The QWR-TENG consists of a quarter-wavelength resonator, an aluminum film with uniformly distributed sound holes, and an FEP film with an ink-printed electrode made of conductive carbon nanotubes. The quarter-wavelength resonator is made of a circular acrylic duct with a length of 60 cm, a diameter of 9.5 cm, and a thickness of 5 mm. For the fabrication of the triboelectric nanogenerator, FEP film was chosen as the electronegative material due to its high dielectric strength and good flexibility, and aluminum was selected as the electropositive material. The thickness of the FEP film is 50 μm, and one side of the film is treated with a corona discharge to facilitate electron transfer with the conductive carbon nanotube electrode. The aluminum film has a thickness of 0.1 mm and is uniformly perforated with one hundred small holes (diameter of 0.5 mm) for contact-separation of the TENG. The FEP membrane and the aluminum film are bonded together with adhesive tape, and then attached to a 3D-printed square ring to form the power generation unit of the TENG. Finally, the power generation unit was connected and fixed to the quarter-wavelength resonator using hot melt adhesive, forming the acoustic energy collector QWR-TENG.

3.2. Equipment of the Acoustic Energy Harvesting System

During the performance testing experiment, the QWR-TENG is installed in an acrylic cover with sound-absorbing cotton on all four walls, which has good sound insulation and shock absorption effects to ensure the accuracy of the experimental data. A loudspeaker, which is driven and adjusted by a function generator (YE1311) with sine waves, is used to produce sound waves. The sound is transmitted through a power amplifier (SA-5016) with an accuracy and resolution of 1.5 and 0.1 dB, respectively. The output signals including open-circuit voltage, short-circuit current, and transferred charge are measured by a Keithley 6514 electrostatic meter.

4. Conclusions

In this paper, a novel acoustic triboelectric nanogenerator QWR-TENG was firstly presented for the efficient collection of low-frequency sound wave energy. QWR-TENG was coupled by a quarter wavelength resonant tube and a contact-separation TENG to achieve acoustic–electrical conversion within a broad bandwidth. Simulation and experimental results showed that QWR-TENG was characterized by two resonance modes in the low-frequency range of 30~250 Hz, which can effectively extend the response bandwidth of the sound energy collector. The effects of structural parameters, including the length and cross-sectional diameter of the quarter-wavelength tube, the thickness of the FEP film, and the power generation area, were experimentally studied to enhance the electrical output performance of QWR-TENG. Results showed that the peak open-circuit voltage, short-circuit current, and transferred charge of the optimized QWR-TENG reached 255 V, 67 μA, and 153 nC, respectively. On this basis, a conical concentrator was introduced to the open end of the QWR to form a composite quarter-wavelength resonator-based triboelectric nanogenerator CQWR-TENG to further improve the electrical output of the acoustic collector. Experimental results showed that compared with QWR-TENG, the peak open-circuit voltage, short-circuit current, and transferred charge of CQWR-TENG increased by 43%, 33%, and 38%, respectively. More specifically, the maximum output power and the power density per unit pressure reached 13.47 mW and 2.27 $WPa^{-1}m^{-2}$. Demonstration experiments on capacitor charging and sensor power supply showed that the proposed QWR/CQWR-TENG has good output performance and the application potential in the field of acoustic energy collection. The response characteristics of being low-frequency and having a wide bandwidth make the QWR/CQWR-TENG promising for sound energy collection in marine and industrial plants, and to power distributed sensor nodes.

Supplementary Materials: The following supporting information can be downloaded at: https://www.mdpi.com/article/10.3390/nano13101676/s1, Figure S1: Effect of power generation unit area on (a) short-circuit current and (b) transferred charge of QWR-TENG; Figure S2: Effect of FEP film thickness on (a) short-circuit current and (b) transferred charge of QWR-TENG; Figure S3: Effect of the quarter-wavelength resonant tube length on (a) the short-circuit current and (b) transferred charge of QWR-TENG; Figure S4: Effect of the quarter-wavelength resonant tube diameter on (a) the short-circuit current and (b) transferred charge of QWR-TENG; Figure S5: Comparison diagram of the output performance of QWR-TENG and CQWR-TENG under the acoustic frequency of 100 Hz and sound pressure level of 95.8 dB. (a) Open-circuit voltage, (b) short circuit current and (c) transferred charge; Figure S6: Demonstration experiment of QWR-TENG collecting low-frequency sound energy in an onshore machinery compartment; Table S1: Acoustic energy harvesting capabilities of different acoustic harvesters; Video S1: The humidity and temperature sensor operates stably under the power supply of QWR-TENG; Video S2: The QWR-TENG powering 196 LEDs.

Author Contributions: Conceptualization, X.X., M.X. and L.L.; methodology, X.X., L.L. and Z.X.; software, L.L., H.Y. and W.L; validation, W.L., Z.X. and C.Z.; formal analysis, C.Z., Y.H. and Q.W.; investigation, Z.X. and Y.H.; resources, M.X. and Y.H.; data curation, Q.W. and W.L.; writing—original draft preparation, L.L. and Z.X.; writing—review and editing, X.X.; supervision, X.X. and M.X.; project administration, M.X.; funding acquisition, X.X. All authors have read and agreed to the published version of the manuscript.

Funding: This research was funded by the National Natural Science Foundation of China (Grant No. 51906029), the China Postdoctoral Science Foundation (2019M661084), and the Open Fund of National Center for International Research of Subsea Engineering Technology and Equipment (No. 3132023354).

Institutional Review Board Statement: Not applicable.

Informed Consent Statement: Not applicable.

Data Availability Statement: Data are available on request from the authors.

Conflicts of Interest: The authors declare no conflict of interest.

References

1. Arslan, H.; Ranjbar, M.; Secgin, E.; Celik, V. Theoretical and experimental investigation of acoustic performance of multi-chamber reactive silencers. *Appl. Acoust.* **2020**, *157*, 106987. [CrossRef]
2. Ding, S.; Tukker, A.; Ward, H. Opportunities and risks of internet of things (IoT) technologies for circular business models: A literature review. *J. Environ. Manag.* **2023**, *336*, 117662. [CrossRef] [PubMed]
3. Motta, R.C.; de Oliveira, K.M.; Travassos, G.H. An evidence-based roadmap for IoT software systems engineering. *J. Syst. Softw.* **2023**, *201*, 111680. [CrossRef]

4. Tripathy, R.R.; Sahoo, R.; Mishra, S.; Das, B.; Balasubramaniam, S.; Ramadoss, A. Fabrication and feasibility study of polymer-based triboelectric nanogenerator towards blue energy harvesting. *Green Energy Resour.* **2023**, *1*, 100006. [CrossRef]
5. Bo, F.; Jiwen, F.; Jiuchun, Z.; Chong, L.; Jia, W.; Mingming, L. Bionic flutter wing piezoelectric-electromagnetic composite energy harvesting system. *Energy Convers. Manag.* **2022**, *271*, 116319. [CrossRef]
6. Carbajo, J.; Ghaffari Mosanenzadeh, S.; Kim, S.; Fang, N.X. Multi-layer perforated panel absorbers with oblique perforations. *Appl. Acoust.* **2020**, *169*, 107496. [CrossRef]
7. Wang, Y.; Zhu, X.; Zhang, T.; Bano, S.; Pan, H.; Qi, L.; Zhang, Z.; Yuan, Y. A renewable low-frequency acoustic energy harvesting noise barrier for high-speed railways using a Helmholtz resonator and a PVDF film. *Appl. Energy* **2018**, *230*, 52–61. [CrossRef]
8. Zhang, Q.; Xi, Z.; Wang, Y.; Liu, L.; Yu, H.; Wang, H.; Xu, M. Multi-Tube Helmholtz Resonator Based Triboelectric Nanogenerator for Broadband Acoustic Energy Harvesting. *Front. Mater.* **2022**, *9*, 896953. [CrossRef]
9. Zuo, J.; Dong, L.; Yang, F.; Guo, Z.; Wang, T.; Zuo, L. Energy harvesting solutions for railway transportation: A comprehensive review. *Renew. Energy* **2023**, *202*, 56–87. [CrossRef]
10. Zhang, Z.; Shao, J.; Nan, Y.; Willatzen, M.; Wang, Z.L. Theory and shape optimization of acoustic driven triboelectric nanogenerators. *Mater. Today Phys.* **2022**, *27*, 100784. [CrossRef]
11. Mālnieks, K.; Kaufelde, P.; Linarts, A.; Lapčinskis, L.; Verners, O.; Šutka, A. Triboelectric laminates from polydimethylsiloxane bilayers for acoustic energy harvesting. *Mater. Lett.* **2022**, *329*, 133188. [CrossRef]
12. Puri, A.; Modak, S.V.; Gupta, K. Global active noise control in vibro-acoustic cavities using acoustic sensing. *J. Sound Vib.* **2019**, *455*, 256–274. [CrossRef]
13. Yang, A.; Li, P.; Wen, Y.; Lu, C.; Peng, X.; He, W.; Zhang, J.; Wang, D.; Yang, F. High-efficiency broadband acoustic energy harvesting using Helmholtz resonator and dual piezoelectric cantilever beams. *Rev. Sci. Instrum.* **2014**, *85*, 066103. [CrossRef] [PubMed]
14. Izhar; Khan, F.U. Electromagnetic-based acoustic energy harvester. In Proceedings of the International Multi Topic Conference, Lahore, Pakistan, 19–20 December 2013.
15. Izhar; Khan, F.U. Electromagnetic based acoustic energy harvester for low power wireless autonomous sensor applications. *Sens. Rev.* **2018**, *38*, 298–310. [CrossRef]
16. Yuan, M.; Cao, Z.; Luo, J.; Zhang, J.; Chang, C. An efficient low-frequency acoustic energy harvester. *Sens. Actuators A Phys.* **2017**, *264*, 84–89. [CrossRef]
17. Qi, S.; Oudich, M.; Li, Y.; Assouar, B. Acoustic energy harvesting based on a planar acoustic metamaterial. *Appl. Phys. Lett.* **2016**, *108*, 263501. [CrossRef]
18. Liu, C.; Qu, G.; Shan, B.; Aranda, R.; Chen, N.; Li, H.; Zhou, Z.; Yu, T.; Wang, C.; Mi, J.; et al. Underwater hybrid energy harvesting based on TENG-MTEG for self-powered marine mammal condition monitoring system. *Mater. Today Sustain.* **2023**, *21*, 100301. [CrossRef]
19. Walden, R.; Aazem, I.; Babu, A.; Pillai, S.C. Textile-Triboelectric nanogenerators (T-TENGs) for wearable energy harvesting devices. *Chem. Eng. J.* **2023**, *451*, 138741. [CrossRef]
20. Yang, M.; Tian, X.; Hua, T. Green and recyclable cellulose based TENG for sustainable energy and human-machine interactive system. *Chem. Eng. J.* **2022**, *442*, 136150. [CrossRef]
21. Liu, C.; Fang, L.; Zou, H.; Wang, Y.; Chi, J.; Che, L.; Zhou, X.; Wang, Z.; Wang, T.; Dong, L.; et al. Theoretical investigation and experimental verification of the self-powered acceleration sensor based on triboelectric nanogenerators (TENGs). *Extrem. Mech. Lett.* **2021**, *42*, 101201. [CrossRef]
22. Xiao, X.; Zhang, X.; Wang, S.; Ouyang, H.; Chen, P.; Song, L.; Yuan, H.; Ji, Y.; Wang, P.; Li, Z.; et al. Honeycomb Structure Inspired Triboelectric Nanogenerator for Highly Effective Vibration Energy Harvesting and Self-Powered Engine Condition Monitoring. *Adv. Energy Mater.* **2019**, *9*, 1902460. [CrossRef]
23. Chen, J.; Guo, H.; Wu, Z.; Xu, G.; Zi, Y.; Hu, C.; Wang, Z.L. Actuation and sensor integrated self-powered cantilever system based on TENG technology. *Nano Energy* **2019**, *64*, 103920. [CrossRef]
24. Xie, Y.; Long, J.; Zhao, P.; Chen, J.; Luo, J.; Zhang, Z.; Li, K.; Han, Y.; Hao, X.; Qu, Z.; et al. A self-powered radio frequency (RF) transmission system based on the combination of triboelectric nanogenerator (TENG) and piezoelectric element for disaster rescue/relief. *Nano Energy* **2018**, *54*, 331–340. [CrossRef]
25. Liao, Y.; Li, W.; Wang, K.; Guo, J.; Shen, Y.; Wang, Q.; Zhang, Y.; Wu, C.; Zhou, X.; Guo, T.; et al. TENG-inspired LED-in-capacitors for smart self-powered high-voltage monitoring and high-sensitivity demodulation of power-line communications. *Nano Energy* **2022**, *102*, 107698. [CrossRef]
26. Zhu, Z.; Li, B.; Zhao, E.; Yu, M. Self-powered silicon PIN neutron detector based on triboelectric nanogenerator. *Nano Energy* **2022**, *102*, 107668. [CrossRef]
27. Yang, J.; Cao, J.; Han, J.; Xiong, Y.; Luo, L.; Dan, X.; Yang, Y.; Li, L.; Sun, J.; Sun, Q. Stretchable multifunctional self-powered systems with Cu-EGaIn liquid metal electrodes. *Nano Energy* **2022**, *101*, 107582. [CrossRef]
28. Zhang, Z.; Gu, G.; Zhang, W.; Gu, G.; Shang, W.; Liu, Y.; Cheng, G.; Du, Z. Double loops power management circuit of pulsed triboelectric nanogenerator with enhanced efficiency at low operating voltage and its application in self-powered flue gas monitoring system. *Nano Energy* **2023**, *110*, 108360. [CrossRef]

29. Wang, D.; Zhang, D.; Tang, M.; Zhang, H.; Sun, T.; Yang, C.; Mao, R.; Li, K.; Wang, J. Ethylene chlorotrifluoroethylene/hydrogel-based liquid-solid triboelectric nanogenerator driven self-powered MXene-based sensor system for marine environmental monitoring. *Nano Energy* **2022**, *100*, 107509. [CrossRef]
30. Yan, W.; Ma, C.; Cai, X.; Sun, Y.; Zhang, G.; Song, W. Self-powered and wireless physiological monitoring system with integrated power supply and sensors. *Nano Energy* **2023**, *108*, 108203. [CrossRef]
31. Lu, S.; Lei, W.; Wang, Q.; Liu, W.; Li, K.; Yuan, P.; Yu, H. A novel approach for weak current signal processing of self-powered sensor based on TENG. *Nano Energy* **2022**, *103*, 107728. [CrossRef]
32. Yang, J.; Chen, J.; Liu, Y.; Yang, W.; Su, Y.; Wang, Z.L. Triboelectrification-based organic film nanogenerator for acoustic energy harvesting and self-powered active acoustic sensing. *ACS Nano* **2014**, *8*, 2649–2657. [CrossRef] [PubMed]
33. Zhao, H.; Xiao, X.; Xu, P.; Zhao, T.; Song, L.; Pan, X.; Mi, J.; Xu, M.; Wang, Z.L. Dual-Tube Helmholtz Resonator-Based Triboelectric Nanogenerator for Highly Efficient Harvesting of Acoustic Energy. *Adv. Energy Mater.* **2019**, *9*, 1902824. [CrossRef]
34. Yuan, H.; Yu, H.; Liu, X.; Zhao, H.; Zhang, Y.; Xi, Z.; Zhang, Q.; Liu, L.; Lin, Y.; Pan, X.; et al. A High-Performance Coniform Helmholtz Resonator-Based Triboelectric Nanogenerator for Acoustic Energy Harvesting. *Nanomaterials* **2021**, *11*, 3431. [CrossRef]
35. Fan, X.; Chen, J.; Yang, J.; Bai, P.; Li, Z.; Wang, Z. Ultrathin, Rollable, Paper-Based Triboelectric Nanogenerator for Acoustic Energy Harvesting and Self-Powered Sound Recording. *ACS Nano* **2015**, *9*, 4236–4243. [CrossRef] [PubMed]
36. Xu, W.; Guo, J.; Wen, H.; Meng, X.; Hong, H.; Yuan, J.; Gao, J.; Liu, D.; Ran, Q.; Wang, Y.; et al. Laminated triboelectric acoustic energy harvester based on electrospun nanofiber towards real-time noise decibel monitoring. *Nano Energy* **2022**, *99*, 107348. [CrossRef]
37. Li, D.; Hu, M.; Wu, F.; Liu, K.; Gao, M.; Ju, Z.; Zhao, J.; Bao, A. Design of tunable low-frequency acoustic energy harvesting barrier for subway tunnel based on an optimized Helmholtz resonator and a PZT circular plate. *Energy Rep.* **2022**, *8*, 8108–8123. [CrossRef]
38. Yuan, M.; Li, C.; Liu, H.; Xu, Q.; Xie, Y. A 3D-printed acoustic triboelectric nanogenerator for quarter-wavelength acoustic energy harvesting and self-powered edge sensing. *Nano Energy* **2021**, *85*, 105962. [CrossRef]
39. Yuan, M.; Cao, Z.; Luo, J.; Pang, Z. Helix structure for low frequency acoustic energy harvesting. *Rev. Sci. Instrum.* **2018**, *89*, 055002. [CrossRef]
40. Gautam, A.; Celik, A.; Azarpeyvand, M. On the acoustic performance of double degree of freedom helmholtz resonator based acoustic liners. *Appl. Acoust.* **2022**, *191*, 108661. [CrossRef]
41. Zhong, R.; Hu, S.; Liu, X.; Qin, B.; Wang, Q.; Shuai, C. Vibro-acoustic analysis of a circumferentially coupled composite laminated annular plate backed by double cylindrical acoustic cavities. *Ocean Eng.* **2022**, *257*, 111584. [CrossRef]
42. Chanlert, P.; Tongyoo, S.; Rordrak, C. Effects of urea–formaldehyde and polyvinyl acetate adhesive on sound absorption coefficient and sound transmission loss of palmyra palm fruit fiber composites. *Appl. Acoust.* **2022**, *198*, 108984. [CrossRef]
43. Gao, N.; Lu, K. An underwater metamaterial for broadband acoustic absorption at low frequency. *Appl. Acoust.* **2020**, *169*, 107500. [CrossRef]
44. Khan, F.; Izhar. Piezoelectric type acoustic energy harvester with a tapered Helmholtz cavity for improved performance. *J. Renew. Sustain. Energy* **2016**, *8*, 705–709. [CrossRef]
45. Cha, S.N.; Seo, J.S.; Kim, S.M.; Kim, H.J.; Park, Y.J.; Kim, S.W.; Kim, J.M. Sound-driven piezoelectric nanowire-based nanogenerators. *Adv. Mater.* **2010**, *22*, 4726–4730. [CrossRef] [PubMed]
46. Chen, F.; Wu, Y.; Ding, Z.; Xia, X.; Li, S.; Zheng, H.; Diao, C.; Yue, G.; Zi, Y. A novel triboelectric nanogenerator based on electrospun polyvinylidene fluoride nanofibers for effective acoustic energy harvesting and self-powered multifunctional sensing. *Nano Energy* **2019**, *56*, 241–251. [CrossRef]
47. Li, B.; You, J.H.; Kim, Y.-J. Low frequency acoustic energy harvesting using PZT piezoelectric plates in a straight tube resonator. *Smart Mater. Struct.* **2013**, *22*, 55013. [CrossRef]
48. Khan, F.U.; Izhar. Electromagnetic energy harvester for harvesting acoustic energy. *Sādhanā* **2016**, *41*, 397–405. [CrossRef]

nanomaterials

Article

Thermally Deposited Sb_2Se_3/CdS-Based Solar Cell: Experimental and Theoretical Analysis

Mamta [1,2,†], Raman Kumari [1,2,†], Chandan Yadav [1,2], Rahul Kumar [1,2], Kamlesh Kumar Maurya [1,2,*] and Vidya Nand Singh [1,2,*]

1 Academy of Scientific and Innovative Research (AcSIR), Ghaziabad 201002, India
2 CSIR-National Physical Laboratory, Dr. K. S. Krishnan Marg, New Delhi 110012, India
* Correspondence: kkmaurya@nplindia.org (K.K.M.); singhvn@nplindia.org (V.N.S.)
† These authors contributed equally to this work.

Abstract: As a promising solar absorber material, antimony selenide (Sb_2Se_3) has gained popularity. However, a lack of knowledge regarding material and device physics has slowed the rapid growth of Sb_2Se_3-based devices. This study compares the experimental and computational analysis of the photovoltaic performance of Sb_2Se_3-/CdS-based solar cells. We construct a specific device that may be produced in any lab using the thermal evaporation technique. Experimentally, efficiency is improved from 0.96% to 1.36% by varying the absorber's thickness. Experimental information on Sb_2Se_3, such as the band gap and thickness, is used in the simulation to check the performance of the device after the optimization of various other parameters, including the series and shunt resistance, and a theoretical maximum efficiency of 4.42% is achieved. Further, the device's efficiency is improved to 11.27% by optimizing the various parameters of the active layer. It thus is demonstrated that the band gap and thickness of active layers strongly affect the overall performance of a photovoltaic device.

Keywords: Sb_2Se_3 thin film; I–V; thickness; SCAPS-1D

check for updates

Citation: Mamta; Kumari, R.; Yadav, C.; Kumar, R.; Maurya, K.K.; Singh, V.N. Thermally Deposited Sb_2Se_3/CdS-Based Solar Cell: Experimental and Theoretical Analysis. *Nanomaterials* **2023**, *13*, 1135. https://doi.org/10.3390/nano13061135

Academic Editors: Qiongfeng Shi and Jianxiong Zhu

Received: 27 February 2023
Revised: 13 March 2023
Accepted: 20 March 2023
Published: 22 March 2023

1. Introduction

Thin-film-based absorber layers are an intriguing option for the development of low-cost, large-scale photovoltaic systems and have already attracted the attention of many researchers. In thin-film-based photovoltaic cells, active layer materials which are several micrometers thick are used compared to silicon-based solar cells. These materials have a higher absorption coefficient than crystalline ones. As a result, only a very thin absorber layer is needed for sunlight absorption. Nowadays, researchers are moving towards chalcogenide thin-film-based devices, which are reaching their commercial stage due to their outstanding performance and higher efficiency. Metal chalcogenides such as cadmium telluride (CdTe) [1], copper gallium diselenide ($CuGaSe_2$) [2], antimony selenide (Sb_2Se_3) [3], etc., provide a range of optical bandgaps, which are beneficial for use in many optical and optoelectronic applications. Sb_2Se_3 is a promising absorber material, having a direct bandgap of 1.2 eV [4–6] and a higher absorption coefficient of ~10^5 cm^{-1} [7]. These are very favourable for use in optoelectronic applications.

The fabrication of Sb_2Se_3 material has been carried out using a variety of deposition processes, such as chemical vapour deposition (CVD) [8], atomic layer deposition (ALD) [9], sputtering [10], spray pyrolysis [11], pulsed laser ablation (PLD) [12], thermal vapour deposition [13], etc. A Sb_2Se_3 thin-film-based solar cell with a 6.06% efficiency was fabricated with a two-step sputtering and post-selenization technique [14]. Another Sb_2Se_3 thin-film-based solar cell achieved an efficiency of 7.4%. In preparing this, the absorber layer was deposited using the vapour transport method and post-selenized [15]. In other research, after selenization, the well-crystallized Sb_2Se_3 thin films were produced with the correct orientation and large crystal grains and an efficiency of 4.86% was attained [16]. Most earlier studies on Sb_2Se_3-based solar cells only used solution processing methods.

Although these procedures can be inexpensive, they frequently contaminate the film or use toxic solvents. Thermal vapour deposition is a vacuum-based deposition technique. It is used widely in the commercial deposition of thin-film-based structures because it is easy to use and reliable. Sb_2Se_3 has a high vapour pressure and low melting point, promoting simple thermal evaporation over magnetron sputtering. The deposited films made using the thermal deposition method are usually categorized as amorphous, and their crystallinity can develop through heat treatment. Until now, no report on a Sb_2Se_3 thin-film solar structure has been entirely fabricated using a vacuum-based process.

CdS is one of the promising buffer layers in photovoltaic cells [17–19]. CdS films must have good transparency and be neither thin nor thick to prevent short circuits or light absorption in the buffer layer [20]. The higher optical band gap of a hexagonal buffer layer makes it preferable over a cubic layer for use in high-efficiency solar cells. The p-Sb_2Se_3-/n-CdS-based solar cells, where the layers are deposited via thermal vapour deposition and then annealed to produce crystallinity, constitute the primary focus of this research. Various characterizations have been analyzed in order to study the properties of active layers. XRD, UV–Vis, and Raman characterizations [21–24] are analyzed to ensure the structure, bandgap, and phase formation of the functional layers, respectively. The material's bandgap and thickness significantly influence the performance of a solar cell. Structural analyses supported the appearance of the orthorhombic and hexagonal phases for Sb_2Se_3 and CdS, respectively (XRD and Raman). I–V characteristics are plotted after the fabrication of the device.

Since Sb_2Se_3 is still in its early stages of development, simulations are crucial to saving the researchers' time and energy while optimizing the properties of the materials and devices. The parameters of p-Sb_2Se_3 obtained from experimental study, such as thickness, bandgap, and absorbance, are used to analyze the efficiency of a Sb_2Se_3-/CdS-based photovoltaic cell. The problems preventing the development of a more efficient cell must be resolved and optimized using proper numerical study. This work is based on the experimental and numerical analysis of Sb_2Se_3-/CdS-based photovoltaic cells. The SCAPS-1D is used for the device's simulation study [25]. The I–V curve provides the solar cell parameters, allowing researchers to study a photovoltaic cell's electrical behaviour.

2. Experimental Study of the Proposed Device Structure

2.1. Material Preparation and Characterization Techniques

Highly pure Sb and Se were placed in a silica tube in a stoichiometry ratio of 2:3. The tube was evacuated at a pressure of ~10^{-5} torr. The silica tube was placed in a furnace and heated to 800 °C. After this, the temperature was maintained for 48 h, as shown in Figure 1. The furnace was turned down, and the tube was left to cool to room temperature. The collected ingots were crushed, and a black powder was obtained. Different characterization techniques can be used to analyze the various properties of the materials [26]. The active layer's structures were ensured by XRD (Cu-Kα, 0.1542 nm). The UV–Vis spectrometer (AvaLight-DH-S-BAL) was used to investigate the optical properties of Sb_2Se_3 in the 200–1000 nm wavelength range. With the help of a stylus profilometer, the film's thickness was measured (Model XP-200). Raman spectroscopy was used to analyze the phase of thin films (Triple Jobin Yvon T 6400).

2.2. Device Fabrication via Thermal Deposition

The structure of the solar device is ITO/CdS/Sb_2Se_3/Ag, as shown in Figure 2a, and the device was fabricated in a superstrate configuration [27]. From the experimental point of view, we successfully fabricated the device based on the different thicknesses of the absorber layer using the thermal vapour deposition method shown in Figure 2b. The deposition chamber was pressurized to a pressure of 10^{-6} torr before evaporation. As the applied current was continuously increased, the powder in the boat melted and evaporated. First, the CdS film was deposited onto ITO glass with a thickness of 65 nm. Then, two different thicknesses (372 nm and 640 nm) of Sb_2Se_3 were deposited onto CdS. Lastly, the silver electrode (Ag) was deposited. The absorber and buffer layers of the device were annealed at 200 °C and 300 °C for 30 and 20 min, respectively, after deposition. The active area of solar cells was ~1 cm^2, and there was no anti-reflective coating.

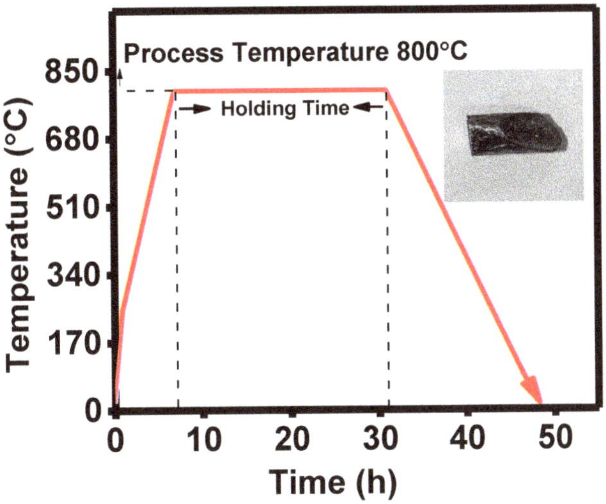

Figure 1. The temperature profile of material inside the vertical tube furnace (inset: single crystal ingot of Sb_2Se_3).

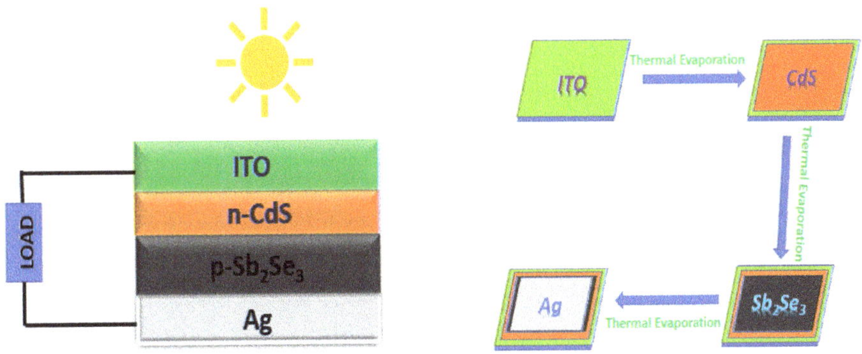

(**a**) (**b**)

Figure 2. Schematic of (**a**) device structure and (**b**) deposition method.

2.3. Characterization Results

The XRD pattern for the Sb_2Se_3-based thin films with two different thicknesses (372 nm and 640 nm) is shown in Figure 3a. There is a sharp peak in the thin film XRD data for both thicknesses (221). When the XRD pattern was compared to Sb_2Se_3's standard JCPDS values, it was concluded that the structure was orthorhombic (JCPDS no.15-0861). The XRD pattern demonstrated sharp peaks for the thin film with a 640 nm thickness. In XRD, sharp peaks analysis indicated that the films were crystalline [28]. The intensity of the peaks likewise increased as the thickness rose. Similarly, in Figure 3b, the XRD for CdS is compared with the standard that of JCPDS no. 89-0019, and the peaks confirm the existence of a hexagonal structure.

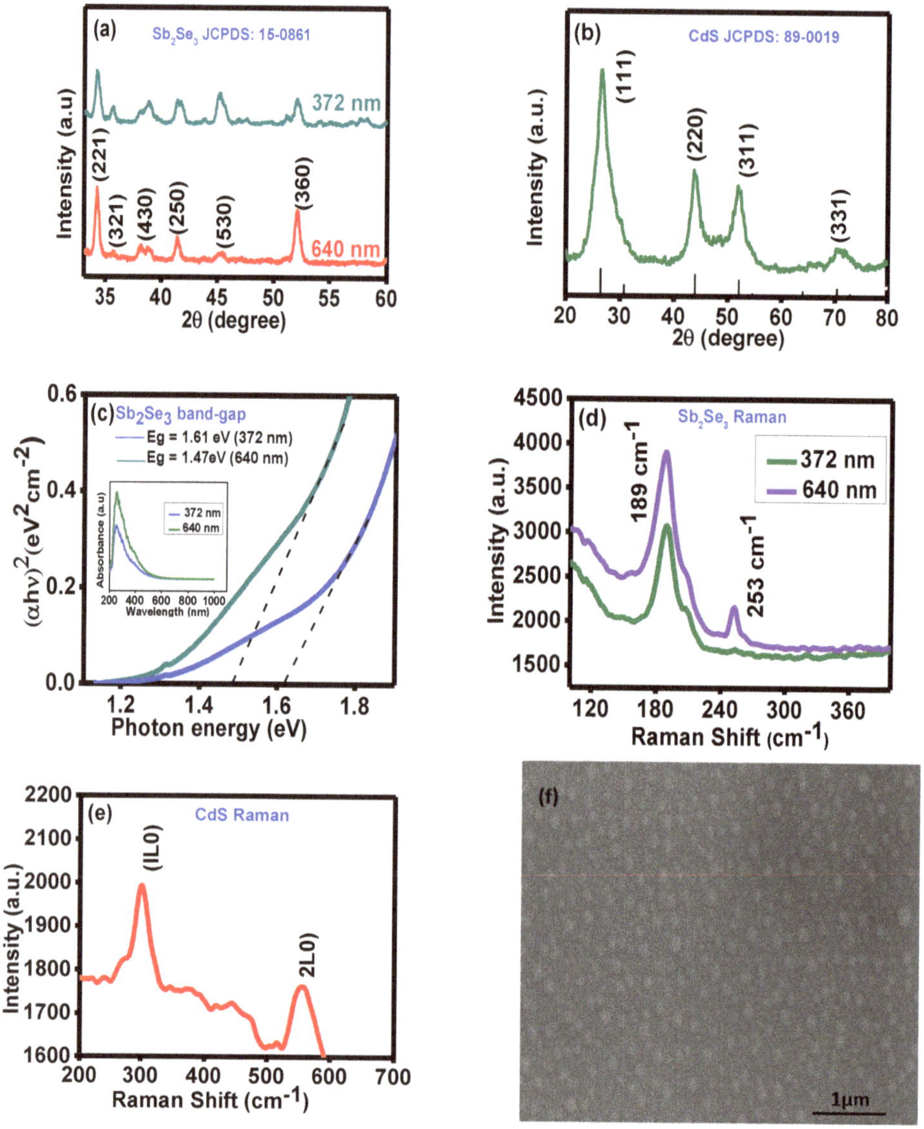

Figure 3. *Cont.*

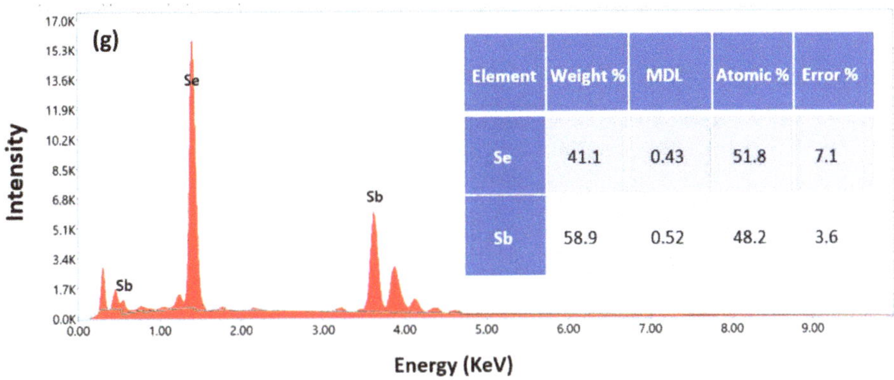

Figure 3. (**a**) XRD analysis for Sb_2Se_3 thin films, (**b**) XRD analysis for CdS (inset: absorbance vs. wavelength spectra), and (**c**) band gap of Sb_2Se_3 thin films at room temperature. Raman analysis for (**d**) Sb_2Se_3 thin films and (**e**) CdS thin films. (**f**) FE-SEM analysis for Sb_2Se_3 thin films, (**g**) EDAX spectroscopy for Sb_2Se_3 thin films, and (**h**) FE-SEM analysis for CdS thin films.

Figure 3c depicts the variation in $(\alpha h\upsilon)^2$ with photon energy ($h\upsilon$) for the direct bandgap of samples with thicknesses 372 nm and the 640 nm, respectively. The absorption spectra are shown in Figure 3c inset, where light absorption in the UV–visible region is enhanced as the Sb_2Se_3 thin film's thickness increases. This work's data agree with those from earlier investigations into Sb_2Se_3 thin films, which were performed using different techniques [29]. In earlier reports, the thin film's absorption coefficient 'α' was determined to be a function of wavelength variation in bandgap resulting from various deposition techniques and synthesis conditions [30]. Raman spectroscopy, a quick and, ideally, non-destructive method, can be used to characterize Sb_2Se_3, CdS and their impurity phases. Raman spectroscopy is based on the inelastic scattering that results from the interaction of monochromatic light with the material. The shift in the frequency of the emitted photons is a property that characterizes numerous frequency modes, including rotational, vibrational, and other low-frequency transitions in the molecules. Raman mode symmetry determines the nature of a material's vibrations. The peaks firmly match those that have already been recorded [31]. The results are shown in Figure 3d. Both films made of Sb_2Se_3 have a vibrational mode at 189 cm^{-1}. For a 640 nm thickness, a new peak (253 cm^{-1}) is observed. The Se-Se bond's stretching vibrations cause the development of a peak at 253 cm^{-1}. The vibrational mode is broader for the film at a lower thickness.

In contrast, the band becomes sharper as thickness increases. The band widens in cases of stress or a structural flaw, suggesting poor lattice structural quality. The intensity of both modes increases with the film thickness to support the increased crystalline nature of the film. Similarly, the Raman spectra of thermally deposited CdS films are shown in Figure 3e, where two peaks which are representative of the first and second longitudinal optical phonon modes, have been observed. The 1LO represents the fundamental and overtone modes (301 cm^{-1}) and 2LO (~580 cm^{-1}) peaks, which were strong and weak, respectively, and almost close to the literature value. Due to the phonon confinement effect, the location of the 2LO mode in the CdS thin films shifted slightly [20,32]. Figure 3f,g depict the field-emission scanning electron microscopy (FE-SEM) and energy-dispersive X-ray (EDAX) analysis for Sb_2Se_3 thin film, which is used to ensure the morphology and chemical composition, respectively. The surface of the thin films was uniform and good.

Similarly, in Figure 3h, FE-SEM of CdS is shown and a surface with small and uniform grains that are free of pinholes is found. These characteristics are in good agreement with the film's high transparency. The Fe-SEM image reveals that the grains were about 15 nm in size.

2.4. Experimental Results

Current–Voltage Characteristics

The I–V measurement for the solar cell was carried out under standard test conditions (100 $mWcm^{-2}$ light illumination). The I–V curves for various absorber layer thicknesses and corresponding device performance parameters are shown in Figure 4. Power output and cell fill factor have been computed from the illuminated I–V characteristics, which measure the short-circuit current and open-circuit voltage.

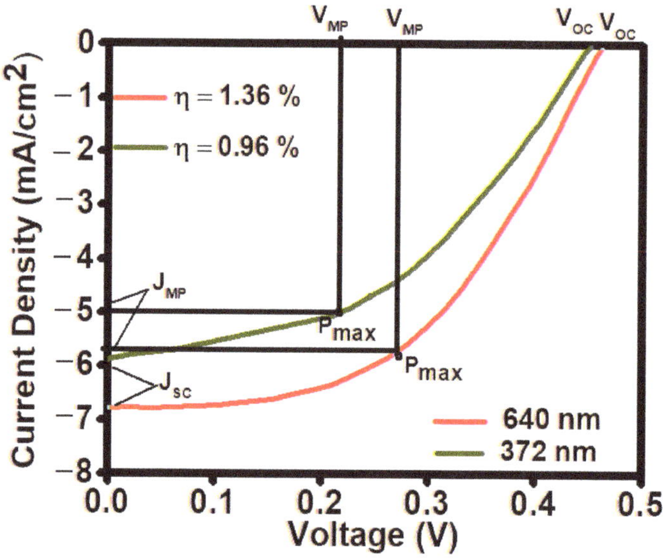

Figure 4. I–V plot for Sb_2Se_3/CdS-based structure with different absorber layer thicknesses.

The performance of solar cells is often described using the fill factor (FF) and efficiency (η), as shown in Equations (1) and (2)

$$FF = \frac{V_{MP} \times J_{MP}}{V_{OC} \times J_{SC}} \quad (1)$$

$$\eta = \frac{P_{max}}{P_{in}} \quad (2)$$

Here, P_{max} is maximum power, P_{in} is incident power, Voc is open-circuit voltage, and Jsc is short-circuited current.

For the 372 nm absorber layer's thickness, η is achieved as 0.96% (Voc = 0.42 V, Jsc = 5.9 mA/cm^2, FF = 39%, Rs = 16.7 Ω cm^2, and Rsh = 342 Ω cm^2) and for 640 nm, η is achieved as 1.36% (Voc = 0.46 V, Jsc = 6.9 mA/cm^2, FF = 48%, Rs = 12.5 Ω cm^2, and Rsh = 386 Ω cm^2). The results show that the device's performance improved for the large thickness of the absorber layer.

3. Numerical Study of the Proposed Device Structure

3.1. Simulation Parameter and Working Conditions

In order to fabricate a complete photovoltaic device, it is necessary to simulate the performance before the deposition of experimentally constructed layers. The Sb_2Se_3-based structure is modelled and simulated with the help of SCAPS-1D software. ITO/n-CdS/p-Sb_2Se_3/Ag is the proposed design of the device employed in this work. ITO and Ag are used as the front and back contacts, respectively. In order to check the device's efficiency, the p-Sb_2Se_3 layer's thickness is varied based on the experimental data. When simulating solar structures numerically, bandgap variation is also taken into account. The other parameters for Sb_2Se_3, CdS, and ITO are chosen from the previous research [33,34]. At its front contact, the solar cell receives 1 (kW/m^2) power from the AM 1.5G solar light spectrum. Table 1 lists the simulation parameters for a solar structure based on Sb_2Se_3, CdS, and ITO materials at 300 K.

Table 1. Simulation parameters for Sb_2Se_3-/CdS-/ITO-based solar structure.

Parameters	**Sb_2Se_3**	**n-CdS**	**ITO**
Thickness	372 nm, 640 nm (Experimental)	65 nm	200 nm
E_g (eV)	1.61, 1.47 (Experimental)	2.4	3.72
χ (eV)	4.04	4.2	4.5
ε_r (relative)	18	10	9.4
μ_n (cm^2/Vs)	15	100	30
μ_p (cm^2/Vs)	5.1	25	5
Donor density (cm^{-3})	0	10^{17}	10^{17}
Acceptor density (cm^{-3})	10^{18}	0	0

3.2. Numerical Results

3.2.1. I–V and QE Characteristic

Figure 5a depicts the I–V characteristics curve of the solar cell of the given Ag-/Sb_2Se_3-/CdS-/ITO-based structure for two different thicknesses (372 nm and 640 nm) of p-Sb_2Se_3 layers, which were achieved from the experimental analysis. We can see that the Sb_2Se_3 layer, which is 640 nm thick, exhibits the highest efficiency from those displayed on this I–V characteristics curve. Efficiencies of 2.59% (Voc = 0.68 V, Jsc = 6.98 mA/cm^2, FF = 53.20%) and 4.42% (Voc = 0.71 V, Jsc = 10.56 mA/cm^2, FF = 58.76%) were achieved for the 372 nm and 640 nm thicknesses of Sb_2Se_3, respectively (Table 2). Figure 5b depicts the external quantum efficiency (EQE) diagram of the Sb_2Se_3 thickness-based solar structure. Quantum efficiency indicates how well a solar cell can absorb carriers from photons of a specific energy when they are incident. A thinner absorber causes less photon absorption to occur at longer wavelengths. Fewer photogenerated electron-hole pairs are found inside the absorber layer as a result. A solar cell's response to various wavelengths is related to quantum efficiency. The device's absorption was in the 300–900 nm wavelength range. A solar cell with a 640 nm thick absorber layer gives higher QE than the 372 nm thick absorber-based cell. Quantum efficiency drops to zero at larger wavelengths because the light is not absorbed below the bandgaps at longer, lower-energy wavelengths. Both simulated devices show a reduction in quantum efficiency at wavelengths below 350 nm because the ITO substrate primarily absorbs the light.

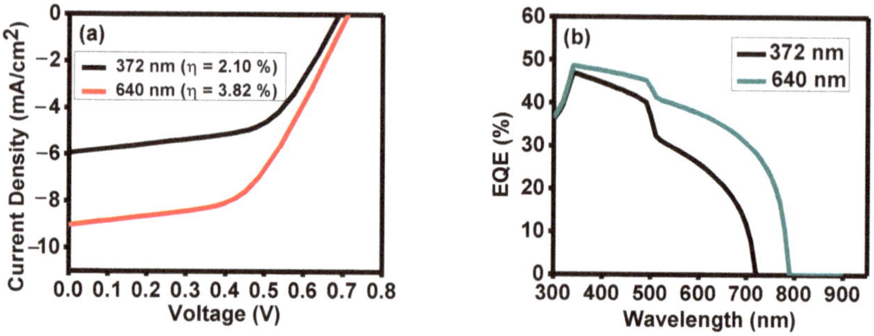

Figure 5. (**a**) I–V and (**b**) QE characteristics for two different thicknesses.

Table 2. Comparison of experimental and theoretical results for two different thicknesses of active Sb_2Se_3 layer.

Thickness (nm)	Voc (V)	Jsc (mA cm^2)	FF (%)	η (%)
372 (Exp.)	0.42	5.9	39	0.96
640 (Exp.)	0.46	6.9	43	1.36
372 (Theo.)	0.68	6.98	53.20	2.56
640 (Theo.)	0.71	10.56	58.76	4.42

3.2.2. The Parasitic Factors: Series and Shunt Resistance

Resistance in a semiconductor's bulk, connections, contacts, etc., is called it's series resistance. The shunt resistance comprises lattice flaws and leakage currents via the edge of the photovoltaic cell. Due to the thinness of the active layer materials, these losses happen when some photons pass through it. Figure 6 represents the influence of series (R_S) and shunt (R_{Sh}) resistance on CdS-/Sb_2Se_3-based cells' performances in ranges of 4–20 Ω-cm^2 and 100–500 Ω-cm^2, respectively, as the R_s and R_{sh} achieved from the experimental study are within these ranges. The solar cells must have low series and high shunt resistances in order to be used in the design and manufacture of high-efficiency PV devices. Here, in the below figures, with an increase in R_S and $R_{Sh,}$ the efficiency decreases and increases, respectively. These simulation results show that the R_S and R_{Sh} significantly impact a designed solar cell's performance.

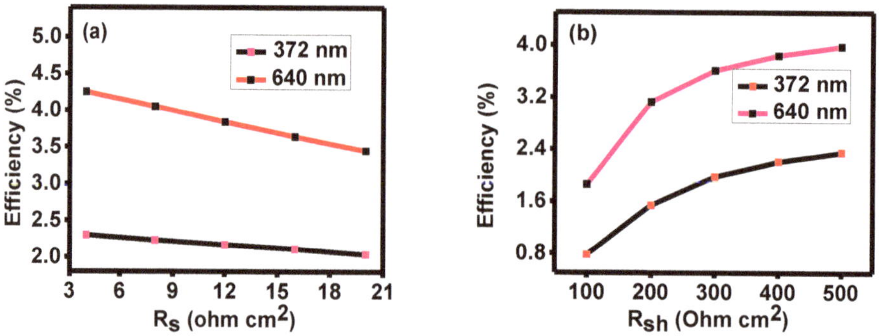

Figure 6. Variation in efficiency with (**a**) series resistance and (**b**) shunt resistance of the device.

In comparison to the simulation results, the experimental fill factor is lower. The resistive parameters primarily affect the fill factor of a solar device. Though the series and shunt resistances are kept the same, this difference can be due to other losses in the experimental study. These limit the performance of solar cells via phenomena such as optical loss due to non-absorption, thermalization, reflection, transmission, and area loss during the fabrication of a device. Additionally, interface defect density can also affect the fill factor.

3.2.3. Improvement in Device Performance

Optical parameters like bandgap and absorption coefficient should allow the solar spectrum to increase the device's performance. Numerous energetic photons will traverse the material if the bandgap is too broad, but they will not form an electron-hole (e–h) pair, and in the case of the small band gap, more energy will be wasted as heat. Therefore, materials are preferred whose band gap is specific. The material's absorption coefficient should also be high because more absorption will occur in that condition, and electrons will excite in the conduction band. The device's performance is also significantly influenced by the active layer thickness. Thus, a highly efficient solar cell can be manufactured with a better knowledge of the above factors. A few factors must be investigated in order to improve the device's performance, which places a substantial load on experimental research. Without fabricating the device, numerical analysis is the better and easiest approach to comprehending the physical mechanisms of a photovoltaic cell. In order to overcome and optimize the difficulties in the direction of a more effective solar cell, proper modeling is necessary. Numerical simulations offer a quick and effective method for identifying the key factors that affect better performance and also produce a valuable and realistic device. The efficiency of the Sb_2Se_3/CdS structure, shown in Figure 7. is improved from 4.42 to 11.27% by simply increasing the thickness and band gap of the absorber layer from 640 nm to 2 μm and 1.47 to 1.2 eV, respectively. Table 3 demonstrates the comparison study for various fabrication methods used to fabricate the Sb_2Se_3-/CdS-based solar cells.

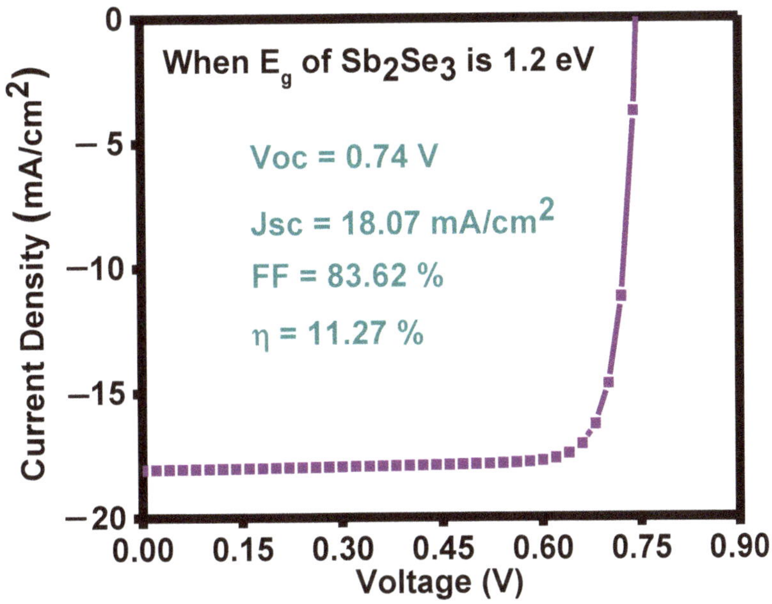

Figure 7. V performance of the optimized solar structure.

Table 3. Comparison of Sb_2Se_3-/CdS-based solar cells fabricated by various methods.

S.no.	Solar Structures	Fabrication Method	Efficiency (%)	References
1.	Mo/Sb_2Se_3/CdS/i-ZnO/ITO	E-beam evaporation/CBD	4.34	[35]
2.	FTO/Sb_2Se_3/CdS/ZnO/ZnO:Al/Au	Thermal/CBD/sputtering	2.1	[36]
3.	Mo/Sb_2Se_3/CdS/i-ZnO/Al-doped ZnO/Ni/Ag	Sputtering/CBD	1.47	[37]
4.	Ag/Sb_2Se_3/CdS/ITO	Thermal evaporation	1.36	This work

4. Conclusions

All the layers of solar cells were thermally deposited by the evaporation technique. These thermally evaporated CdS/Sb_2Se_3 solar cells showed open circuit voltages of 0.42 V and 0.46 V for the two different absorber layer's thicknesses, 372 nm, and 640 nm, respectively. These compare favourably to the reported values for a similar structure. Furthermore, experimentally, the highest efficiencies of 0.96% and 1.36% are achieved for these thicknesses of the absorber. Experimental work is then compared with theoretical study with the help of simulation, and the results state that the efficiency (2.56% and 4.42%) increases with the increase in absorber layer thickness. Series and shunt resistance, obtained from the experimental results, are also used in the simulation work, and their impact on the device's efficiency can also be noticed. In the simulation study, by further improving the absorber's band gap from 1.47 eV to 1.2 eV, a maximum efficiency of 11.27% is obtained. This study shows that a one-step thermal deposition technique and a proper numerical approach can produce a highly efficient solar cell.

Author Contributions: M. and R.K. (Raman Kumari): Conceptualization, Methodology, Data curation, Writing—original draft, C.Y.: Data curation, R.K. (Rahul Kumar): Methodology, K.K.M.: Supervision, Writing—review & editing, and V.N.S.: Supervision, Writing—review & editing. All authors have read and agreed to the published version of the manuscript.

Funding: This research received no external funding.

Data Availability Statement: Data will be available upon valid request. The raw/processed data required to reproduce these findings cannot be shared as the data also forms part of an ongoing study.

Acknowledgments: The authors highly acknowledge the Council for Scientific and Industrial Research (CSIR) and INDIA for the Senior Research Fellowship (SRF) grant. The authors would like to thank their institute for its support.

Conflicts of Interest: The authors declare no conflict of interest.

References

1. Houshmand, M.; Zandi, M.H.; Gorji, N.E. Degradation and device physics modeling of SWCNT/CdTe thin film photovoltaics. *Superlattices Microstruct.* **2015**, *88*, 365–370. [CrossRef]
2. Romanyuk, Y.E.; Hagendorfer, H.; Stücheli, P.; Fuchs, P.; Uhl, A.R.; Sutter-Fella, C.M.; Werner, M.; Haass, S.; Stückelberger, J.; Broussillou, C.; et al. All solution-processed chalcogenide solar cells–from single functional layers towards a 13.8% efficient CIGS device. *Adv. Funct. Mater.* **2015**, *25*, 12–27. [CrossRef]
3. Leng, M.; Luo, M.; Chen, C.; Qin, S.; Chen, J.; Zhong, J.; Tang, J. Selenization of Sb_2Se_3 absorber layer: An efficient step to improve device performance of CdS/Sb_2Se_3 solar cells. *Appl. Phys. Lett.* **2014**, *105*, 083905. [CrossRef]
4. Chen, C.; Li, W.; Zhou, Y.; Chen, C.; Luo, M.; Liu, X.; Zeng, K.; Yang, B.; Zhang, C.; Han, J.; et al. Optical properties of amorphous and polycrystalline Sb_2Se_3 thin films prepared by thermal evaporation. *Appl. Phys. Lett.* **2015**, *107*, 043905. [CrossRef]
5. Vadapoo, R.; Krishnan, S.; Yilmaz, H.; Marin, C. Electronic structure of antimony selenide (Sb_2Se_3) from GW calculations. *Phys. Status Solidi B* **2010**, *248*, 700–705. [CrossRef]
6. El-Shair, H.; Ibrahim, A.; El-Wahabb, E.A.; Afify, M.; El-Salam, F.A. Optical properties of Sb_2Se_3 thin films. *Vacuum* **1991**, *42*, 911. [CrossRef]
7. Kamruzzaman, M.; Liu, C.; Farid Ul Islam, A.K.M.; Zapien, J.A. A comparative study on the electronic and optical properties of Sb_2Se_3 thin film. *Semiconductors* **2017**, *51*, 1615–1624. [CrossRef]

8. Ma, Z.; Chai, S.; Feng, Q.; Li, L.; Li, X.; Huang, L.; Liu, D.; Sun, J.; Jiang, R.; Zhai, T.; et al. Chemical vapor deposition growth of high crystallinity Sb_2Se_3 nanowire with strong anisotropy for near-infrared photodetectors. *Small* **2019**, *15*, 1805307. [CrossRef]
9. Mahuli, N.; Halder, D.; Paul, A.; Sarkar, S.K. Atomic layer deposition of a Sb_2Se_3 Photo-absorber Layer using Selenium Dimethyldithiocarbamate as new Se precursor. *Chem. Mater.* **2019**, *31*, 7434–7442. [CrossRef]
10. Tang, R.; Chen, X.Y.; Liang, G.X.; Su, Z.H.; Luo, J.T.; Fan, P. Magnetron sputtering deposition and selenization of Sb_2Se_3 thin film for substrate Sb_2Se_3/CdS solar cells. *Surf. Coat. Technol.* **2019**, *360*, 68–72. [CrossRef]
11. Rajpure, K.; Bhosale, C. Preparation and characterization of spray-deposited photoactive Sb_2S_3 and Sb_2Se_3 thin films using aqueous and non-aqueous media. *Mater. Chem. Phys.* **2002**, *73*, 6–12. [CrossRef]
12. Virt, I.S.; Rudyj, I.O.; Kurilo, I.V.; Lopatynskyi, I.Y.; Linnik, L.F.; Tetyorkin, V.V.; Luka, G. Properties of Sb_2S_3 and Sb_2Se_3 thin films obtained by pulsed laser ablation. *Semiconductors* **2013**, *47*, 1003–1007. [CrossRef]
13. Li, Y.Z.; Li, F.; Liang, G.X.; Zheng, W.L.; Xu, Y.M.; Zheng, Z.H.; Fan, P. Sb_2Se_3 thin films fabricated by thermal evaporation deposition using the powder prepared via mechanical alloying. *Surf. Coat. Technol.* **2019**, *358*, 1013–1016. [CrossRef]
14. Tang, R.; Zheng, Z.; Su, Z.; Li, X.; Wei, Y.; Zhang, X.; Fu, Y.; Luo, Y.; Fan, P.; Liang, G. Highly efficient and stable planar hetero-junction solar cell based on sputtered and post-selenized Sb_2Se_3 thin film. *Nano Energy* **2019**, *64*, 103929. [CrossRef]
15. Liang, G.; Chen, M.; Ishaq, M.; Li, X.; Tang, R.; Zheng, Z.; Su, Z.; Fan, P.; Zhang, X.; Chen, S. Crystal growth promotion and defects healing enable minimum open-circuit voltage deficit in antimony selenide solar cells. *Adv. Sci.* **2022**, *9*, 2105142. [CrossRef]
16. Fan, P.; Chen, G.; Chen, S.; Zheng, Z.; Azam, M.; Ahmad, N.; Su, Z.; Liang, G.; Zhang, X.; Chen, Z. Quasi-vertically oriented Sb_2Se_3 thin film solar cells with open-circuit voltage exceeding 500 mV prepared via close-space sublimation and selenization. *ACS Appl. Mater. Interfaces* **2021**, *13*, 46671–46680. [CrossRef] [PubMed]
17. Komilov, A.G. Influence of CdS buffer layer thickness on the photovoltaic parameters of solar cells. *Appl. Sol. Energy* **2018**, *54*, 308–309. [CrossRef]
18. Caballero, R.; Kaufmann, C.A.; Cwil, M.; Kelch, C.; Schweigert, D.; Unold, T.; Rusu, M.; Schock, H.W.; Siebentritt, S. The role of the CdS buffer layer in $CuGaSe_2$-based solar cells. *J. Phys. Condens. Matter* **2007**, *19*, 356222. [CrossRef]
19. Hu, X.; Tao, J.; Wang, Y.; Xue, J.; Weng, G.; Zhang, C.; Chen, S.; Zhu, Z.; Chu, J. 5.91%-efficient Sb_2Se_3 solar cells with a radio-frequency magnetron-sputtered CdS buffer layer. *Appl. Mater. Today* **2019**, *16*, 367–374. [CrossRef]
20. Ashok, A.; Regmi, G.; Romero-Núñez, A.; Solis-López, M.; Velumani, S.; Castaneda, H. Comparative studies of CdS thin films by chemical bath deposition techniques as a buffer layer for solar cell applications. *J. Mater. Sci. Mater. Electron.* **2020**, *31*, 7499–7518. [CrossRef]
21. Hobson, T.D.; Hutter, O.S.; Birkett, M.; Veal, T.D.; Durose, K. 2018, June. Growth and characterization of Sb_2Se_3 single crystals for fundamental studies. In Proceedings of the 2018 IEEE 7th World Conference on Photovoltaic Energy Conversion (WCPEC) (A Joint Conference of 45th IEEE PVSC, 28th PVSEC & 34th EU PVSEC), Waikoloa Village, HI, USA, 10–15 June 2018; pp. 0818–0822.
22. Abhijit, B.K.; Jayaraman, A.; Molli, M. Electronic band structure and optical properties of antimony selenide under pressure. *Dae Solid State Phys. Symp.* **2015**, *1731*, 90020. [CrossRef]
23. Herrmann, M.G.; Stoffel, R.P.; Sergueev, I.; Wille, H.-C.; Leupold, O.; Haddouch, M.A.; Friese, K. Lattice dynamics of Sb_2Se_3 from inelastic neutron and X-ray scattering. *Phys. Status Solidi* **2020**, *257*, 2000063. [CrossRef]
24. Dong, Z.; Xu, H.; Liang, F.; Luo, C.; Wang, C.; Cao, Z.Y.; Chen, X.J.; Zhang, J.; Wu, X. Raman characterization on two-dimensional materials-based thermoelectricity. *Molecules* **2018**, *24*, 88. [CrossRef]
25. Burgelman, M.; Verschraegen, J.; Degrave, S.; Nollet, P. Modeling thin-film PV devices, progress in photovoltaics: Research and applications. *Proj. Photo Volt Res. Appl.* **2004**, *12*, 143–153. [CrossRef]
26. Mamta; Singh, Y.; Maurya, K.K.; Singh, V. A review on properties, applications, and deposition techniques of antimony selenide. *Sol. Energy Mater. Sol. Cells* **2021**, *230*, 111223. [CrossRef]
27. Ferekides, C.S.; Balasubramanian, U.; Mamazza, R.; Viswanathan, V.; Zhao, H.; Morel, D.L. CdTe thin film solar cells: Device and technology issues. *Sol. Energy* **2004**, *77*, 823–830. [CrossRef]
28. Ghahramanifard, F.; Rouhollahi, A.; Fazlolahzadeh, O. Electrodeposition of Cu-doped p-type ZnO nanorods; effect of Cu doping on structural, optical, and photo electrocatalytic properties of ZnO nanostructure. *Superlattices Microstruct.* **2018**, *114*, 1–14. [CrossRef]
29. Liu, C.; Yuan, Y.; Cheng, L.; Su, J.; Zhang, X.; Li, X.; Li, J. A study on optical properties of Sb_2Se_3 thin films and resistive switching behavior in Ag/Sb_2Se_3/W heterojunctions. *Results Phys.* **2019**, *13*, 102228. [CrossRef]
30. Mamta; Singh, Y.; Maurya, K.K.; Singh, V.N. n-Si/p-Sb_2Se_3 structure-based simple solar cell device. *Mater. Today Sustain.* **2022**, *18*, 100148. [CrossRef]
31. Tao, R.; Tan, T.; Zhang, H.; Meng, Q.; Zha, G. Sb_2Se_3 solar cells fabricated via close-space sublimation. *J. Vac. Sci. Technol. B* **2021**, *39*, 052203. [CrossRef]
32. Sathishkumar, R.; Devakirubai, E.; David, A.; Tamilselvan, S.; Nithiyanantham, S. Structural and optical studies of cadmium sulfide (CdS) thin film by chemical bath deposition (CBD). *Mater. Focus* **2017**, *6*, 41–46. [CrossRef]
33. Li, Z.-Q.; Ni, M.; Feng, X.-D. Simulation of the Sb_2Se_3 solar cell with a hole transport layer. *Mater. Res. Express* **2020**, *7*, 016416. [CrossRef]
34. Ullah, H.; Ullah, S.; Soucase, B.M. The baseline of numerical simulations for ZnTe-based thin-film solar cells. In Proceedings of the 2014 International Conference on Energy Systems and Policies (ICESP), Islamabad, Pakistan, 24–26 November 2014; pp. 1–6. [CrossRef]

35. Tiwari, K.J.; Neuschitzer, M.; Espíndola-Rodriguez, M.; Sanchez, Y.; Jehl, Z.; Vidal-Fuentes, P.; Saucedo, E.; Malar, P. Efficient Sb_2Se_3/CdS planar heterojunction solar cells in substrate configuration with (hk0) oriented Sb_2Se_3 thin films. *Sol. Energy Mater. Sol. Cells* **2020**, *215*, 110603. [CrossRef]
36. Liu, X.; Chen, J.; Luo, M.; Leng, M.; Xia, Z.; Zhou, Y.; Qin, S.; Xue, D.J.; Lv, L.; Huang, H.; et al. Thermal evaporation and characterization of Sb_2Se_3 thin film for substrate Sb_2Se_3/CdS solar cells. *ACS Appl. Mater. Interfaces* **2014**, *6*, 10687–10695. [CrossRef]
37. Lee, D.; Cho, J.Y.; Heo, J. Improved efficiency of Sb_2Se_3/CdS thin-film solar cells: The effect of low-temperature pre-annealing of the absorbers. *Sol. Energy* **2018**, *173*, 1073–1079. [CrossRef]

nanomaterials

MDPI

Communication

Schottky-Diode Design for Future High-Speed Telecommunications

Chi-Ho Wong [1,2], Leung-Yuk Frank Lam [3], Xijun Hu [3], Chi-Pong Tsui [1,*] and Anatoly Fedorovich Zatsepin [4,*]

1 Department of Industrial and Systems Engineering, The Hong Kong Polytechnic University, Hong Kong 999077, China; roych.wong@polyu.edu.hk
2 Research Institute for Advanced Manufacturing, The Hong Kong Polytechnic University, Hong Kong 999077, China
3 Department of Chemical and Biological Engineering, The Hong Kong University of Science and Technology, Hong Kong 999077, China
4 Institute of Physics and Technology, Ural Federal University, Yekaterinburg 620002, Russia
* Correspondence: gary.c.p.tsui@polyu.edu.hk (C.-P.T.); a.f.zatsepin@urfu.ru (A.F.Z.)

Abstract: The impact of 5G communication is expected to be widespread and transformative. It promises to provide faster mobile broadband speeds, lower latency, improved network reliability and capacity, and more efficient use of wireless technologies. The Schottky diode, a BN/GaN layered composite contacting bulk aluminum, is theoretically plausible to harvest wireless energy above X-band. According to our first principle calculation, the insertion of GaN layers dramatically influences the optical properties of the layered composite. The relative dielectric constant of BN/GaN layered composite as a function of layer-to-layer separation is investigated where the optimized dielectric constant is ~2.5. To push the dielectric constant approaching ~1 for high-speed telecommunication, we upgrade our BN-based Schottky diode via nanostructuring, and we find that the relative dielectric constant of BN monolayer (semiconductor side) can be minimized to ~1.5 only if it is deposited on an aluminum monolayer (metal side). It is rare to find a semiconductor with a dielectric constant close to 1, and our findings may push the cut-off frequency of the Al/BN-based rectenna to the high-band 5G network.

Keywords: dielectric properties; 2D materials; Schottky diode; energy harvesting system

check for updates

Citation: Wong, C.-H.; Lam, L.-Y.F.; Hu, X.; Tsui, C.-P.; Zatsepin, A.F. Schottky-Diode Design for Future High-Speed Telecommunications. *Nanomaterials* **2023**, *13*, 1448. https://doi.org/10.3390/nano13091448

Academic Editor: Albert Chin

Received: 30 March 2023
Revised: 14 April 2023
Accepted: 20 April 2023
Published: 24 April 2023

1. Introduction

5G communication promises to revolutionize wireless technology, enabling faster download and upload speeds, lower latency, more robust and reliable connections, and many more innovative remote applications [1], such as medical surgery and automobiles. With the potential for improved connectivity, 5G will enable more efficient, creative, and cost-effective solutions for businesses and consumers alike. It could also create opportunities for developing countries to gain similar access to high-speed internet as those in developed nations, providing a better quality of wireless communication. 5G wireless communication is based on harvesting wireless energy with the help of a local antenna array and a low-power automated transceiver [1]. The automated transceiver assigned the optimum frequency channels where the local antennas are connected to the LTE network. The frequency of the low-band 5G is close to 700 MHz, and the transmission of the mid-band 5G is about 2.5–3.7 GHz [2]. The high-band 5G frequencies of 25–40 GHz should provide the fastest internet speed of a few gigabits per second [3]. Although a frontier company such as Samsung has been investigating the feasibility of mobile communications at frequencies of 28 and 38 GHz [3], the current state-of-the-art Schottky diode harvests wireless energy up to 8 GHz–12 GHz (X band) only [4]. This X-band Schottky diode is made up of two-dimensional MoS_2 sheets and bulk palladium[4], where the relative dielectric constant of MoS_2 sheets is above six [5]. Harvesting wireless energy beyond X-band requires the clearance of electric charges in the depletion region rapidly, where the cut-off frequency of the Schottky diode is inversely proportional to the dielectric constant [6]. A high dielectric constant increases the capacitance of Schottky diodes and eventually requires a longer transient time to clear the electric charges [7].

Designing a Schottky diode that is capable of harvesting wireless energy at the high-band 5G signals has proven to be challenging because of the issue of finding semiconductors that have a relative dielectric constant [7] close to 1. This is because of the problem of having high electric susceptibilities in semiconductors. For example, the dielectric constants of 3D silicon, 3D germanium, and 2D MoS_2 sheets are 11, 16, and 6, respectively [7,8]. As the dielectric constant depends on size and pressure [9], we have decided to alter the nanostructure of a Schottky diode in order to create an ultra-low dielectric constant, which would better prepare for the 5G world. The dielectric constants of GaN and BN films are ~8 and ~3.5, respectively. We are going to study whether the heterostructure made up of these two materials will yield a much lower dielectric constant. In this paper, two Schottky diodes will be proposed: (BN-GaN/Al and BN monolayer/Al monolayer). We will apply the ab-initio simulations to design high-speed Schottky diodes with a key feature of ultralow dielectric constant.

2. Materials and Methods

The BN layers and GaN layers are stacked alternatingly, as shown in Figure 1. We apply geometric optimization under a boundary condition that the layer-to-layer distance remains at 0.5 nm. The geometric optimization is implemented under spin-restricted GGA-PBE, functional [10,11] under the CASTEP platform. The convergence tolerance is that the maximum displacement and the maximum force of atoms are 0.002 Å and 0.05 eV/Å, respectively. We calculate the band structure and the dielectric constant using the same DFT functional. The maximum SCF cycle is 100, and the SCF tolerance is 2×10^{-6} eV/atom. The internal of k-points is 1/Å. The ultrasoft pseudopotential is applied. The bulk aluminum is attached to BN/GaN layered composite side-by-side with specific crystallographic directions, i.e., the Al [100] and BN/GaN [1120] axes are parallel (abbreviated as Schottky diode A).

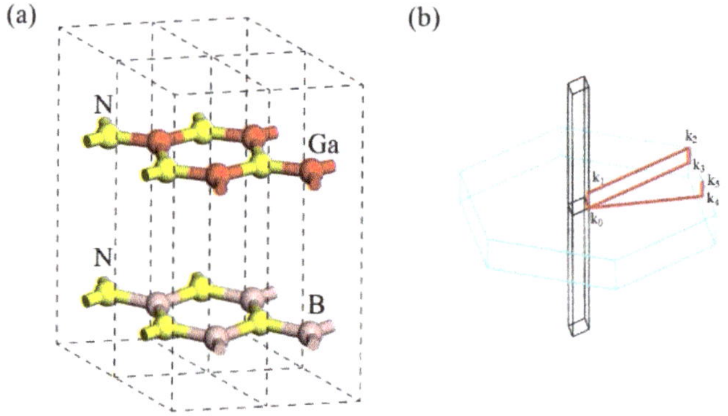

Figure 1. (**a**) The repeated unit of BN/GaN layered composite. (**b**) The real (black) and reciprocal (light blue) spaces. The red lines label the direction of k space.

On the other hand, we examine the semiconducting properties of Schottky diode B, in which the boron nitride monolayer is deposited on top of an aluminum monolayer. After implementing geometric optimization in the presence of an aluminum monolayer under the same DFT method, we proceed to calculate the band gap and the dielectric constant (parallel to the surface) of the BN monolayer.

3. Results and Discussions

The band structure of the BN-GaN layered composite is plotted in Figure 2. Despite the band gap of the isolated BN sheet being as large as 6 eV, the direct band gap of the BN/GaN layered composite is reduced to 1.6 eV because of the proximity effect across the layers. The bigger gallium atoms raise the atomic density in each GaN layer so that a stronger internal pressure [12,13] is created in the composite. The insertion of GaN layers also strengthens electronic overlaps along the lateral direction [12,13]. The wavefunction of electrons overlapped more effectively in the BN/GaN layered composite, and hence a narrower band gap is observed. The theoretical band gaps simulated by the GGA functional like PBE or PW91 are usually underestimated [14–17]. However, there is no universal way to calibrate the band gap calculated by the ab-initio simulations unless the experimental band gap of the BN-GaN layered composite is known in combination with the parametric tests in the ab-initio simulations semi-empirically. Though our calculated band gap is slightly underestimated, the band gap of the BN/GaN layered composite still reaches 1.6 eV, in which the value of the band gap is suitable to be a part of the Schottky diode.

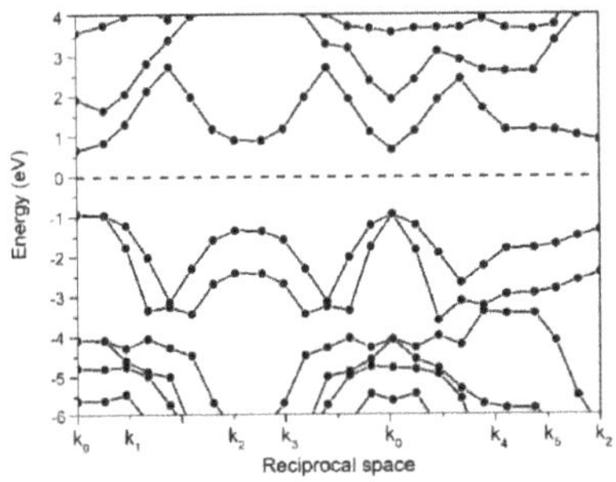

Figure 2. The band structure of BN/GaN layered composite. The layer-to-layer distance is 0.5 nm. The Fermi level is shifted to 0 eV for convenience.

To identify if the BN/GaN layered composite belongs to a p-type or n-type semiconductor, we estimate the Fermi level (E_F) relative to the conduction band (E_c) and the valence band (E_v), respectively. Figure 2 shows that the Fermi level of the BN/GaN layered composite is located closer to the conduction band, and we conclude that the BN-GaN layered composite is an n-type semiconductor, where the energy levels before the bulk Al and BN-GaN join together, drawn in Figure 3. The Fermi level is located nearer to the conduction band because the bigger gallium atom decreases the positive charge density of the composite [18].

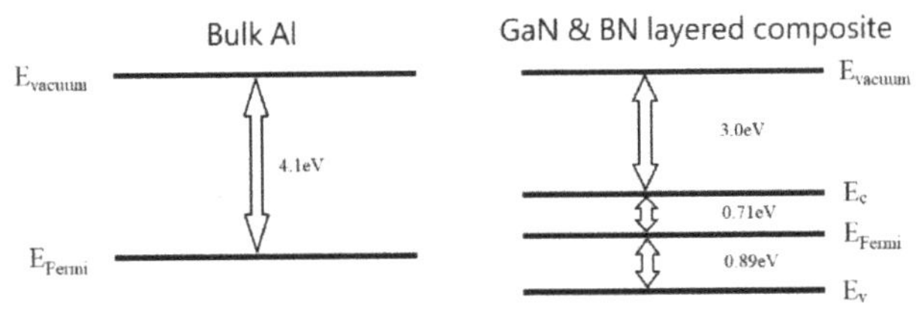

Figure 3. The energy levels of the bulk Al and the BN-GaN layered composite before they join together to form the Schottky diode A. The work function is the energy difference between the vacuum level E_{vaccum} and Fermi level E_{Fermi}, while the band gap is the energy difference between the conduction E_c and valence bands E_v. The electron affinity is the energy difference between the vacuum level E_{vaccum} and conduction band E_c. Subtracting the work function from the electron affinity yields the Schottky barrier.

Figure 3 displays the simplified energy diagrams of Schottky diode A. The work function of aluminum is 4.1 eV, and the electron affinity of the BN/GaN layered composite is 3 eV. The Schottky barrier refers to the offset between the work function of the metal and the electron affinity of the semiconductor [7,18]. The Schottky barrier of most silicon-based diodes is around 0.7 eV [19]. By aligning the energy states relative to the vacuum level, the Schottky barrier of Schottky diode A is 1.1 eV. In other words, the driven voltage of Schottky diode A for forward-bias operation is comparable to the modern silicon-based diodes [7].

The dielectric function describes the response of the material subjected to a time-dependent electromagnetic field [18]. The excitations induced by the interaction of incident electrons with materials refer to the real and imaginary parts of the dielectric function [20]. We show the relative dielectric constant of the BN/GaN layered composite in Table 1. The relative dielectric constant of the BN/GaN layered composite is as low as 2.5. In comparison with the modern X-band rectenna consisting of two-dimensional MoS_2 sheets and palladium [4], the dielectric constant of BN/GaN layered composite is at least 50% lower than the dielectric constant of two-dimensional MoS_2 sheets [5]. With the use of the BN/GaN layered composite as a part of the Schottky diode, it is possible to push the cut-off frequency far beyond X-band because the cut-off frequency is inversely proportional to the capacitance across the depletion region. The non-regular atomic layers across the metal-semiconductor interface make the theoretical prediction of junction resistance inaccurate. However, the electrical resistance of aluminum is ~4 times smaller than that of palladium, and hence a low junction resistance is likely expected. The imaginary parts of the dielectric constants of Schottky diodes A and B are close to zero, and therefore their phase lags are likely negligible [20].

Table 1. The dielectric constants of Schottky diodes A and B.

	Material (s)	Dielectric Constant (s)
Schottky diode A	Bulk Al and BN/GaN layered composite	2.5–3.1 (d: 0.5 nm–0.8 nm)
Schottky diode B	Al monolayer and BN monolayer	1.5

The dielectric constant is directly proportional to the permittivity of the medium [18]. If the layer-to-layer distance of the BN/GaN layered composite increases to 0.8 nm, the relative dielectric constant can drop to 2.6, but the band gap rises to 2.5 eV, as shown in Figure 4. Materials with low dielectric constants generally have weaker intermolecular interactions, meaning their molecules are less likely to hold onto, and may facilitate the flow of, the electric charge. Hence, weakening the layer-to-layer coupling decreases the effectiveness of electric polarization in the medium [20] and presumably drops the permittivity of the substance. While softening the layer-to-layer coupling may yield a low dielectric constant, we proceed to study the dielectric properties of the isolated BN monolayer. We observe that the relative dielectric constant of the isolated BN monolayer can drop even further to 2.4.

The BN layer is used as a building block because the dielectric constant of pristine hexagonal BN film is already small (~3.5). The relationship between the dielectric constant ε_r and imposed E-field[6] is $\varepsilon_r = 1 + \frac{2\pi B\gamma^{1/2}}{\xi^2} \times f(Aiya)$, where B is related to the optical absorption in the absence of static field. $\gamma \propto \frac{1}{F_{field}}$, where F_{field} is the force under an imposed E-field [6]. $f(Airy)$ is the integral related to Airy functions [6]. To impose an E-field ($\gamma \propto \frac{1}{F_{field}}$) can drop the dielectric constant of semiconductors [6], we need to insert another thin-film semiconducting material with the same hexagonal structure to create an internal E-field perpendicular to the plane, where the GaN layer is preferable. By introducing a dissimilar size of cation (Ga and B) between the adjacent layers, the dissimilar layers contribute an internal E-field perpendicular to the plane. Hence, the closely packed BN and GaN layered composite yields a lower dielectric constant than either BN or GaN layers where the positive charge density of B ion is higher than that of Ga ion in the layers, and meanwhile, the B ion induces negative charges on the nearest GaN layers under an induced internal E-field [6]. The electric field usually drops the dielectric constant by reducing the number of effective charge carriers in the material. This happens because the polar molecules in the material are oriented and shifted in response to the external E field. This impedes materials from concentrating the electrostatic flux. As a result, the induced negative charges on GaN layers owing to the induced E-field reduce the dielectric constant [6] in the entire composite. The band gap is increased from 1.6 eV to 2.5 eV because the wider layer-to-layer separation reduces the internal pressure of the composite along the lateral plane [12,18].

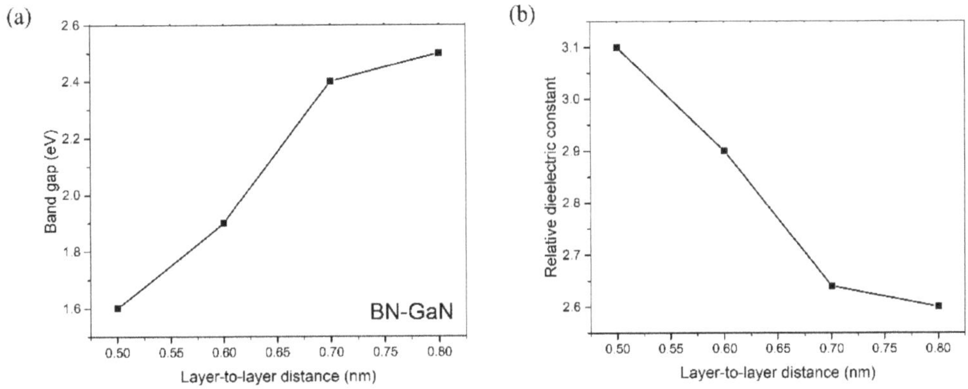

Figure 4. (**a**) The band gap and (**b**) the dielectric constant of the BN-GaN composite depend on the layer-to-layer separation d between BN and GaN layers.

Motivated by the benefits of the internal E-field between the layers [6], we decided to examine if the application of an internal E-field between an aluminum monolayer and a BN monolayer can drop the dielectric constant of the BN monolayer below 2.0. After the BN monolayer is relaxed in the presence of an aluminum monolayer, the B-N-B bond angle changes from 120 degrees to 116.9 degrees, and the N-B bond length changes from 1.45 Å to 1.48 Å. The Al ions in the aluminum monolayer create a high positive charge density and induce negative charges on the BN monolayer, which sets up an internal E-field. The induced E-field drops the dielectric constant[6] of the BN monolayer below 2.0 successfully. As listed in Table 1, the boron nitride monolayer in Schottky diode B shows the relative dielectric constant of 1.5 only, and the energy levels of the components are listed in Figure 5. In comparison with bulk aluminum, the work function of the aluminum monolayer is only 2.5 eV because the size-dependent work function is always expected [21]. We draw a parallel between the function parameters of Schottky diode B and the MoS_2-based Schottky diode in Figure 6. Although the energy barrier of Schottky diode B is increased to 1.41 V, its ultralow dielectric constant is still able to harvest wireless energy beyond X-band communication if the driven voltage is above the forward-bias voltage [7]. The dielectric constant of the BN monolayer in Schottky diode B is ~4 times smaller than the dielectric constant of two-dimensional MoS_2 sheets in the modern X-band Schottky diode [4]. As the cut-off frequency of the Schottky diode is inversely proportional to the capacitance across the depletion region [7], the ultralow dielectric constant of Schottky diode B may be able to boost the cut-off-frequency towards the high-band 5G signals. Moreover, the wide band gap in the Schottky diode B is expected to operate at higher temperatures than the MoS_2-based Schottky diodes because a wide band gap in a semiconductor allows for greater resistance to thermal breakdown, as the material is able to withstand higher temperatures without carrier excitation. This increases the maximum operating temperature of the Schottky diode, allowing it to operate more efficiently at higher temperatures. Concerning electric conductivity, aluminum is more metallic than palladium, which allows a low ohmic resistance value (minimizes power dissipation) on the metallic side.

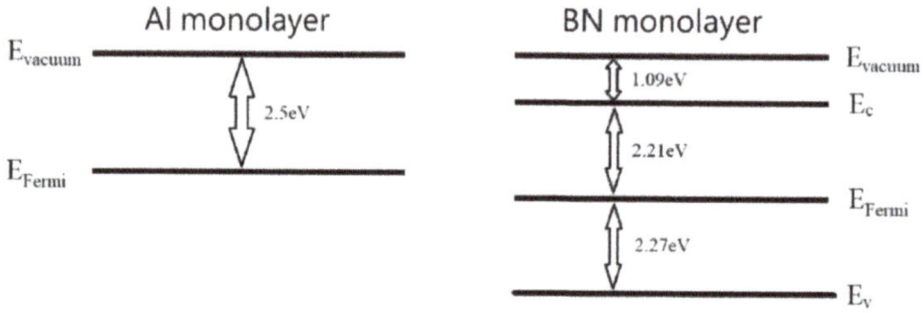

Figure 5. The energy levels of the Al monolayer and BN monolayer before they join together to form Schottky diode B. **Left panel**: The work function is defined by E_{vaccum}—E_{Fermi}; **Right panel**: The band gap is defined by E_c—E_V and the electron affinity is defined by E_{vaccum}—E_C; The work function subtracted by the electron affinity is called the Schottky barrier.

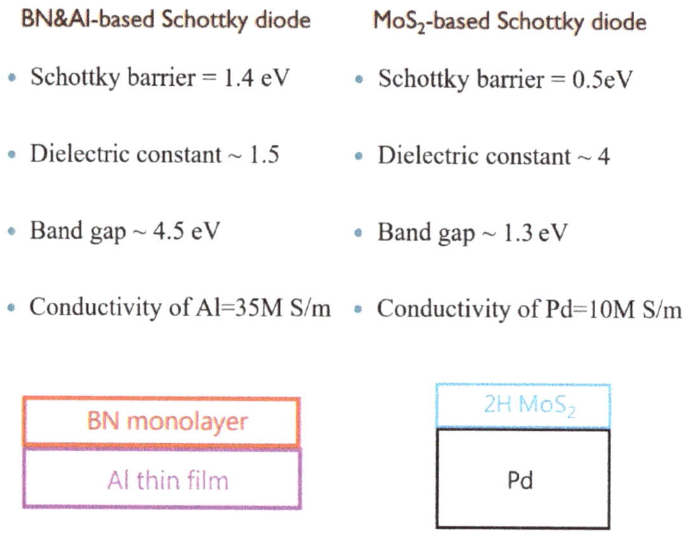

Figure 6. Comparison between Schottky diode B and the MoS_2-based Schottky diode [4]. The electric conductivity refers to 293 K.

4. Conclusions

Our research shows the potential for utilizing a BN/GaN layered composite and aluminum to build Schottky diodes for creating high-frequency wireless energy-harvesting devices. By tuning the induced electric field of the interface, we identify that the dielectric constant of the boron nitride monolayer on top of the aluminum monolayer is as low as 1.5, suggesting its suitability for harvesting high-band 5G signals. The implication of this finding is that the proposed heterostructures are promising to construct devices capable of harvesting higher frequency signals, such as those produced by 5G networks. Such a device would enable the capture and conversion of these high-frequency signals into energy sources, creating an efficient form of energy transfer and high-speed telecommunications.

Author Contributions: Conceptualization: C.-H.W.; Formal analysis: C.-H.W. and A.F.Z.; Investigation: C.-H.W., L.-Y.F.L., X.H., C.-P.T. and A.F.Z.; Methodology: C.-H.W.; Validation: C.-H.W.; Supervision: C.-H.W. and A.F.Z.; Visualization: C.-H.W., L.-Y.F.L., X.H., C.-P.T. and A.F.Z.; Writing—original draft: C.-H.W.; Writing—review and editing: C.-H.W., L.-Y.F.L., X.H., C.-P.T. and A.F.Z.; Resources: C.-P.T. and A.F.Z.; Funding acquisition: C.-P.T. and A.F.Z. All authors have read and agreed to the published version of the manuscript.

Funding: This research was funded by Russian Federation for support (Ural Federal University Program of Development within the Priority-2030 Program) and the Department of Industrial and Systems Engineering, The Hong Kong Polytechnic University. A.F.Z. thanks the Ministry of Science and Higher Education of the Russian Federation for support (Ural Federal University Program of Development within the Priority-2030 Program, project. 4.38). C.H.W thanks Research Institute for Advanced Manufacturing, The Hong Kong Polytechnic University. The authors would like to thank for the financial support from the Department of Industrial and Systems Engineering, The Hong Kong Polytechnic University. The APC was funded by Department of Industrial and Systems Engineering, The Hong Kong Polytechnic University.

Institutional Review Board Statement: Not applicable.

Informed Consent Statement: Not applicable.

Data Availability Statement: Data are sharable under a reasonable request.

Acknowledgments: We thank the Department of Industrial and Systems Engineering and Research Institute for Advanced Manufacturing at The Hong Kong Polytechnic University. We also thank the Ministry of Science and Higher Education of the Russian Federation for support.

Conflicts of Interest: The authors declare no conflict of interest.

References

1. Luo, Y.; Pu, L.; Wang, G.; Zhao, Y. RF Energy Harvesting Wireless Communications: RF Environment, Device Hardware and Practical Issues. *Sensors* **2019**, *19*, 3010. [CrossRef] [PubMed]
2. Trinh, L.H.; Ferrero, F.; Lizzi, L.; Staraj, R.; Ribero, J.-M. Reconfigurable Antenna for Future Spectrum Reallocations in 5G Communications. *IEEE Antennas Wirel. Propag. Lett.* **2015**, *15*, 1297–1300. [CrossRef]
3. Wang, Y.; Li, J.; Huang, L.; Jing, Y.; Georgakopoulos, A.; Demestichas, P. 5G Mobile: Spectrum Broadening to Higher-Frequency Bands to Support High Data Rates. *IEEE Veh. Technol. Mag.* **2014**, *9*, 39–46. [CrossRef]
4. Zhang, X.; Grajal, J.; Vazquez-Roy, J.L.; Radhakrishna, U.; Wang, X. Two-dimensional MoS2-enabled flexible rectenna for Wi-Fi-band wireless energy harvesting. *Nature* **2019**, *566*, 368–372. [CrossRef] [PubMed]
5. Laturia, A.; Van de Put, M.L.; Vandenberghe, W.G. Dielectric properties of hexagonal boron nitride and transition metal dichalcogenides: From monolayer to bulk. *NPJ 2D Mater. Appl.* **2018**, *2*, 6. [CrossRef]
6. Viswanathan, K.S.; Callaway, J. Dielectric Constant of a Semiconductor in an External Electric Field. *Phys. Rev.* **1966**, *143*, 564. [CrossRef]
7. McCowen, A.; Shaari, S.B.H.; Board, K. Transient analysis of Schottky-barrier diodes. *IEE Proc. Solid-State Electron Devices* **1988**, *135*, 71–75. [CrossRef]
8. Kar, S. *Data Tables of Semiconductor, High-k Dielectrics, and Metal Constants*; Springer Verlag: Berlin/Heidelberg, Germany, 2013.
9. Chan, R.K.; Davidson, D.W.; Whalley, E. Effect of Pressure on the Dielectric Properties of Ice I. *J. Chem. Phys.* **1965**, *43*, 2376. [CrossRef]
10. Burke, K.; Perdew, J.P.; Wang, Y. Derivation of a Generalized Gradient Approximation: The PW91 Density Functional. In *Electronic Density Functional Theory*; Springer: New York, NY, USA, 1998; pp. 81–111.
11. Perdew, J.P.; Burke, K.; Ernzerhof, M. Generalized Gradient Approximation Made Simple. *Phys. Rev. Lett.* **1996**, *77*, 18. [CrossRef] [PubMed]
12. Zhang, L.; Liu, C.; Wang, L.; Liu, C.; Wang, K.; Zou, B. Pressure-Induced Emission Enhancement, Band-Gap Narrowing, and Metallization of Halide Perovskite Cs3Bi2I9. *Angew. Chem. Int. Ed.* **2018**, *57*, 35.
13. Yang, Z.; Hu, K.; Meng, X.; Tao, Q.; Dong, J.; Liu, B.; Lu, Q.; Zhang, H.; Sundqvist, B.; Zhu, P.; et al. Tuning the band gap and the nitrogen content in carbon nitride materials by high temperature treatment at high pressure. *Carbon* **2018**, *130*, 170–177. [CrossRef]
14. Sánchez, P.M.; Cohen, A.J.; Yang, W. Localization and Delocalization Errors in Density Functional Theory and Implications for Band-Gap Prediction. *Phys. Rev. Lett.* **2008**, *100*, 146401. [CrossRef] [PubMed]
15. Becke, A.D. Density-functional Exchange-Energy Approximation with Correct Asymptotic Behavior. *Phys. Rev. A* **1998**, *38*, 3098. [CrossRef] [PubMed]
16. Lee, C.; Yang, W.; Parr, R.G. Development of the Colle-Salvetti Correlation-Energy Formula into a Functional of the Electron Density. *Phys. Rev. B* **2018**, *37*, 785. [CrossRef] [PubMed]

17. Crowley, J.M.; Kheli, J.T.; William, A. Resolution of the Band Gap Prediction Problem for Materials Design. *J. Phys. Chem. Lett.* **2016**, *7*, 1198–1203. [CrossRef] [PubMed]
18. Kittel, C. *Introduction to Solid State Physics*, 8th ed.; Wiley: Hoboken, NJ, USA, 2004; ISBN 978-0471415268.
19. Milnes, A.G. Metal-Semiconductor Schottky-Barrier Diodes. In *Semiconductor Devices and Integrated Electronics*; Springer: Dordrecht, The Netherlands, 1980.
20. Petzelt, J.; Rychetský, I. Dielectric Function. In *Encyclopedia of Condensed Matter Physics*; Elsevier: Amsterdam, The Netherlands, 2005; pp. 426–429.
21. Khoa, N.T.; Kim, S.W.; Yoo, D.-H.; Kim, E.J.; Hahn, S.H. Size-dependent work function and catalytic performance of gold nanoparticles decorated graphene oxide sheets. *Appl. Catal. A Gen.* **2014**, *469*, 159–164. [CrossRef]

nanomaterials

MDPI

Review

Triboelectric and Piezoelectric Nanogenerators for Self-Powered Healthcare Monitoring Devices: Operating Principles, Challenges, and Perspectives

Enrique Delgado-Alvarado [1,*], Jaime Martínez-Castillo [1], Luis Zamora-Peredo [1], Jose Amir Gonzalez-Calderon [2], Ricardo López-Esparza [3], Muhammad Waseem Ashraf [4], Shahzadi Tayyaba [5] and Agustín L. Herrera-May [1,6,*]

1 Micro and Nanotechnology Research Center, Universidad Veracruzana, Boca del Río 94294, Veracruz, Mexico
2 Cátedras CONACYT-Institute of Physic, Universidad Autónoma de San Luis Potosí, San Luis Potosí 78290, San Luis Potosí, Mexico
3 Departamento de Física, Universidad de Sonora, Hermosillo 83000, Sonora, Mexico
4 Department of Physics, Government College University Lahore, Lahore 54000, Pakistan
5 Department of Computer Engineering, The University of Lahore, Lahore 54000, Pakistan
6 Maestría en Ingeniería Aplicada, Facultad de Ingeniería de la Construcción y el Hábitat, Universidad Veracruzana, Boca del Río 94294, Veracruz, Mexico
* Correspondence: endelgado@uv.mx (E.D.-A.); leherrera@uv.mx (A.L.H.-M.); Tel.: +52-2297752000 (E.D.-A.)

Citation: Delgado-Alvarado, E.; Martínez-Castillo, J.; Zamora-Peredo, L.; Gonzalez-Calderon, J.A.; López-Esparza, R.; Ashraf, M.W.; Tayyaba, S.; Herrera-May, A.L. Triboelectric and Piezoelectric Nanogenerators for Self-Powered Healthcare Monitoring Devices: Operating Principles, Challenges, and Perspectives. *Nanomaterials* **2022**, *12*, 4403. https://doi.org/10.3390/nano12244403

Academic Editors: Qiongfeng Shi and Jianxiong Zhu

Received: 2 October 2022
Accepted: 8 December 2022
Published: 9 December 2022

Abstract: The internet of medical things (IoMT) is used for the acquisition, processing, transmission, and storage of medical data of patients. The medical information of each patient can be monitored by hospitals, family members, or medical centers, providing real-time data on the health condition of patients. However, the IoMT requires monitoring healthcare devices with features such as being lightweight, having a long lifetime, wearability, flexibility, safe behavior, and a stable electrical performance. For the continuous monitoring of the medical signals of patients, these devices need energy sources with a long lifetime and stable response. For this challenge, conventional batteries have disadvantages due to their limited-service time, considerable weight, and toxic materials. A replacement alternative to conventional batteries can be achieved for piezoelectric and triboelectric nanogenerators. These nanogenerators can convert green energy from various environmental sources (e.g., biomechanical energy, wind, and mechanical vibrations) into electrical energy. Generally, these nanogenerators have simple transduction mechanisms, uncomplicated manufacturing processes, are lightweight, have a long lifetime, and provide high output electrical performance. Thus, the piezoelectric and triboelectric nanogenerators could power future medical devices that monitor and process vital signs of patients. Herein, we review the working principle, materials, fabrication processes, and signal processing components of piezoelectric and triboelectric nanogenerators with potential medical applications. In addition, we discuss the main components and output electrical performance of various nanogenerators applied to the medical sector. Finally, the challenges and perspectives of the design, materials and fabrication process, signal processing, and reliability of nanogenerators are included.

Keywords: energy harvesting; green energy; internet of medical things; monitoring healthcare devices; piezoelectric nanogenerator; triboelectric nanogenerator

1. Introduction

Due to the health problems caused by COVID-19, the World Health Organization (OMS) has implemented strategies for monitoring, controlling, and treating chronically ill patients [1,2]. An interesting strategy is digital home hospitalization, which could be used in future health contingencies [3–5]. This strategy is justified by the lack of hospital beds and the increase in the world population of older adults [6,7]. The world population could reach 8.5 billion and 9.7 billion in 2030 and 2050, respectively [8]. In addition, digital home hospitalization is suitable for people with chronic diseases who live in isolated areas [9,10]. Thus, patients could be cared for from their homes, preserving their integrity with the help of the implementation of a wireless patient monitoring tool.

In recent years, the internet of medical things (IoMT) and the internet of things (IoT) have registered great interest from the scientific and medical communities. The IoMT is a subarea developed and applied within the IoT to focus on different current medical applications. This process uses the acquisition, processing, transmission, and storage of medical information through specific devices and the security of patient data [11–15]. The IoMT processes and contains patients' confidential data. Thus, a critical challenge of the IoMT is achieving data security [16,17]. The IoMT requires devices with soft, elastic, and biocompatible materials, safe performance, and stable electrical response [18–20]. In addition, these devices require a great capacity for monitoring different patient signals using electrical energy sources of small size and long lifetime. These devices must provide real-time data on the health status of patients, informing the patients, family members, medical centers, or doctors. In addition, the IoMT devices must alert possible health risks of patients.

Wireless and portable electronics have increased with the gradual technological development of 5th generation mobile networks around the world [21–25]. Using portable electronic devices and wireless connections, the IoMT technology may allow the possibility to control and assist patients. Thus, this technology can help patients' health by providing and controlling medical treatments [26–29]. However, the future IoMT portable devices will require energy sources that replace conventional batteries, which have limitations such as limited lifetimes, large volume and weight, and toxic materials that can contaminate the environment. To improve the performance and lifetime of energy storage devices, recent investigations [30–35] have developed rechargeable metal-air batteries and metal-ion batteries, as well as supercapacitors. Another alternative to substitute conventional batteries is the fabrication of nanogenerators, which can harvest green energy from the environment or the human body. The nanogenerators offer advantages such as a long lifetime, small volume, light weight, a non-expensive fabrication process, and high output power density [36–40]. The nanogenerators can use the triboelectric and piezoelectric effects for harvesting biomechanical energy into electricity [41–49]. The piezoelectric nanogenerators can convert the mechanical deformations of the piezoelectric materials used in their structures into electrical voltages. On the other hand, triboelectric nanogenerators use both contact electrification and electrostatic induction between two moving triboelectric layers to generate an electrical potential difference. Table 1 depicts several advantages and disadvantages of piezoelectric and triboelectric nanogenerators, metal-air batteries, and supercapacitors.

Nanogenerators can harvest biomechanical energy due to the movement of the hands and fingers [50–57], wrists [58–66], knee joints [67–74], and feet [75–84]. In addition, this energy is generated by external agents, including the movement or friction of clothing, backpacks, and the airflow interacting with the human body when running, riding a motorcycle or bicycle, skating, and so on. Nanogenerators could also power future implantable medical devices. Implantable medical devices can maintain and monitor vital signs of patients, such as heart rate monitors, pacemakers, defibrillators, and brain stimulation [85–87]. However, these devices are still powered using conventional batteries, which have limited-service life. Thus, nanogenerators could harvest the biomechanical energy of patients for powering implantable medical devices [88–102]. Figure 1 illustrates the application of nanogenerators to convert biomechanical motion into electrical energy to power potential IoMT electronic devices.

Table 1. Comparison of advantages and disadvantages of different devices used in the energy harvesting and storage process.

Device Type	Advantages	Disadvantages	Ref.
Metal-air batteries	High energy density, low cost, flat discharge voltage, and high safety	Performance is affected by environmental conditions, dendrite formation on the anode, carbonation of alkaline electrolyte, and limited range of operating temperature	[30,32]
Supercapacitor	High energy density, high specific surface area, long cycle life, and good conductivity and stability	Capacitance and charge storage depend on the employed electrode materials and non-simple fabrication process	[33–35]
Piezoelectric nanogenerator	Simple structure and easy fabrication process, good electromechanical stability, and non-complex signal processing system	Performance depends on the properties and structural configuration of the piezoelectric material	[36]
Triboelectric nanogenerator	High electrical performance, compact structure, simple working principle, low-cost materials, and good electrical stability	Wear of triboelectric material by friction, and performance depends on the properties and working mode of the triboelectric film	[36]

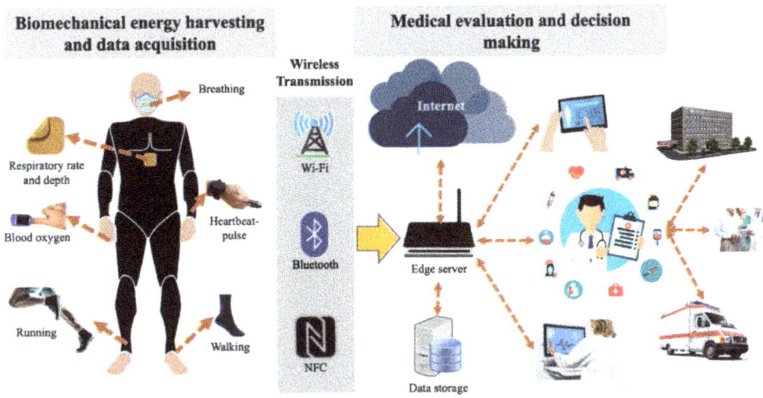

Figure 1. Potential application of piezoelectric and triboelectric nanogenerators to convert biomechanical energy into electrical energy, which could be used to power IoMT electronic devices.

We reviewed piezoelectric and triboelectric nanogenerators that can harvest green energy from environmental sources such as human body motion, mechanical vibrations, and wind. The harvested energy can be converted into electrical energy to power potential medical devices that monitor and process signals of different health parameters of patients. These signals could be transmitted to hospitals or medical centers using the IoMT. The operating principle, materials, and manufacturing processes of nanogenerators are discussed. Future medical applications of piezoelectric and triboelectric nanogenerators are reported, including materials, signal processing components, and packaging. Finally, we proposed the challenges and perspectives on the output performance of nanogenerators, taking into account the topics of design, materials and fabrication processes, signal processing, and reliability.

2. Operating Principle, Materials, and Fabrication

Triboelectric and piezoelectric nanogenerators have potential applications in the medical sector due to their high electrical output performance, simple structure and operation principle, and cost-effective fabrication process.

In 2017, Chen et al. [103] reported a flexible hybrid nanogenerator that can be attached on soft surfaces, such as human skin, for body motion harvesting and monitoring physiological signals. A single-electrode TENG and a PENG are integrated into the nanogenerator to improve its electromechanical performance (Figure 2). The PENG module used poly(vinylidenefluoride-co-trifluoroethylene) P(VDF-TrFE) nanofiber mats as the piezoelectric structure, and polydimethylsiloxane (PDMS) as the protective layer. Thermoplastic polyurethane (PU) nanofibers coated with carbon nanotubes (CNT) and silver nanowires (AgNWs) were employed in the electrodes. Thus, the PU electrodes sandwiched the piezoelectric layer. The PENG part was separated from the TENG part by a PDMS isolation film to produce piezoelectric and triboelectric outputs. Another PDMS layer operated as triboelectric material, and the PU film as an electrode for the single-electrode TENG. Based on the triboelectric and piezoelectric mechanisms, this hybrid nanogenerator registered a maximum peak power up to 84 mW/m^2 and 0.11 mW/m^2 for the TENG and PENG modules under compressive stress (Figure 3). This nanogenerator can be attached to any soft hybrid to convert tapping and pressing energy into electricity. The hybrid nanogenerator can be placed on the skin for real-time monitoring of human physiological signals, such as respiratory information and radial artery pulse. Figure 4 shows the hybrid nanogenerator attached to the back of the hand to collect the body motion energy of closing the fist and punching. This nanogenerator can be mounted on the abdomen for monitoring the human respiratory rate and depth. Thus, the nanogenerator could be employed for monitoring the respiratory condition of patients. Also, this nanogenerator has potential applications for self-powered healthcare monitoring systems and e-skins.

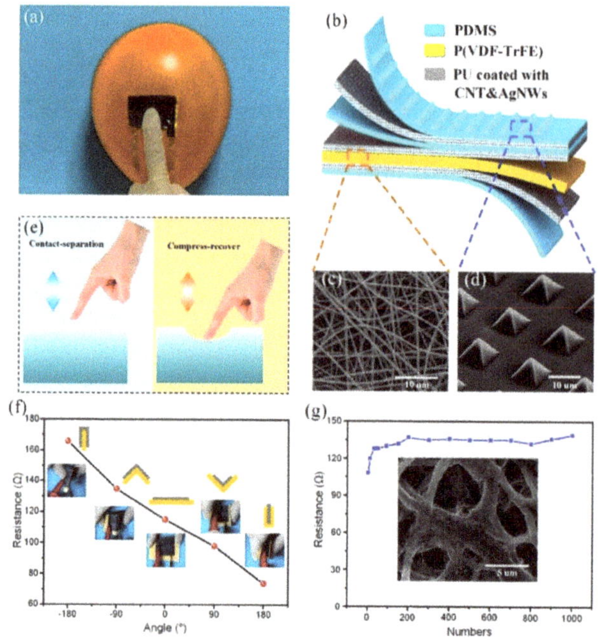

Figure 2. Flexible hybrid nanogenerator reported by Chen et al. [103]. (**a**) Nanogenerator placed on the surface of a balloon. (**b**) Schematic view of the different materials of the nanogenerator. (**c**) SEM image of the electrospun P(VDF-TrFE) nanofiber mat. (**d**) SEM image of the micro-patterned PDMS layer. (**e**) Process of contact-separation and compress-recover on a soft surface. (**f**) Measurements of resistance of the conductive fiber mat as a function of its bending angle. (**g**) Stability results on the resistance variation of the fiber-based electrode under fold/unfold deformation for 1000 cycles. SEM image of the fiber-based electrode. Reprinted with permission from [103]. Copyright ©2017, Elsevier Ltd.

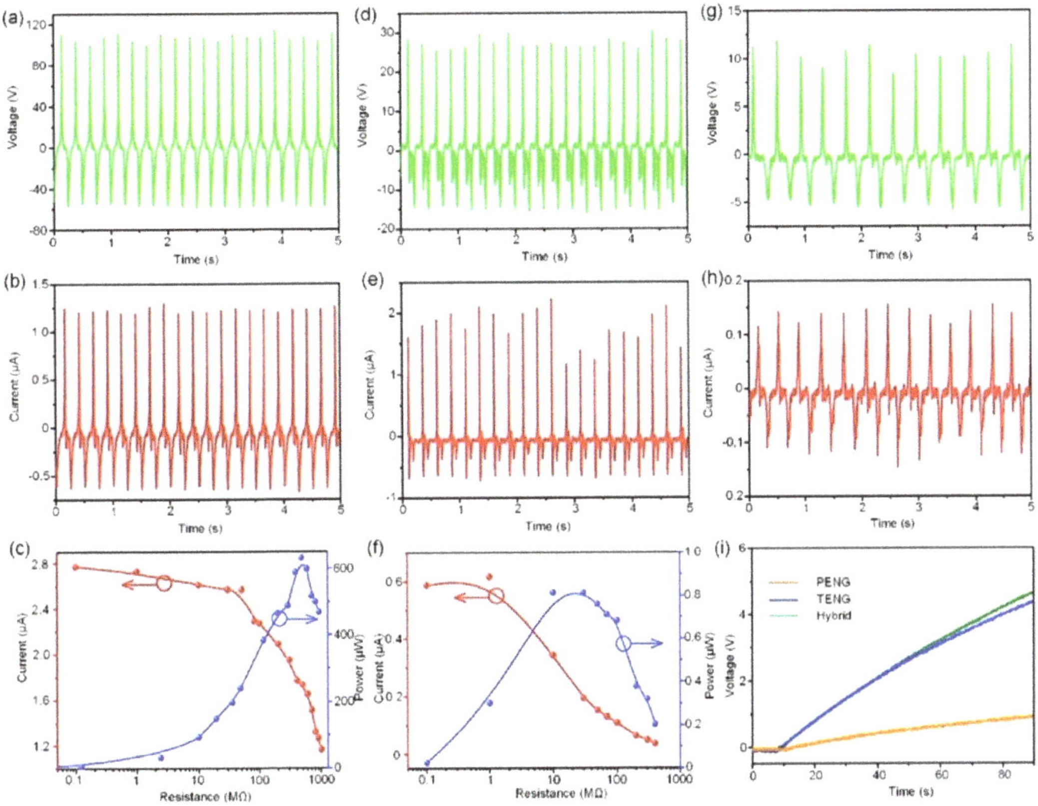

Figure 3. Electrical output performance of the flexible hybrid nanogenerator reported by Chen et al. [103]. (**a**) The output voltage, (**b**) current, and (**c**) maximum peak power as a function of resistance load of the TENG module. (**d**) The output voltage, (**e**) current, and (**f**) maximum peak power versus resistance load of the PENG module under compression. (**g**) The output voltage and (**h**) current of the PENG module under bending mode. (**i**) The charge voltage of a capacitor of 1 μF using the hybrid nanogenerator. Reprinted with permission from [103]. Copyright ©2017, Elsevier Ltd.

In 2018, Sun et al. [104] designed a biocompatible triboelectric–piezoelectric–pyroelectric hybrid nanogenerator with high flexibility. This nanogenerator has a transparent structure formed by polyvinylidene fluoride (PVDF), PDMS, and Ag nanowires. The PVDF material acts as piezoelectric and pyroelectric film and the PDMS acts as a triboelectric film. The Ag nanowires are used as high-performance transparent electrodes (TEs) and can be distributed into a network like a leaf venation (LV), inspired by optimized LV (Figure 5). Thus, the working mechanism of the hybrid nanogenerator is integrated by a triboelectric nanogenerator, a piezoelectric nanogenerator (TENG-PiENG), and a pyroelectric nanogenerator (pyENG). This nanogenerator can be mounted on different body parts to collect mechanical and thermal energy from the human body. This nanogenerator has an environmentally friendly design, which the structure of LV can reuse many times. The nanogenerator structure registered a minimum sheet resistance of 1.4 Ωsq^{-1} with 82% transmission and sheet resistance of 68.2 Ωsq^{-1} with an ultra-high transmission of up to 99%. The hybrid nanogenerator can harvest a maximum open-circuit output voltage of 55 V and 86 V from mechanical and thermal energy with the three coupled transduction mechanisms (Figure 6). The nanogenerator has a potential application for assessing the patient's health status with a cold, using frequency, coughing, and breathing through the amplitude of the voltage (Figure 7). The hybrid nanogenerator can monitor various human physiological signals, such as heartbeat, swallowing, and neck tilting. This nanogenerator could be used for cost-effective medical diagnostics and prognosis of cardiovascular, Parkinson, and esophagus diseases. Also, the nanogenerator integrated with a thermochromic liquid crystal display (LCD) could operate as a thermometer for medical diagnostics.

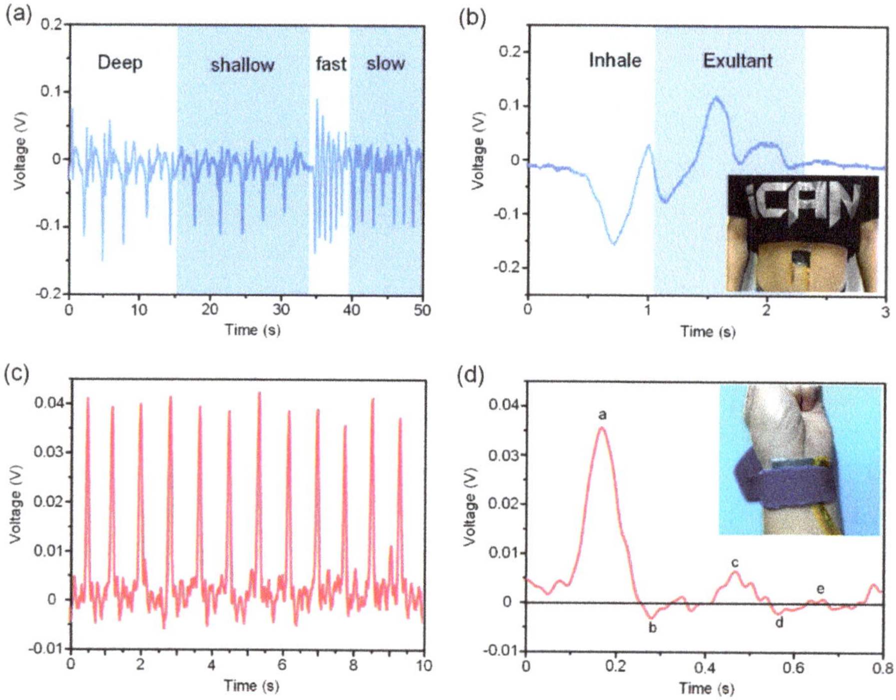

Figure 4. Results of real-time physiological monitoring using the flexible hybrid nanogenerator developed by Chen et al. [103]. (**a**) Respiration signal during 50 s with four different breathing modes such as deep, shallow, fast, and slow. (**b**) The enlarged signal in one breath cycle of a patient, which was measured using the hybrid nanogenerator attached to the belly. (**c**) The real-time artery pulse signal and (**d**) its enlarged signal of output voltage in one cycle that was measured with the hybrid nanogenerator. Reprinted with permission from [103]. Copyright ©2017, Elsevier Ltd.

In 2019, Zhu et al. [105] investigated a self-powered and self-functional cotton sock (S^2-sock) based on a poly(3,4-ethylenedioxythiophene) polystyrene sulfonate (PEDOT:PSS)-coated fabric TENG and lead zirconate titanate (PZT) piezoelectric device (Figure 8). The PEDOT:PSS is a polymer with high conductivity and mechanical stability. A Cu substrate was bonded to the piezoelectric device to improve the ductility and optimize the mechanical behavior. This sock can produce the featured waveforms for walking pattern recognition and motion tracking of patients under home care. The experimental tests show that in-shoe situation, the TENG part of the sock can generate the output power of 66 μW and 137 μW at 1 Hz walking and 2 Hz jumping with a load resistance of 10 MΩ, respectively (Figure 9). For indoor use, the TENG part of the sock stepping on a polytetrafluoroethylene (PTFE) film, which is mounted on the ground, registers maximum powers of 1.17 and 1.71 mW under a load resistance of 59.7 MΩ. This allows a power density of 11 $\mu W/cm^2$ considering a contact area close to 150 cm^2. For the piezoelectric device (size of 5 cm × 5 cm), the maximum output power was 32 μW under load resistance of 0.4 MΩ and 12 N at 1 Hz. The maximum power density of the PZT device is 128 $\mu W/cm^2$, which offers a good performance for smart textiles. For the indoor case using bare shocks at home, the pattern recognition and motion tracking can be measured for home care applications of the IoMT. The S^2-shock was used to measure the mimetic motions of a Parkinson's disease patient regarding normal motion, loss of stride, and freezing of gait (FOG). Moreover, this sock can collect energy from body motion and sense diversified physiological signals for healthcare and home sports. In addition, this cotton sock can be enhanced by incorporating wireless transmission modules and integrated circuits.

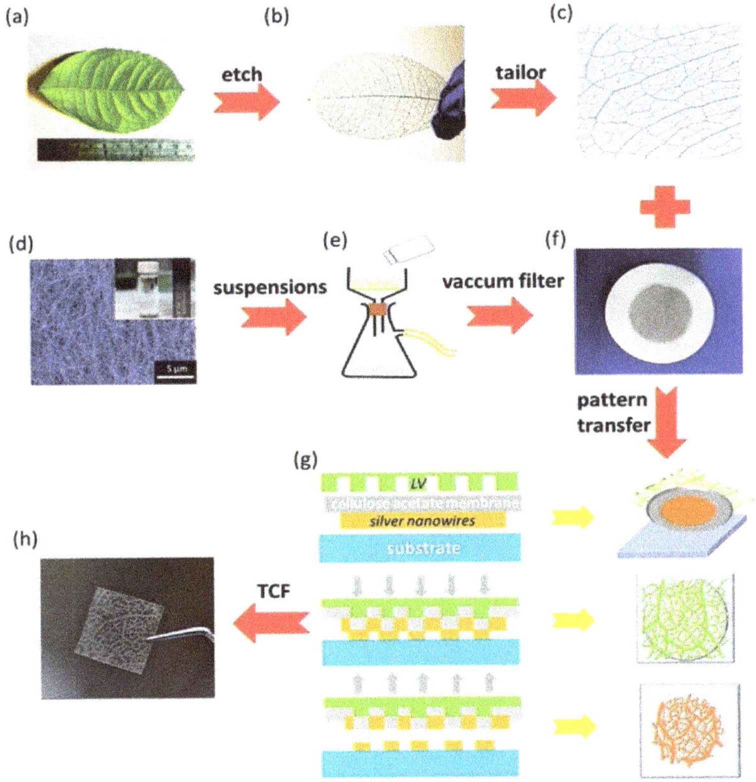

Figure 5. Network of LV-like Ag nanowires of the hybrid nanogenerator fabricated by Sun et al. [104]. Images of (**a**) the structure of a ramified leaf used as a mold to fabricate the patterns of TEs. (**b**) The skeleton of the leaf venation and (**c**) LV network were prepared in a square shape to be used as a mold. (**d**) SEM image of Ag nanowires. These nanowires were synthesized in a small bottle. (**e**) Schematic of vacuum filtration of the Ag nanowires. (**f**) Image of Ag nanowires film filtered using a cellulose acetate membrane. (**g**) Schematic of a modified dry transfer printing technique to fabricate the LV-like Ag nanowires network. (**h**) Image of the LV-like Ag network on a PDMS substrate (size of 3.5 cm × 3 cm). Reprinted with permission from [105]. Copyright ©2018, Elsevier Ltd.

In 2019, Li et al. [106] reported a hybrid nanogenerator based on cellulose nanofibril (CNF) as a triboelectric layer and bacterial cellulose (BC) with nanoparticles of $BaTiO_3$ and a multi-walled carbon nanotube (MWCNT) as piezoelectric layer. The CNF fibers were treated by employing a mixed solution of HNO_3 and H_2SO_4. In addition, a -NO_3 group is applied to enhance the electronegativity of the CNF fibers. This hybrid nanogenerator is formed by two composite layers. The top arched layer acts as PENG, using a functionalized BC paper with two Ni electrode films. The PENG has an arched shape to improve the strength of the effective strain applied to the bacterial cellulose film. The bottom layer operates as TENG integrated by a Nitro-CNF paper and a bottom Cu electrode film. The working mechanism of the PENG consists of the strain of the piezoelectric layer caused by external forces on its top surface, which generates a voltage variation between the two Ni electrodes. On the other hand, TENG operates with contact electrification and electrostatic induction. The triboelectric and piezoelectric parts have high output electrical performance, achieving short-circuit current density and open-circuit voltage of 1.23 $\mu A/cm^2$, 37 V, 220 nA/cm^2, and 22 V, respectively. With a rectifier circuit to integrate the outputs of both triboelectric and piezoelectric parts, the hybrid nanogenerator has a short-circuit current density and open-circuit voltage of 1.6 $\mu A/cm^2$ and 18 V, respectively. This cellulose-composed nanogenerator has an environmentally friendly simple structure, light weight, and cost-effective fabrication. Thus, this nanogenerator could apply to wearable or implantable devices and self-powered electronic sensors.

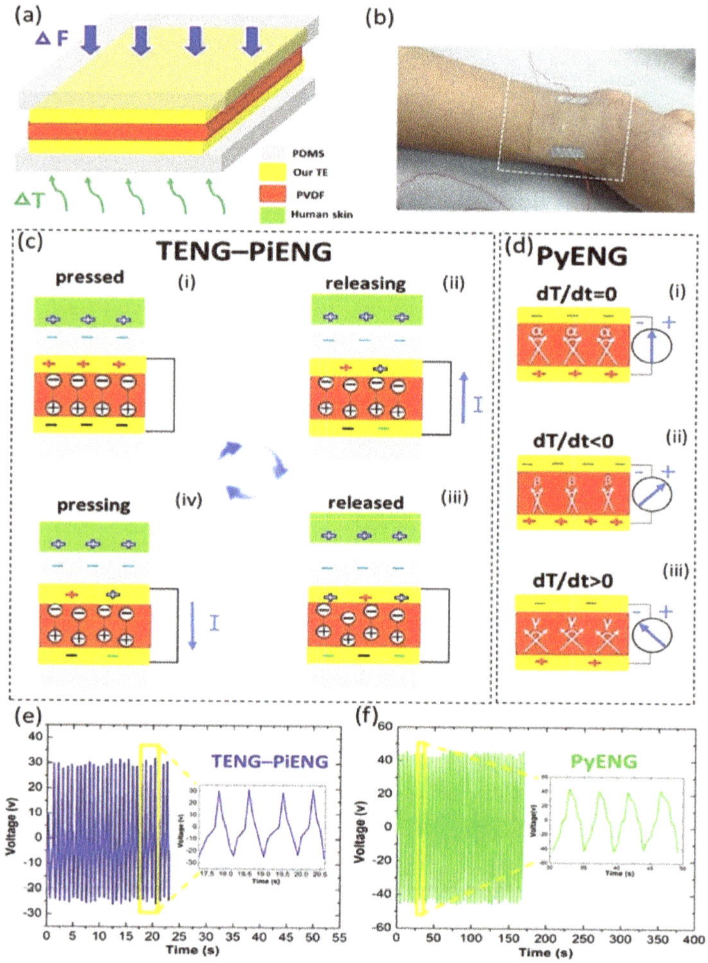

Figure 6. Working mechanism of the flexible hybrid nanogenerator designed by Sun et al. [104]. (**a**) Schematic view of the components of the hybrid nanogenerator. (**b**) Image of the hybrid nanogenerator placed on the wrist. Working mechanism of the (**c**) TENG-PiENG and (**d**) PyENG of the hybrid nanogenerator. Open-circuit voltage of the (**e**) TENG-PiENG and (**f**) PyENG. Reprinted with permission from [104]. Copyright ©2018, Elsevier Ltd.

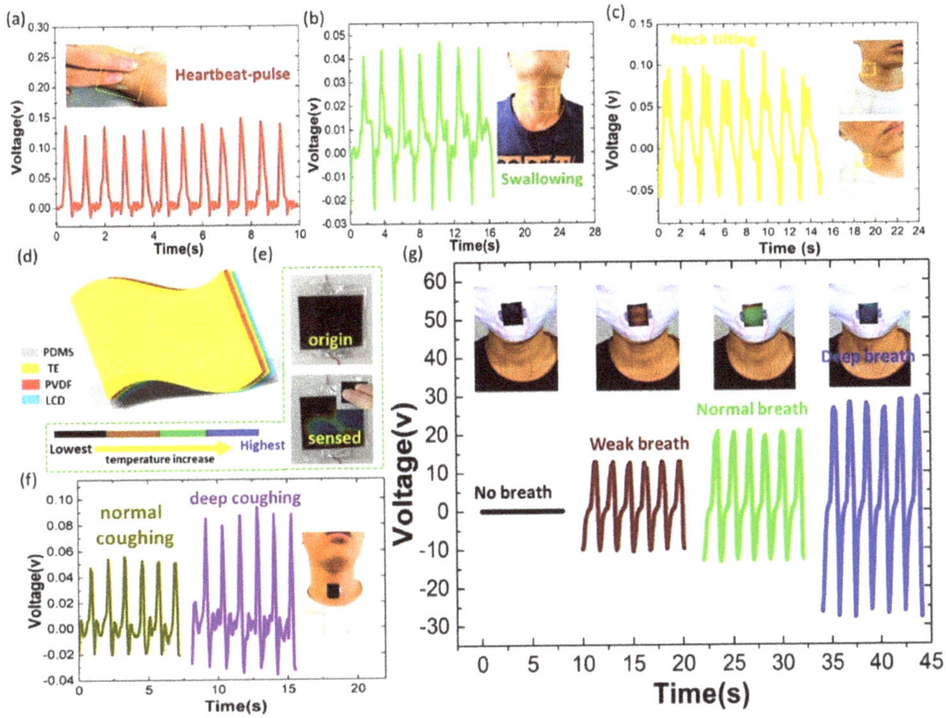

Figure 7. Potential applications of the flexible hybrid nanogenerator reported by Sun et al. [104]. Voltage output of the hybrid nanogenerator for assessing (**a**) heartbeat pulse, (**b**) swallowing, and (**c**) neck tilting. (**d**) Schematic view of a smart device composed of transparent hybrid nanogenerator and LCD film. (**e**) Smart device for temperature monitoring. (**f**) Voltage output of the smart device for sensing different degrees of coughing. (**g**) Voltage output of the smart device for monitoring three types of breathing conditions. Reprinted with permission from [104]. Copyright ©2018, Elsevier Ltd.

In 2020, Syu et al. [107] reported a biomimetic and flexible hybrid self-powered sensor (BHSS) that was formed by triboelectric and piezoelectric elements. The triboelectric structure is based on the shell of Mytilidae and is fabricated with PDMS film. The piezoelectric component contains porous PVDF fibers deposited on a printed circuit board (PCB) substrate. To develop the biomimetic Mytilidae surface, Mytilidae nano-structured patterns on a PDMS layer were applied using a soft transfer molding process. This hybrid sensor registered an open-circuit voltage and short-circuit current of 15 V and 115 nA, respectively. For a load resistance of 10 MΩ, the hybrid sensor achieved a maximum average power density of 675 $\mu W/m^2$. The hybrid sensor was tested. This self-powered sensor was attached on the thumb, index, and middle fingers of a latex glove for monitoring bottle-holding actions. For this experiment, its electrical signals under three plastic bottles with different water volumes (loading weight of 0.65 kg, 1.3 kg, and 2.06) were examined. The electrical signals of the sensors were studied for various contact positions along the bottles. This self-powered sensor could be used for wearable electronic devices and monitoring biomechanical motion, and human gesture recognition, using a machine learning algorithm of long short-term memory.

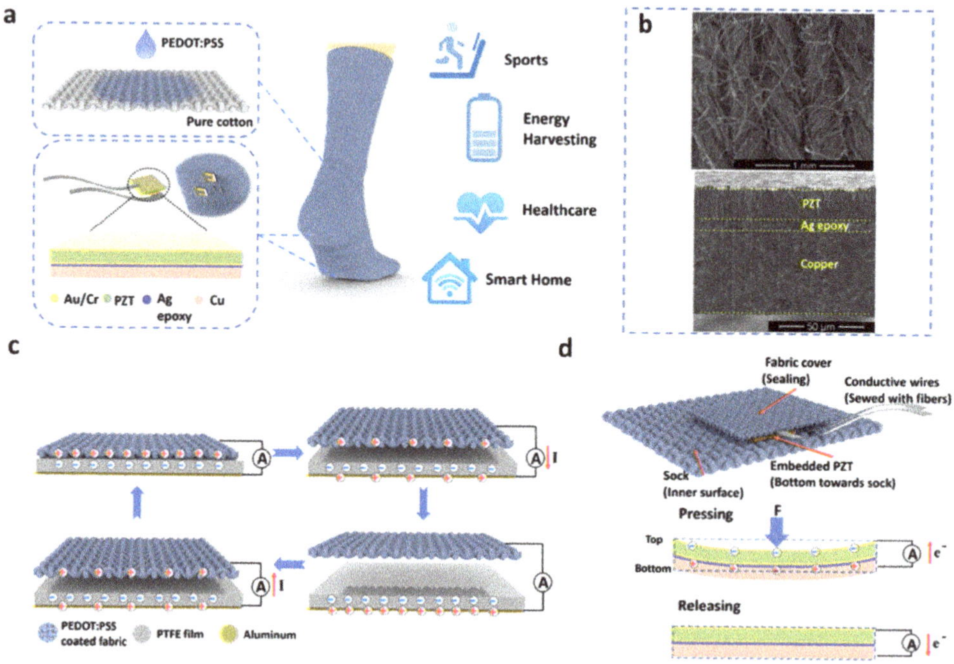

Figure 8. TENG and piezoelectric device integrated in a S^2-sock manufactured by Zhu et al. [105]. (**a**) Schematic of the materials used in the triboelectric S^2-sock coupled with piezoelectric devices for various applications, including energy harvesting, healthcare, smart home, and sports. (**b**) (Top) SEM image (top view) of PEDOT:PSS-coated textile. SEM image of a cross-section view of piezoelectric device. (**c**) Working mechanism of TENG sock using contact-separation condition. (**d**) Working mechanism of piezoelectric device and integration design for S^2-sock. Reprinted with permission from [105]. Copyright ©2019, American Chemical Society.

Huang et al. [108] fabricated a flexible biocompatible triboelectric-piezoelectric nanogenerator by employing recombinant spider silk and poly(ethylene terephthalate) (PET)/PVDF films. Graphene was incorporated into PVDF to improve the piezo-PVDF performance. Thus, the PET/PVDF-graphene composite film was added with genetically engineered spider silk. The electron-cloud-potential-well model was used to describe the charge transfer between spider silk and PET, with or without the PVDF film. The electrons in molecular orbits of spider silk protein can be transferred to the empty orbits of PET film. This transference of electrons depends on the difference in potential-well depths (difference of the surface electron potential) between both materials. PVDF, under mechanical strain, can alter the surface electron potential of the coupling material. Due to this, PVDF was added to PET film to enhance its surface potential. Thus, the improved surface-potential difference between PET/PVDF and spider silk can increase their electrons transfer and amount of harvested energy. Figure 10 depicts the output electrical signals and stability results of the hybrid nanogenerator. This hybrid nanogenerator (area of 6.25 cm^2) registered a maximum output power density of 4016 mW/m^2 under a load resistance of 8 MΩ. For this resistance, the nanogenerator reached the output voltage, output short-circuit current, and output power close to 200 V, 12 μA, and 2.51 mW, respectively. This nanogenerator could be used as wearable hand-gesture sensors, body motion sensors, bioenergy harvesters, and implantable organ monitors (Figure 11). For instance, this nanogenerator could be included in implantable sensors for monitoring signals of the heart, stomach, chest, and bladder.

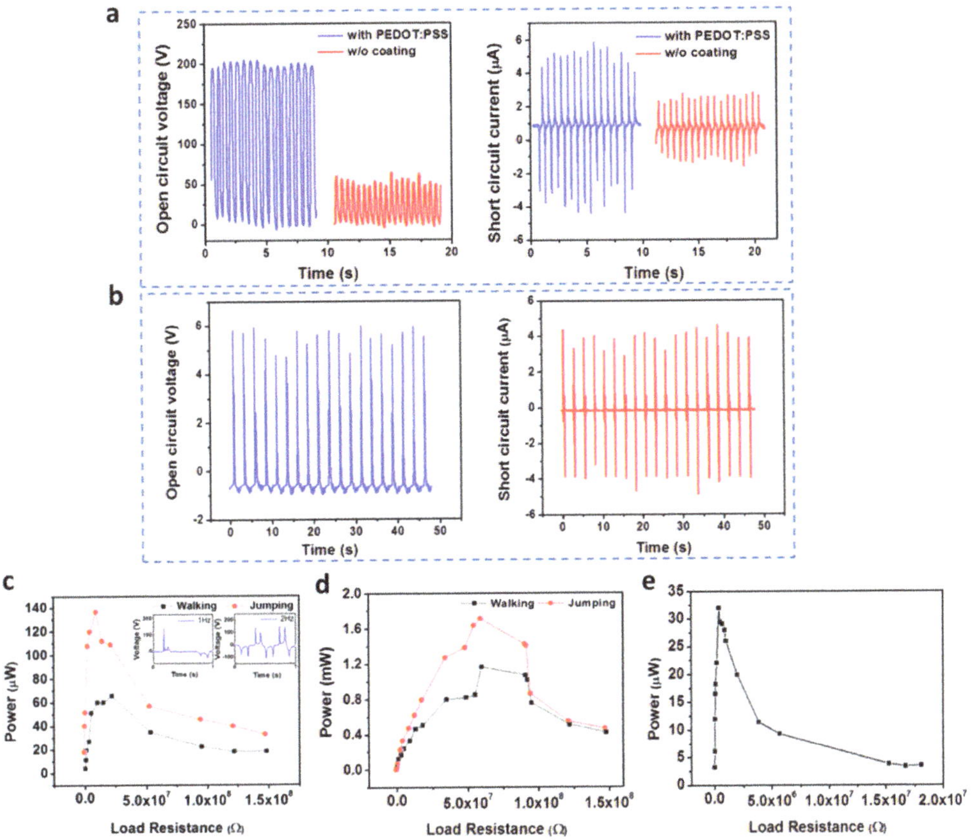

Figure 9. Output electrical performance of the TENG and piezoelectric device integrated in a S^2-sock designed by Zhu et al. [105]. (**a**) Results of open-circuit voltage and short circuit current of the sock with and without PEDOT:PSS under contact-separation operation with respect to PFTE film placed on the ground (Figure 9c). (**b**) Results of open-circuit voltage and short-circuit current of piezoelectric device integrated to the sock (Figure 9d). (**c**,**d**) Output power as a function of the external load resistance with and without shoes. (**e**) Output power of the piezoelectric device as a function of external load resistance. Reprinted with permission from [105]. Copyright ©2019, American Chemical Society.

In 2022, Du et al. [109] designed a shoe insole hybrid nanogenerator (IHN) composed of a multilayered TENG and arched PENG. This nanogenerator can convert the mechanical energy of footsteps into electrical energy. Moreover, this nanogenerator can identify three types of motion states: walking, stepping, and jumping. This nanogenerator has high performance of electrical stability and durability, which is suitable to collect biomechanical energy of feet to power wearable electronic sensors. In addition, this nanogenerator has potential application for self-powered biosensor systems in the fields of sports and medicine. Figure 12 illustrates an application of the IHN integrated with a dorsalis pedis artery sensor and a processor module. The sensor is placed inside the tongue of the shoe and the processor module is mounted on the outside of the shoe. The sensor contains a PVDF film that is packaged with PTFE and Kapton films. The PVDF film harvests the vibration signal of the pulse and converts it into an electrical signal. The hindfoot and forefoot of the IHN are formed by the arched PENG and TENG, respectively. The TENG part has three triboelectric films (PTFE-Al-PTFE), in which two sponge layers with through-holes are added to separate the Al sheet from both PTFE films. Both PTFE films and the Al layer are contacted and separated due to the motion (e.g., walking, stepping, and jumping) of the foot. On the back of PTFE is attached the Cu electrode that operates as a source for the charge transference caused by electrostatic induction. In addition, the Al sheet can act as both an electrode and a triboelectric layer. On the other hand, a surface area (2 cm × 7 cm) of the sponge is extracted to incorporate the arched PENG. This PENG is composed of a PVDF film that is packaged by an arched PE and Kapton film. Furthermore, the IHN was encapsulated using a cotton cloth and PE film, which avoids the effect of sweat on the electrical performance of the hybrid nanogenerator.

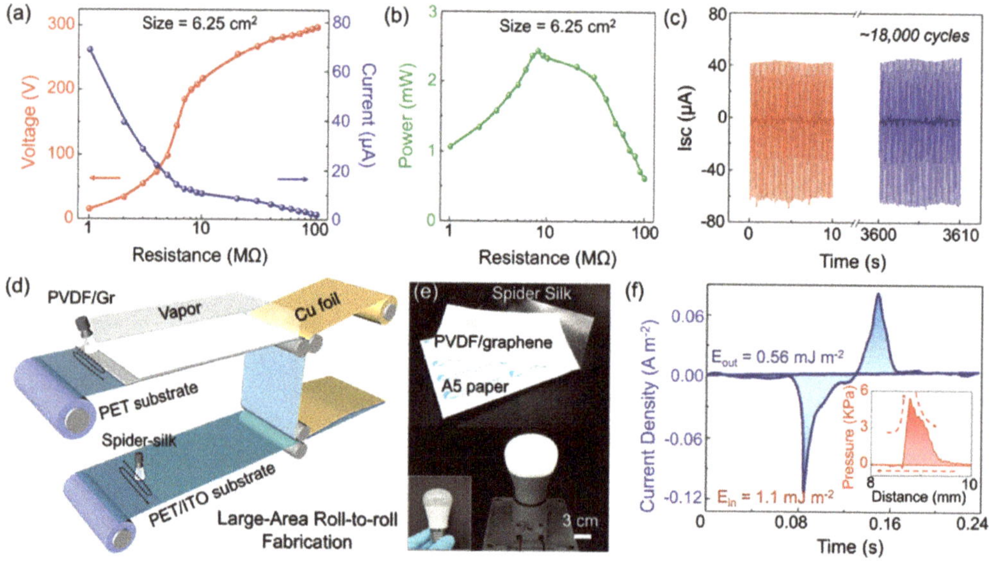

Figure 10. Performance of the hybrid nanogenerator developed by Huang et al. [108]. (**a**) Output open-circuit voltage and short-circuit current of the hybrid nanogenerator as a function of load resistance. (**b**) The output power of the hybrid nanogenerator as a function of load resistance. (**c**) Results of stability and durability tests of the hybrid nanogenerator. Short-circuit current measured over 18,000 cycles. (**d**) Schematic of the prototype for the large-scale fabrication of the hybrid nanogenerator. (**e**) Photographs of the spider silk and PVDF/graphene films (both films with area equal to A5 paper), and white LED light (3 W). This LED was lightened using a hybrid nanogenerator with an area of 48 cm^2. (**f**) The output current density of the hybrid nanogenerator during one cycle of hand tapping. The inset illustrates the input pressure generated by hand-tapping under a distance range. Reprinted with permission from [108]. Copyright ©2020, WILEY-VCH Verlag GmbH & Co. KGaA, Weinheim.

Figure 13 shows the working mechanism of the IHM reported by Du et al. [109] during a walking cycle. Due to the pressure of the heel of the shoe in a walking cycle, the PENG shape is altered from arch to plane. This deformation generates a potential difference between both electrodes of the PENG. When the forefoot falls on the ground, the PTFE and Al layers are contacted in the through-hole of the sponge. The increment of the pressure of the forefoot increases the contact area between both PTFE and A layers. After the contact of both layers, a surface charge transfer induces between them a potential difference in the back Cu electrode (Figure 13b). This potential difference generates the electrons flow from the Cu electrode to the Al electrode using the external circuit. The PENG returns to arch shape after the heel lifts, generating a current opposite to that when the hindfoot falls (Figure 13c). When the forefoot is lifted, the PTFE and Al layers of the TENG are separated.

This causes the electrons flow from the Al electrode to the Cu electrode in the external circuit, which has an opposite direction to that when the forefoot falls (Figure 13d). This sequence of four states occurs during one cycle of walking. Due to the parallel design of TENG and PENG, both nanogenerators have the same electron flow direction. It allows the composite signal output of the IHN to be higher than those of the TENG and PENG. From human body motions, the IHN generated a maximum open circuit voltage of 150 V and a short-circuit current of 4.5 μA, respectively. Figure 14g shows the pulse signal of a dorsal pedis monitoring system that is supplied by the IHN.

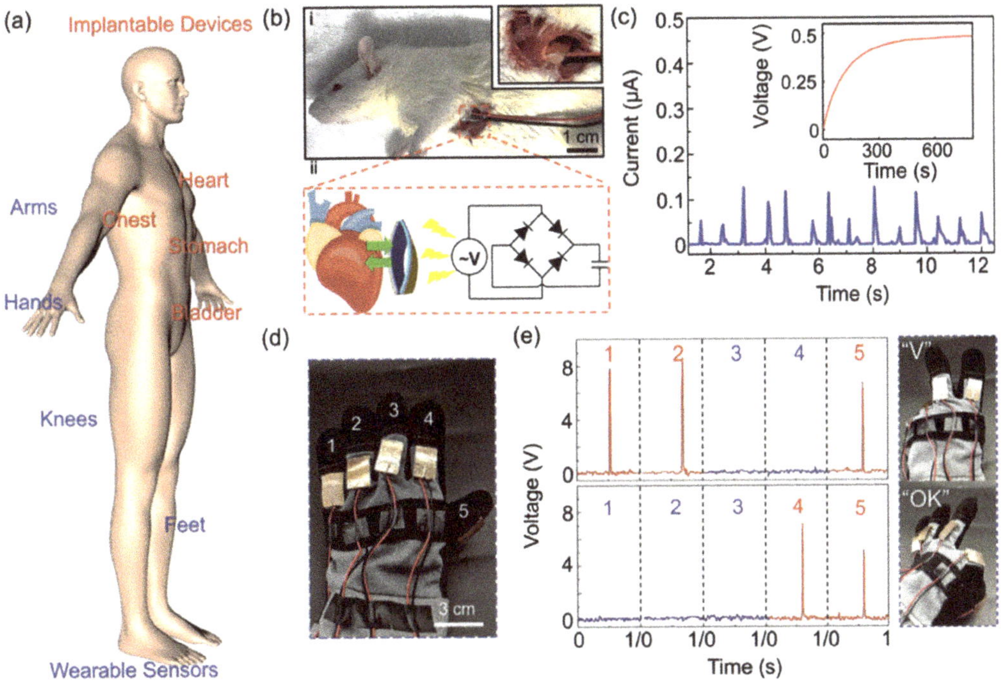

Figure 11. Potential applications of the hybrid nanogenerator proposed by Huang et al. [108]. (**a**) Schematic of a hybrid nanogenerator with potential applications in implantable devices and wearable sensors. (**b**) (i) Photograph of a hybrid nanogenerator placed in the subdermal chest of a Sprague-Dawley rat; (ii) Schematic of the working mechanism of the nanogenerator for energy harvesting from a beating rat heart. For this, a rectifier circuit and a capacitor (4.7 μF) were implemented. (**c**) Current signals of an implanted hybrid nanogenerator related to the beating heart. Inset shows the storage voltage of a capacitor (4.7 μF). (**d**) Photograph of a hybrid nanogenerator adhered on the glove for hand-gesture monitoring. (**e**) Output voltage signals of nanogenerators placed on the five fingers for "Victory" and "OK" hand gestures. Reprinted with permission from [108]. Copyright ©2020, WILEY-VCH Verlag GmbH & Co. KGaA, Weinheim.

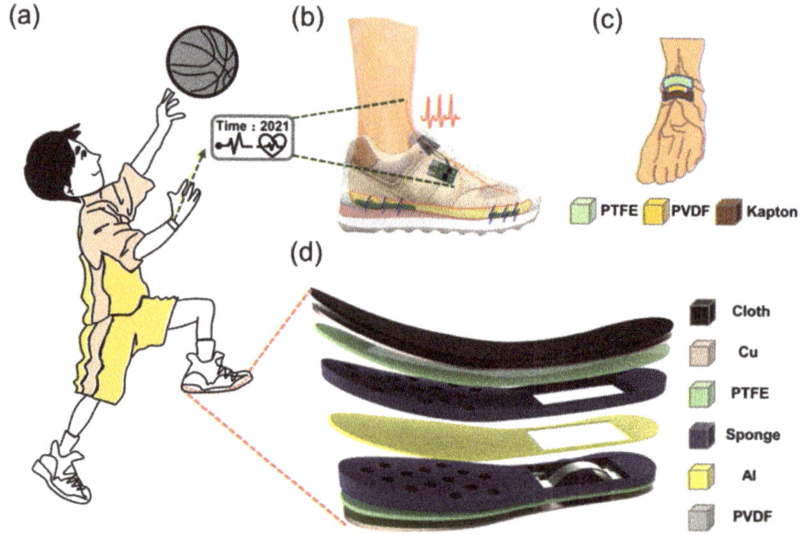

Figure 12. Schematic of the main components of the isolate hybrid nanogenerator presented by Du et al. [109]. (**a**) IHN applied to a sports shoe for monitoring the footstep types. (**b**) Schematic of the self-powered system for dorsalis pedis artery monitoring. (**c**) Schematic of the signal acquisition of the dorsalis pedis artery sensor. (**d**) Schematic of the materials and elements for the TENG and PENG modules of the IHN. Reprinted with permission from [109]. Copyright ©2022, John Wiley and Sons.

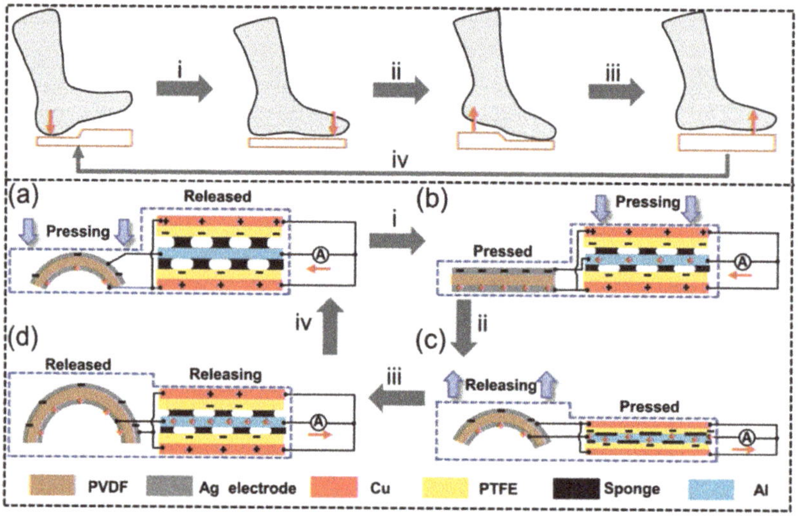

Figure 13. Working mechanisms of the PENG and TENG parts of the IHN during one walking cycle [109]. Schematic of the charge transfer process of the PENG and TENG parts when (**a**) the hindfoot falls, (**b**) the forefoot falls, (**c**) the hindfoot lifts, and (**d**) the forefoot lifts, respectively. Reprinted with permission from [109]. Copyright ©2022, John Wiley and Sons.

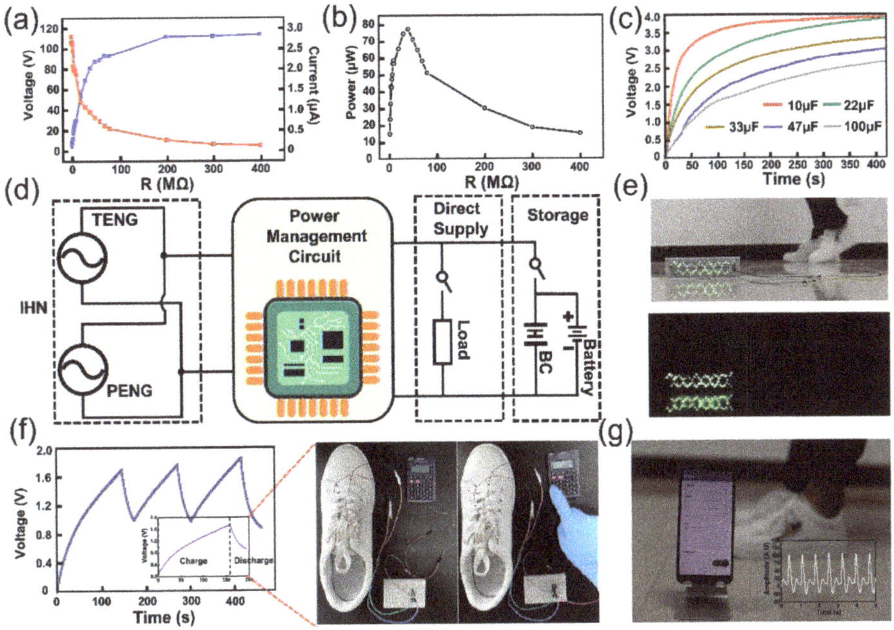

Figure 14. Electrical output performance and applications of IHN proposed by Du et al. [109]. (**a**) Output voltage and current as a function of external load resistance. (**b**) Output power as a function of external load resistance. (**c**) Charge response of four capacitors powered by IHN. (**d**) Schematic view of the electrical circuit for the IHN direct power supply and energy storage system. (**e**) LEDs lighted using the IHN. (**f**) Output voltage and calculator powered by IHN. (**g**) Pulse signal of the dorsalis pedis monitoring system, which was sent to a mobile phone. Reprinted with permission from [109]. Copyright ©2020, WILEY-VCH Verlag GmbH & Co. KGaA, Weinheim.

3. Potential Applications

This section describes the performance of potential biomedical applications of triboelectric or piezoelectric nanogenerators.

The PENGs and TENGs can be used in the modulation of neuronal activity, diagnostics for cardiovascular treatment, heart implants, drug delivery, and abdominal implants, among others. Ouyang et al. [110] reported an implanted symbiotic pacemaker integrated by a triboelectric nanogenerator, a power management unit (PMU), and pacemaker unit. The implantable triboelectric nanogenerator (iTENG) can harvest and store energy from cardiac motion. The energy collected by the iTENG is stored in the capacitor of the PMU. Next, this stored energy can be used to drive the pacemaker unit, generating pacing electrical pulses, and controlling the rate of cardiac motion (Figure 15). Figure 16 shows the operating principle of the PTFE-based iTENG. Figure 17 depicts the electrical performance of the iTENG implanted in the chest of a Yorkshire porcine. For this case, the iTENG was located between the heart and pericardium of the pig. In addition, the PTFE component was placed on the left ventricular surface. Figure 18 illustrates the symbiotic cardiac pacemaker system in vivo using a wireless passive trigger. The pacemaker unit produces electrical pulses that can induce myocardial contraction and adjust the heart rate using pacing electrodes. The electrical pulses registered the output voltage and duration of 3 V and 0.5 ms, respectively. The symbiotic pacemaker could correct sinus arrhythmia and prevent damage. The iTENG achieved an open circuit voltage of 65.2 V and harvested energy from each cardiac motion cycle up to 0.495 mJ. This energy harvested is higher than that required for endocardial pacing (0.377 mJ). This iTENG could be employed for implantable medical electronic devices due to its high output power density and suitable stability. Thus, conventional batteries could be substituted.

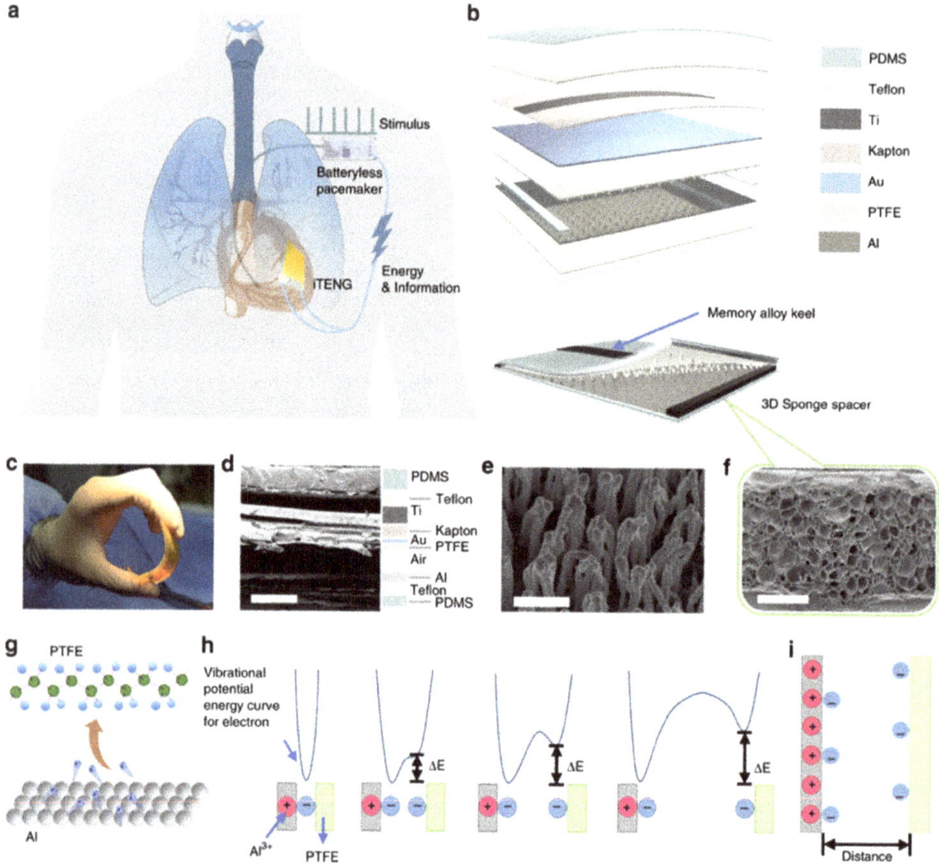

Figure 15. Schematic view of main components of symbiotic pacemaker system developed by Ouyang et al. [110]. (**a**) Schematic representation of different elements used in the symbiotic cardiac pacemaker system. (**b**) Illustration of the materials used in the iTENG. (**c**) Image of the iTENG structure under bending stress. (**d**) SEM image of different layers of an iTENG section (scale bar: 500 mm). (**e**) SEM image of the nanostructure on the PTFE layer (scale bar: 1 mm) of the iTENG. (**f**) SEM image of a 3D elastic sponge structure (scale bar: 500 mm). (**g**,**h**) Schematic view of the mechanism of charge transfer. (**i**) Model of the charge separation caused by the charge transfer. Reprinted with permission from [110]. Copyright ©2019, Springer Nature.

Ouyang et al. [111] fabricated a flexible, self-powered, ultrasensitive pulse sensor (SUPS) formed by a triboelectric active sensor, which has an output voltage of 1.52 V, a peak signal–noise ratio of 45 dB, and good stability close to 107 cycles, as well as no expensive components. The SUPS structure (Figure 19) is composed of a nanostructured Kapton (n-Kapton) triboelectric film (100 mm thick) with an electrode Cu film (50 nm thick) deposited on its back side, a second Cu film (50 nm thick) deposited on other Kapton film to form a nanostructured Cu (n-Cu) film, and a spacer. These components are encapsulated using an elastomer. The working mechanism of the SUPS depends on the coupling of contact electrification and electrostatic induction (Figure 19d). The concave structures of the device allow the contact electrification and electrostatic induction to occur simultaneously.

This characteristic of the SUPS structure achieves a good relationship between applied external force and the electrical response of the device. Thus, the electrical signal of the SUPS can be modulated by the applied external force. Based on four different structures of triboelectric films, the electrical characterization of the SUPS was measured by applying a vertically compressive force (close to 50 N) with a linear motor on the triboelectric films (Figure 20a). The triboelectric layers integrated by n-Kapton and n-Cu films registered the best electrical output performance of the SUPS. For this case, the maximum output voltage, current, and transferred charge of the device were approximately 109 V, 2.73 mA, and 7.6 nC, respectively. Furthermore, the electrical performance of the device was measured on the radial arteria of a man (24 years old), as shown in Figure 20b. For this second case, the maximum output voltage, current, and transferred charge of the structures of n-Kapton and n-Cu were 1.52 V, 5.4 nA, and 1.08 nC, respectively. Also, the SUPS performance was tested to detect the mechanical oscillations of honeybee wings, which were converted to electrical signals (Figure 20c).

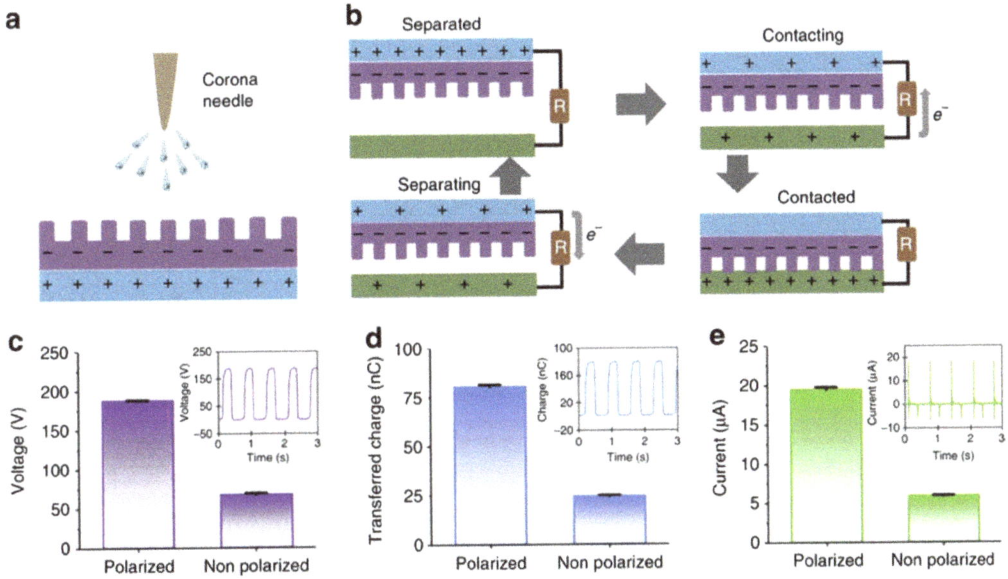

Figure 16. Working mechanism of the PTFE-based iTENG designed by Ouyang et al. [110]. (**a**) Schematic view of a corona discharge system. (**b**) Working mechanism of the iTENG. (**c**–**e**) Results of electrical performance (output voltage, transferred charge, and current) of the polarized and non-polarized PTFE film considering a linear motor. Reprinted with permission from [110]. Copyright ©2019, Springer Nature.

Two SUPS devices placed at distinct locations on the radial artery of the wrist of a man (24 years old) were used to measure the pulse wave velocity (PWV), as shown in Figure 21a,b. The PWV variations of this male patient before and after exercise consisting of 500 m jogs were recorded as shown in Figure 21c. Figure 21d depicts a schematic view of the main elements of the wireless pulse sensor system. This system includes the SUPS, analog-digital conversion, storage unit (8M-bit data), and Bluetooth interface. The SUPS output signals can be digitized using the ADC and wirelessly transmitted to a smart phone or laptop with the Bluetooth interface (Figure 21h). Moreover, the SUPS can be adjusted to different human body parts for monitoring the pulse signals of the ankle artery, the finger, the radial artery, the brachial artery, and carotid artery zones, as shown in Figure 21e. Also, the SUPS could detect variations of the heart rates of patients due to different diary movement activities. In the future, this SUPS could be used in self-powered and wearable mobile diagnosis devices for cardiovascular diseases.

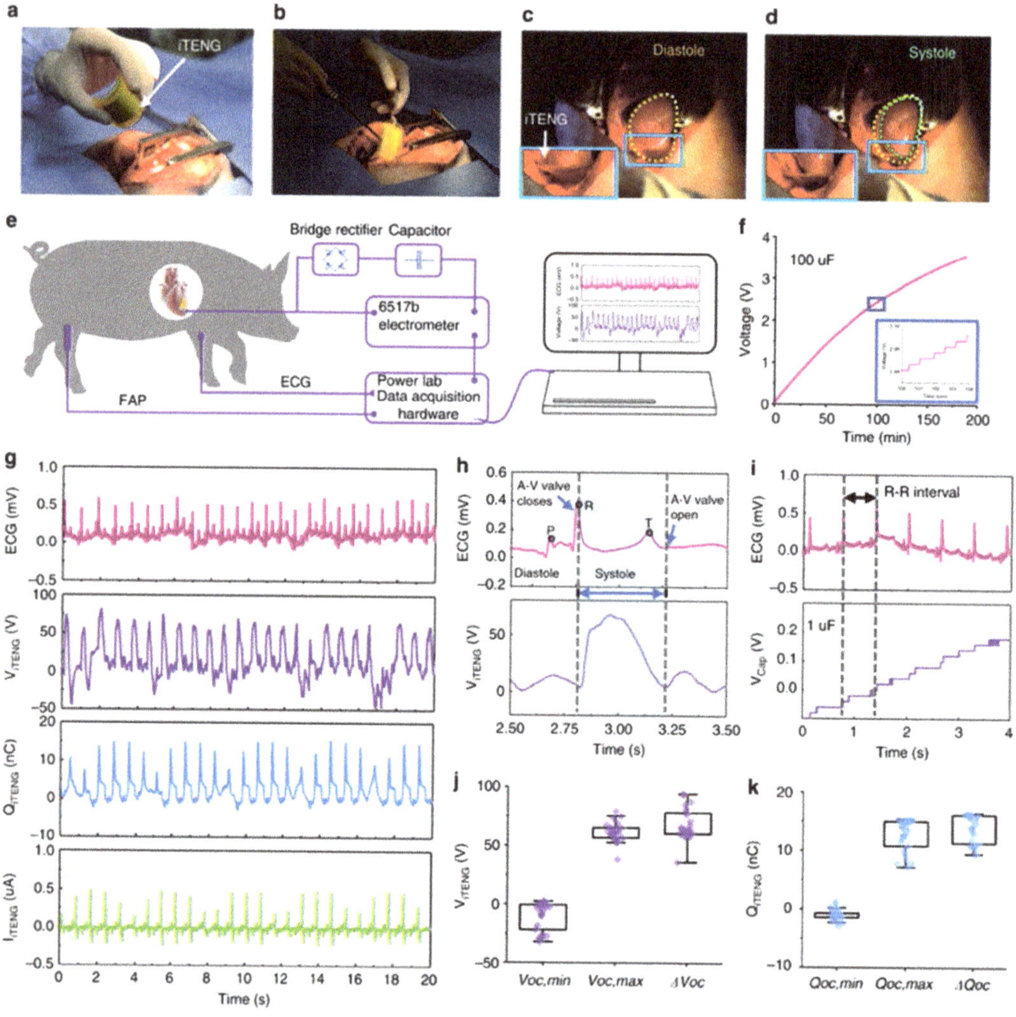

Figure 17. Electrical performance of the iTENG implanted in an animal specimen [110]. (**a**,**b**) Photographs of the iTENG implanted between the heart and pericardium of a Yorkshire porcine. (**c**,**d**) Photographs of the iTENG driven by the cardiac motion of a Yorkshire porcine. (**e**) Schematic view of the electrical performance test of the iTENG. (**f**) Charging response of a capacitor (100 μF) that is charged by iTENG. (**g**) In vivo electrical tests of the short-circuit current, transferred charge, and open-circuit voltage of the iTENG, as well as simultaneously recorded ECG signals. (**h**) Output open-circuit voltage of the iTENG and simultaneously recorded ECG results. (**i**) Charging signal of a capacitor (1 μF) charged by iTENG and ECG curve. (**j**) Statistical results of voltage shift, maximum voltage, and minimum voltage of the iTENG. (**k**) Statistical results of the transferred charge shift, maximum transferred charge, and minimum transferred charge of the iTENG. Reprinted with permission from [110]. Copyright ©2019, Springer Nature.

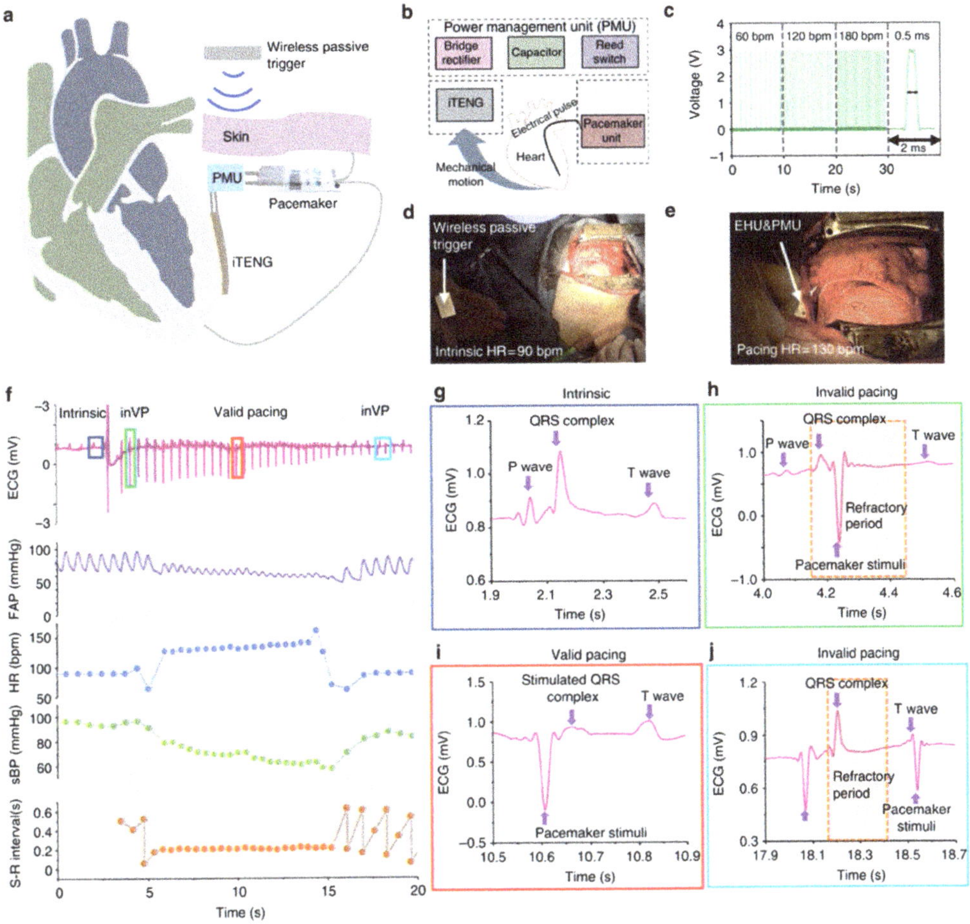

Figure 18. Performance of the symbiotic cardiac pacemaker designed by Ouyang et al. [110]. (**a**) Representation of the pacemaker powered by iTENG that harvests energy from the cardiac motion of a Yorkshire porcine. (**b**) Illustration of the main components of the symbiotic cardiac pacemaker system. (**c**) Stimulation voltage pulse under different frequencies produced by a pacemaker unit. (**d**,**e**) Photograph of the symbiotic cardiac pacemaker system switched on by wireless passive trigger in a pig. (**f**) Results of the stimulus-R wave intervals (S-R), systolic blood pressure (sBP), heart rate (HR), femoral artery pressure (FAP), and ECG during the stimulation stage of the symbiotic pacemaker unit. (**g**) ECG signal of the heart rate considering a normal systolic blood pressure. (**h**) ECG signal under a pacing stimulus in the refractory period, considering a normal systolic blood pressure. (**i**) ECG signal of the heart successfully placed by the symbiotic pacemaker unit, with decreased systolic blood pressure. (**j**) ECG signal under failed pacing stimuli, regarding a restored systolic blood pressure. Reprinted with permission from [110]. Copyright ©2019, Springer Nature.

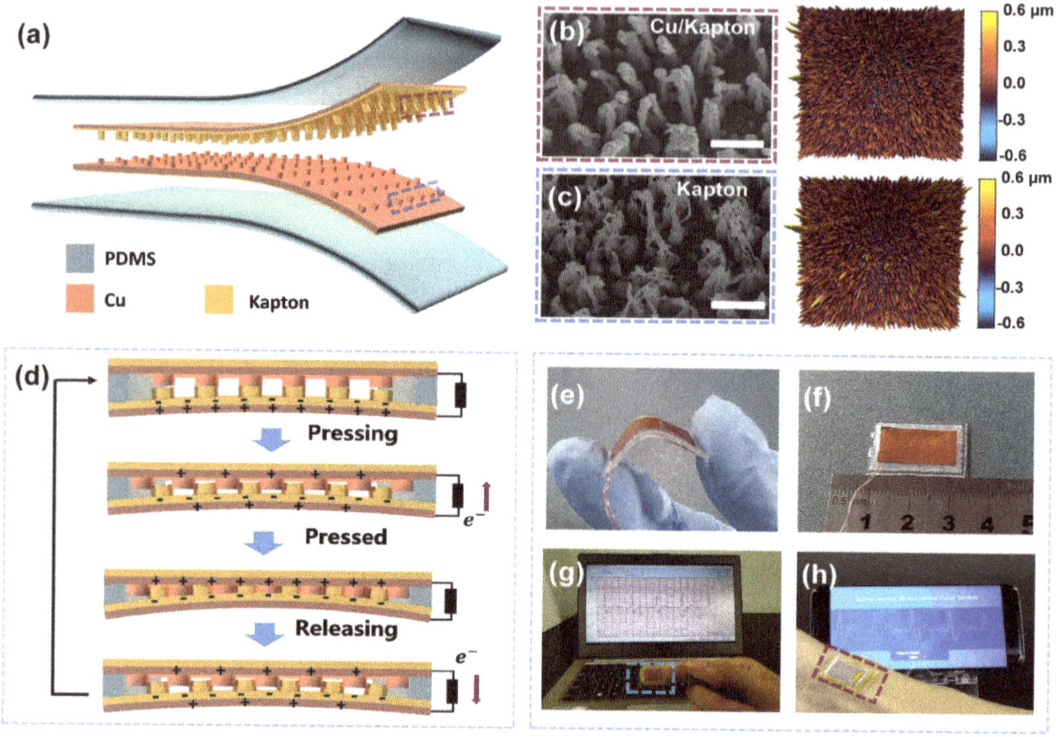

Figure 19. Components and operating mechanism of the SUPS developed by Ouyang et al. [111]. (**a**) Schematic view of the structure of the triboelectric nanogenerator of the SUPS. (**b**,**c**) SEM and atomic force microscopy (AFM) images of the nanostructured Cu and Kapton film, respectively. (**d**) Schematic view of the operating principle of the triboelectric nanogenerator. (**e**) Illustration of the good flexibility of the SUPS. (**f**) Image of the size of the SUPS structure (2 cm × 1 cm). (**g**,**h**) Experimental results of the signal outputs of SUPS pressed on the finger and radial artery region, respectively. Reprinted with permission from [111]. Copyright ©2017, WILEY-VCH Verlag GMbH & Co. KGaA, Weinheim.

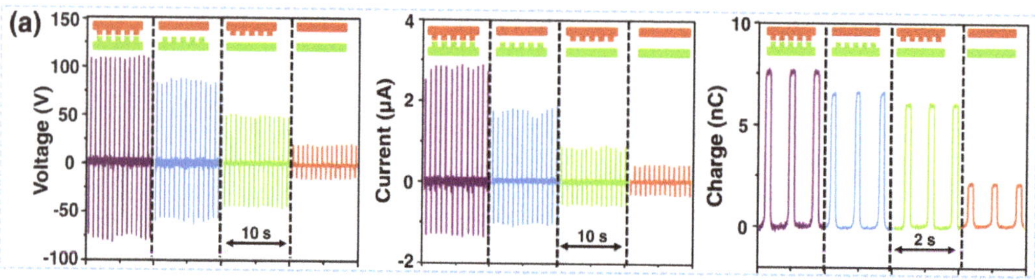

Figure 20. *Cont.*

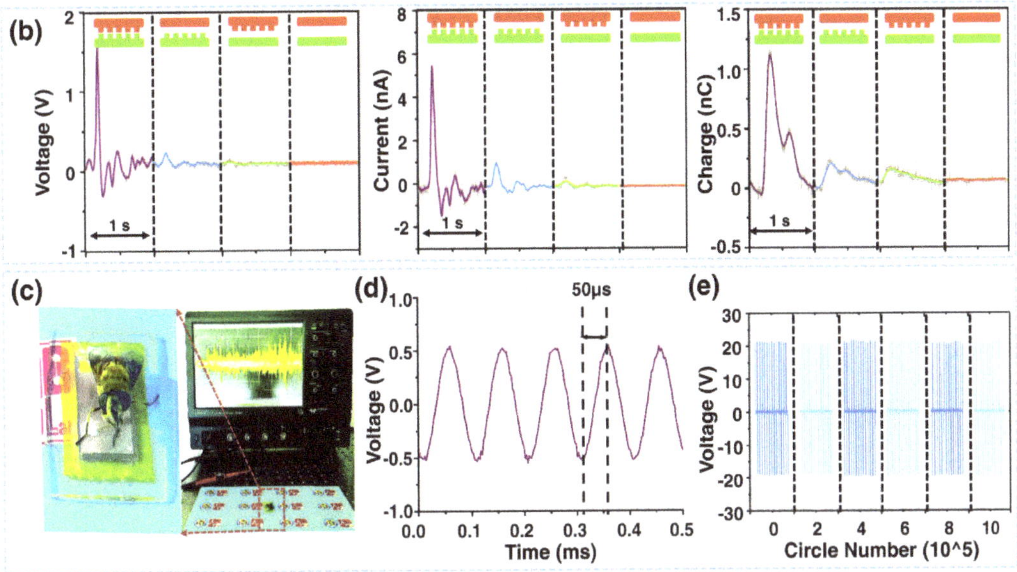

Figure 20. Electrical output performance of the SUPS with four different surface structures of the triboelectric films [111]. (**a**) Results of the output voltage, current, and transferred charge of SUPS when a lineal motor applied a compressive force of 50 N over four distinct structural configurations of triboelectric films. (**b**) Results of the output voltage, current, and transferred charge of SUPS placed on the radial artery and considering four different structural configurations of triboelectric films. (**c**) Real-time output voltage of SUPS caused by the oscillations (frequency close to 200 Hz) of the wings of a honeybee. (**d**) Output voltage of SUPS due to mechanical vibrations signals (10 kHz) generated by a loudspeaker. (**e**) Stability tests of the open-circuit voltage of SUPS during one million operating cycles. For this test, external forces of 30 N were applied on SUPS using a linear motor. Reprinted with permission from [111]. Copyright ©2017, WILEY-VCH Verlag GMbH & Co. KGaA, Weinheim.

Chu et al. [112] proposed an active pulse sensing system for monitoring the vibration signals of the human radial artery. This system includes a sandwich-structure piezoelectret with high equivalent piezoelectricity. This structure is composed of fluorinated ethylene propylene (FEP)/Ecoflex/FEP layers. For the initial stage of this structure, the two electrodes (Al and Cu layers) have the same electrical potential. When the structure is deformed and compressed by external pressure, its dipole moment suffers a variation that generates a positive current in the external circuit. The reported system with wireless transmission and big data analyses could operate as a wearable m-Health system (Figure 22a). Figure 23 depicts the results of pulse wave intervals and corresponding Poincare plots measured from ten volunteers. In addition, Chu et al. [112] developed a three-channel pulse sensor array to measure the pulse waves at the Cun, Guan, and Chi pulse regions on the radial artery of the wrist considering the traditional Chinese medicine (TCM) technique (Figure 24). Thus, a pulse-sensing system could detect and collect a large amount of the pulse signals in these three regions for potential applications of big data analyses and diagnoses. The high precision and stability of the proposed system are suitable for potential application as medical assessments, considering the identification of common heart diseases and measurement of blood pressure.

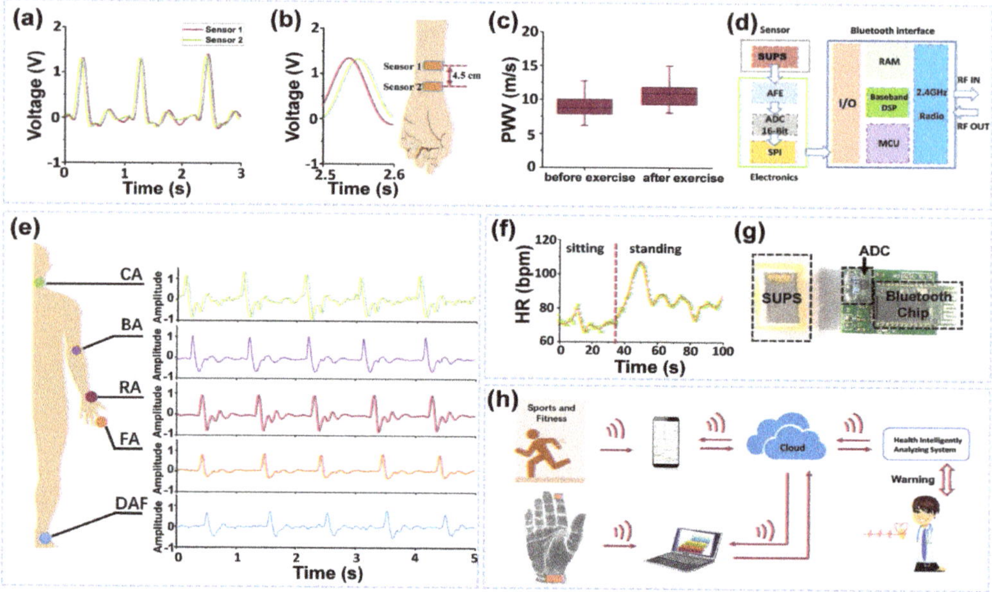

Figure 21. Electrical output variations on different artery regions from the human body and exercise activities, and the elements with big data and cloud of wireless pulse sensor system [111]. (**a**) Output voltage of two SUPS placed on different radial artery regions of a wrist. (**b**,**c**) Output voltage of two SUPS due to radial artery signals before and after an exercise activity. (**d**) Schematic representation of the components in the wireless pulse sensor system. (**e**) Output voltage of SUPS due to signals of different artery regions from a human body. (**f**) Variations of the heart rate during the activity of sitting and standing up. (**g**) Photograph of SUPS, ADC components, and Bluetooth interface of the wireless health monitoring system. (**h**) Schematic representation for online monitoring of cardiovascular signals of patients using the intelligent mobile diagnosis system. Reprinted with permission from [111]. Copyright ©2017, WILEY-VCH Verlag GMbH & Co. KGaA, Weinheim.

Liu et al. [113] designed a self-powered photodynamic therapy (s-PDT) system for potential application in cancer treatment. This s-PDT system is composed of a wearable twinning structure piezoelectric nanogenerator (ts-PENG), a power management unit (PMU), a miniaturized LED (m-LED), and a photosensitizer (PS) (Figure 25). This nanogenerator converts biomechanical energy into electrical energy to power a m-LED. This m-LED is used to stimulate light over tumor tissues, which inhibits tumor growth. The performance of this LED is controlled by the PMU with dimensions of 1.7 cm × 4.8 cm × 1.3 cm. This PMU can drive two irradiation modes: intermittent continuous light stimulation (ICLS) and pulsed light stimulation (PLS). For the ICLS mode, the tumor tissue can suffer strong radiation intermittently, which renders it dead. For the PLS case, small residual tumors can be destroyed by applying continuous low-dose irradiation.

The ts-PENG structure developed by Liu et al. [113] is formed by a Parylene-C/PET packaging layer, a PET substrate, and a double piezoelectric layer of PVDF and Ag electrodes. The ts-PENG achieves the open-circuit voltage (V_{oc}) and short-circuit transferred charge (Q_{sc}) close to 200 V and 0.46 μC (Figure 26b). Furthermore, the ts-PENG reaches 340 mW/m^2 under load resistances between 10 kΩ and 10 GΩ (Figure 26d). Moreover, the wearable ts-PENG was tested on human and rat models (Figure 27). First, the ts-PENG was attached on the knee joint and the elbow joint of a human body. The maximum values of the V_{oc} and Q_{sc} of the ts-PENG placed on knee joint and the elbow joint achieve 200 V, 220 V, 0.5 μC, and 0.65 μC, respectively. Also, an LED can be lighted using the biomechanical energy harvested by the wearable ts-PENG. In future applications, the PMU of the ts-PENG could drive the LED to apply both PLS and ICLS irradiation modes to tumors of the human body. On the other hand, the ts-PENG was positioned on the leg of a rat to convert its motion into electrical energy. A linear motor was used to pull the leg and simulate its mechanical motion. For this rat model, the maximum results of the V_{oc} and Q_{sc} of the ts-PENG are close to 8 V and 10.5 nC (Figure 27d). On the other hand, the s-PDT system was used for inhibiting tumors in vivo with intermittent continuous light stimulation (Figure 28). Finally, the ts-PENG offers advantages such as simple structural configuration, high flexibility, good stability, light weight, and long service life. These ts-PENG characteristics are suitable for self-powered wearable medical devices.

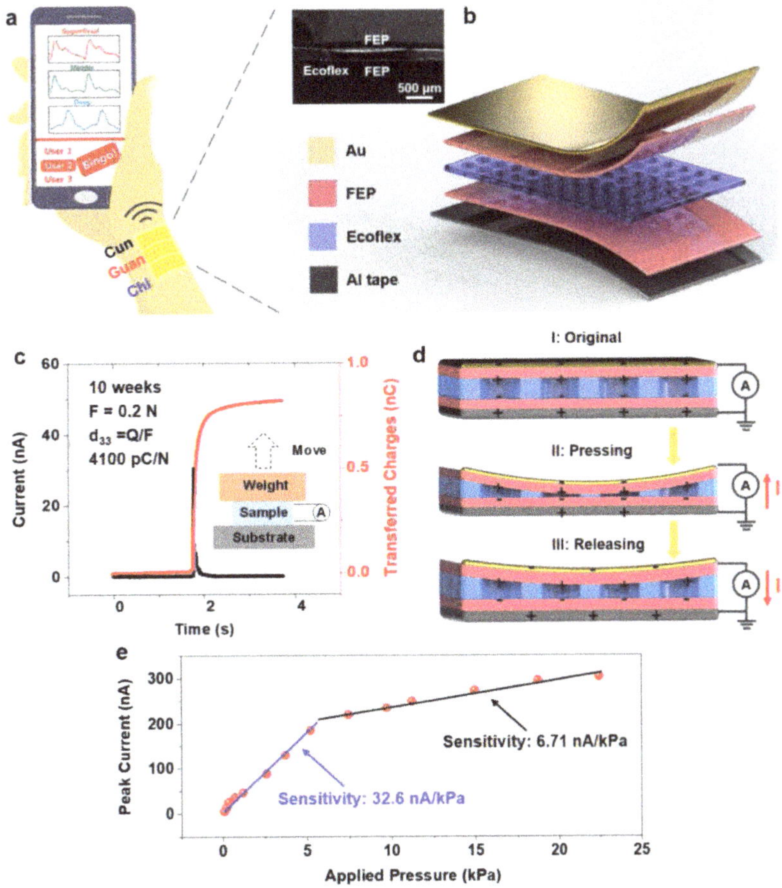

Figure 22. Representation of the sandwich-structure piezoelectret to supply a pulse-sensing system fabricated by Chu et al. [112]. (**a**) Schematic representation of the pulse-sensing system for monitoring pulse signals at the Cun, Guan, and Chi regions of the wrist. These signals could be transmitted to a smart phone. (**b**) Structural configuration of the pulse sensor integrated by the FEP/Ecoflex/FEP piezoelectret layer. (**c**) Experimental results of the equivalent piezoelectric coefficient (d_{33}) of the piezoelectret structure of the pulse sensor. (**d**) Working mechanism of the pulse sensor into its stages of pressing (I-II) and releasing (III-I), which induces the flow of electrical currents. (**e**) Experimental results of the sensitivity of peak short-circuit currents generated by a pulse sensor when an external pressure range is applied to it with a frequency of 1.5 Hz. Reprinted with permission from [112].

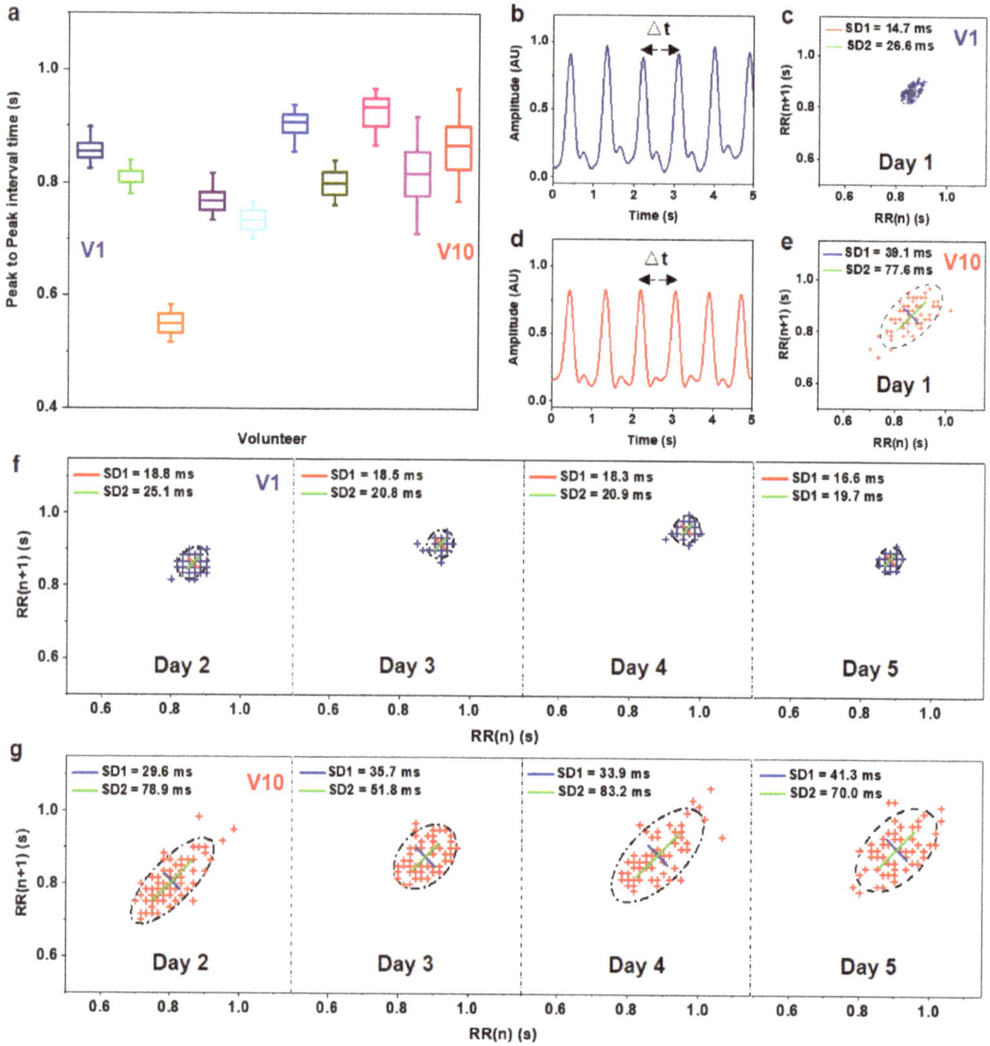

Figure 23. Results of the pulse-sensing system measured from ten volunteers [112]. (**a**) Statistical data of the peak-to-peak pulse wave intervals for the ten volunteers. The ninth and tenth volunteers (V9 and V10) have pulse intervals with larger dispersion compared to the other eight volunteers. It is due to the ninth and tenth volunteers having arrhythmia that was previously diagnosed using ECG tests in hospitals. (**b**) Amplitude of pulse waves and (**c**) corresponding Poincare plots for the first volunteer (V1), who was not diagnosed with arrhythmia. (**d**) Amplitude of pulse waves and (**e**) corresponding Poincare plots for the tenth volunteer, who has arrhythmia. Poincare plots for the (**f**) first and tenth (**g**) volunteers. The volunteer V10 has larger dispersion of pulse wave intervals. Large dispersion of the pulse wave intervals could be related to arrythmia disease. Reprinted with permission from [112]. Copyright ©2018, WILEY-VCH Verlag GMbH & Co. KGaA, Weinheim.

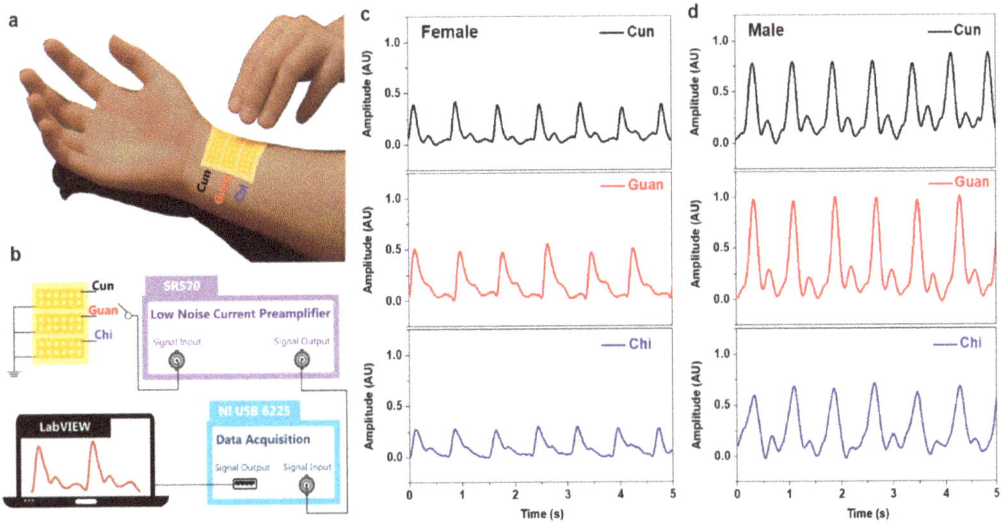

Figure 24. Results of the pulse waves at the Cun, Guan, and Chi regions of the radial artery of the wrist from a human body that were measured employing a three-channel pulse sensing array [112]. (**a**) Representation of the pulse-sensing array placed at the Cun, Guan, and Chi regions. (**b**) Signal processing of the pulse-sensing array. Results of normalized Cun, Guan, and Chi pulse signals for (**c**) a 26-year-old female volunteer, and (**d**) a 28-year-old male volunteer. Reprinted with permission from [112]. Copyright ©2018, WILEY-VCH Verlag GMbH & Co. KGaA, Weinheim.

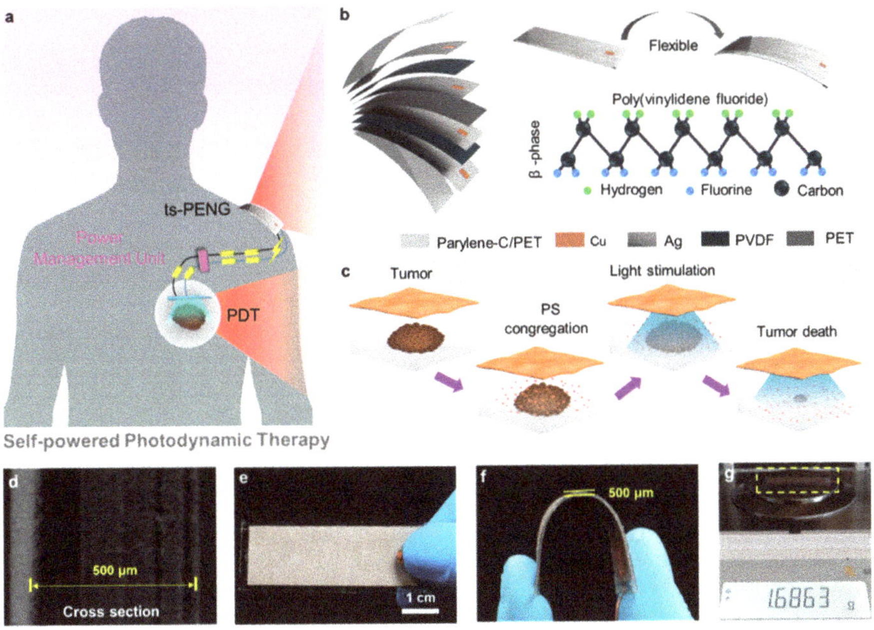

Figure 25. *Cont.*

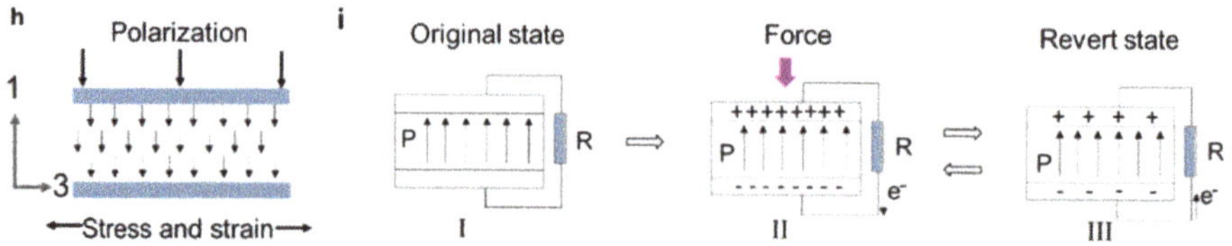

Figure 25. Representation of a self-powered photodynamic therapy system proposed by Liu et al. [113]. (**a**) This system includes a ts-PENG, a PMU, an m-LED, and a photosensitizer. (**b**) Materials of the different layers of the ts-PENG considering the packaging and substrate. (**c**) Representation of the s-PDT used in the apoptotic process of subcutaneous tumor tissue. (**d**) SEM image of the cross section of the st-PENG. (**e**) Image of the initial state of the ts-PENG. (**f**) Image of the bending state of the ts-PENG. (**g**). Weight of the ts-PENG. (**h**) Schematic view of the d_{31} operating mode of the PVDF film. (**i**) Operating principle of the PVDF film. Reprinted with permission from [113]. Copyright ©2020, American Chemical Society.

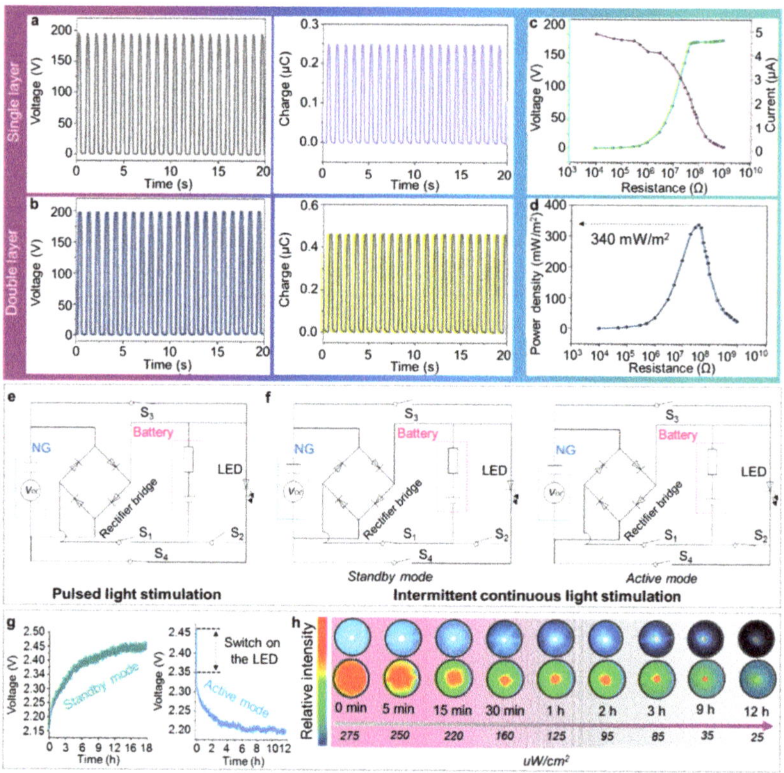

Figure 26. Performance of the ts-PENG designed by Liu et al. [113]. The open-circuit voltage and short-circuit transferred charge of the PENG with (**a**) single and (**b**) double PVDF film. (**c**) Voltage and current and (**d**) power density of the ts-PENG under different load resistances between 10 kΩ and 10 GΩ. (**e**,**f**) Schematic representation of the irradiation modes of PLS and ICLS. (**g**) Charging and discharging voltage of the button cell by the ts-PENG and LED, respectively. (**h**). Results of the light intensity of LED as a function of time. Reprinted with permission from [113]. Copyright ©2020, American Chemical Society.

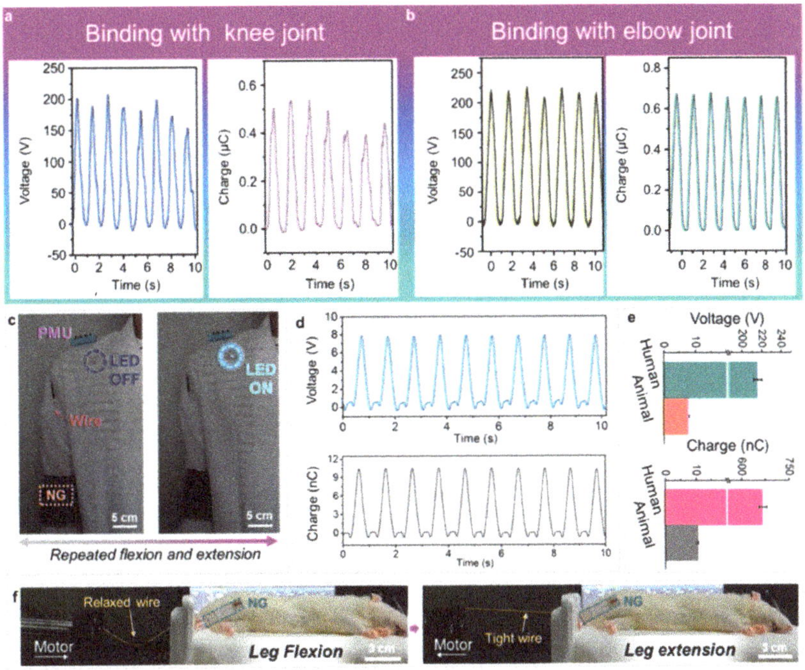

Figure 27. Experimental results of the output performance of the ts-PENG tested in human and rat models [113]. The open-circuit voltage and short-circuit transferred charge of the ts-PENG placed on (**a**) the knee joint and (**b**) elbow joint. (**c**) Test of the ts-PENG system to light an LED. (**d**) The open-circuit voltage and short-circuit transferred charge of the ts-PENG attached to the leg of a rat model. (**e**) Average results of the open-circuit voltage and short-circuit transferred charge of the ts-PENG tested on the human and rat models. (**f**). Image of the ts-PENG tested on the leg of the rat model. Reprinted with permission from [113]. Copyright ©2020, American Chemical Society.

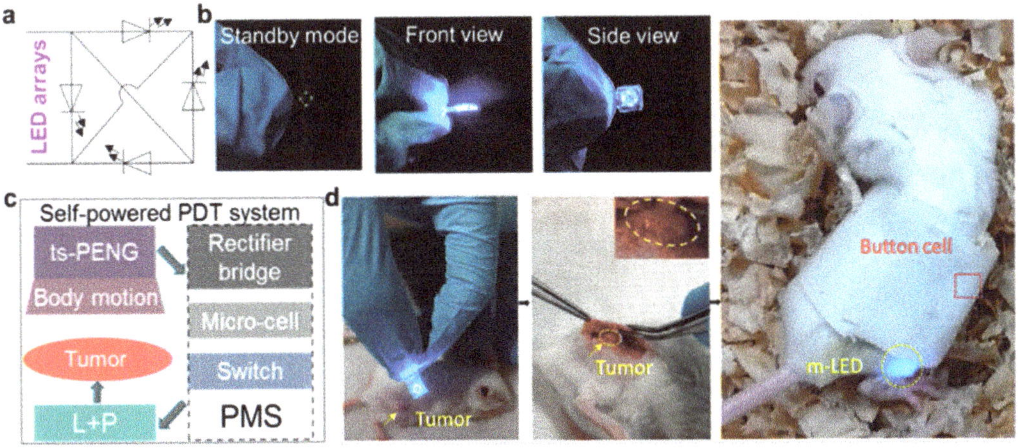

Figure 28. *Cont.*

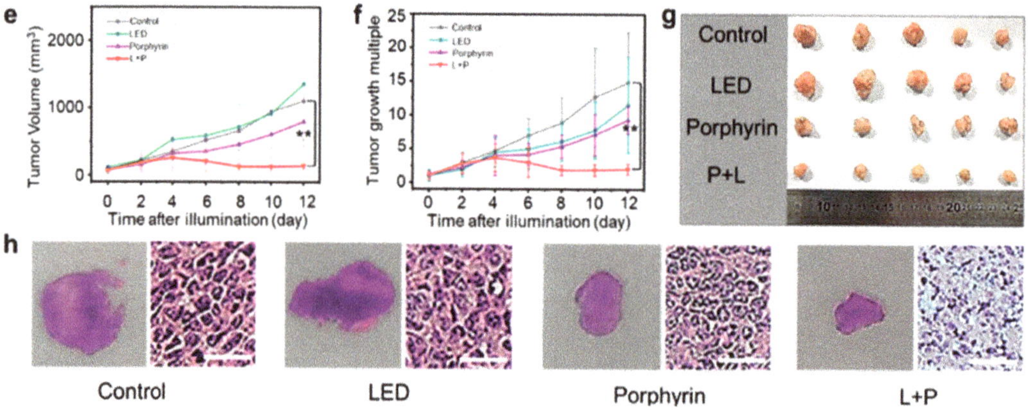

Figure 28. Experimental results of intermittent continuous light stimulation applied to a tumor in vivo [113]. (**a**) Electrical diagram of LED array. (**b**) Different views and states of the LED array packaging by PDMS. (**c**) Schematic diagram of the self-powered PDT system. (**d**) Application of the self-powered PDT system to inhibit a tumor model on a rat. (**e**) Variation of the tumor volume using a control group without any treatment and experimental groups cultured with porphyrin, LED, and LED + porphyrin. (**f**). Results of tumor growth multiplication during 12 days of a control group without any treatment and experimental groups cultured with porphyrin, LED, and LED + porphyrin. (**g**,**h**) Photographs of the tumors and pathological section (50 μm of scale bar) obtained from different experimental groups. These results were measured after 12 days. Reprinted with permission from [113]. Copyright ©2020, American Chemical Society.

4. Challenges and Perspectives

This section describes various challenges and perspectives of piezoelectric and triboelectric nanogenerators to power future healthcare monitoring devices, including the topics of design, materials and fabrication processes, signal processing, output performance, and reliability.

4.1. Design

The nanogenerator design process is key to developing better piezoelectric and triboelectric structural configurations that optimize energy harvesting to power healthcare monitoring devices. For each potential medical device, the design of a nanogenerator must examine the main performance requirements and limitations of this device, optimal materials and manufacturing process, surrounding conditions (e.g., temperature, relative humidity, pressure, dust, and so on), encapsulated type, signal processing components, and the characteristics of the original energy sources (e.g., biomechanical, wind, mechanical vibrations, and so on). For instance, nanogenerators can harvest biomechanical motion to power medical devices. However, optimal energy harvesting from human body motion has great challenges. In order to develop efficient and stable nanogenerators attached to the human body, the design of nanogenerators must consider scalable, wearable, durable, and stretchable materials. In the coming years, smart textile-based nanogenerators could be incorporated into clothing using piezoelectric or triboelectric materials with special characteristics such as being wearable, stretchable, durable, and washable. These nanogenerator types could power smart sensors that monitor real-time signals on the health status of patients. By using the IoMT, these signals could be transmitted to medical hospitals or doctors. In addition, future nanogenerators could be attached to different parts of the human body, which will require designs of electromechanical components composed of materials with high flexibility, high reproducibility, easy fabrication, and friendly performance. Future implantable self-powered medical devices will require nanogenerators with biocompatible packaging materials.

To achieve optimal designs for the output performance of nanogenerators, multi-optimization techniques can be implemented, including the various objective functions and constraints. For instance, the output power density and cost of a nanogenerator can be the objective functions to be maximized and minimized. The constraints equations of a nanogenerator can consider its size, service time, reliability parameters, and so on. For these optimization models, the mechanical and physical properties of the materials of the nanogenerator must be known.

4.2. Materials and Fabrication Processes

The output performance and reliability of piezoelectric and triboelectric nanogenerators for monitoring healthcare devices depend on the suitable selection of their materials and manufacturing processes. These materials should have electromechanical properties that allow for the development of nanogenerators with the benefits of cost-effectiveness, wearability, flexibility, easy fabrication, stability, and robust performance. Furthermore, the optimal selection of materials can be useful to fabricate smart nanogenerators with both functions for energy scavenging and working as active/self-powered sensors, which reduce their size and weight. Moreover, smart nanogenerators can be eco-friendly to the environment, eliminating the use of harmful materials from conventional batteries.

Recent investigations of nanogenerators have taken into account organic or waste materials, including chitin, spider silk, tomato, rice paper, peanut, walnut, pistachio, almond, fish gelatin, fish bladder, eggshell, sunflower husks, garbage soda cans, silk fibroin, and so on [114–121]. For these materials, two research challenges are the analytical modeling and experimental measurement of their piezoelectric and triboelectric transduction mechanisms. In addition, future investigations must consider standard rules and tests to measure the performance and reliability of nanogenerators. Another challenge is the development of scalable manufacturing processes for nanogenerators, which decrease costs and satisfy the future commercial market.

4.3. Signal Processing

Generally, the output voltage and current of triboelectric nanogenerators are not direct current (DC) signals. Due to this limitation, triboelectric nanogenerators need rectifier circuits to convert their AC output signals into DC signals. This DC voltage can be stored using capacitors or batteries to power the monitoring healthcare devices. A research challenge in this topic is the development of cost-effective energy storage systems coupled with nanogenerators.

Recent investigations [122–124] on nanogenerators have included novel electrical interfaces of high efficiency and minimum power consumption. The electrical interfaces could be self-powered and include cold-start circuit architectures [125,126]. In addition, the electrical interfaces could incorporate a small footprint to decrease the size of the nanogenerators. For this, the Application-Specific Integrated Circuit (ASIC) could be implemented.

4.4. Reliability

The implementation of reliability standard tests and rules is required to measure the safe and stable performance of nanogenerators for medical applications. The reliability analyses of the electromechanical performance of nanogenerators under different environments and operating conditions are required to study their stability, durability, and working capability. These reliability tests can provide the main electromechanical failures of the nanogenerators due to various factors, such as high relative humidity, large temperature variations, dust, mechanical impact, wear, crack growth, fatigue, and so on. Furthermore, the different packaging types and signal-processing components of nanogenerators must be included in the reliability tests. In order to improve the reliability of nanogenerators, the designers could consider simple operating mechanisms, minimum electrical and structural components, robust and durable materials, safe packaging, and so on. Figure 29 depicts various challenges of the piezoelectric and triboelectric nanogenerators for their application in IoMT electronic devices.

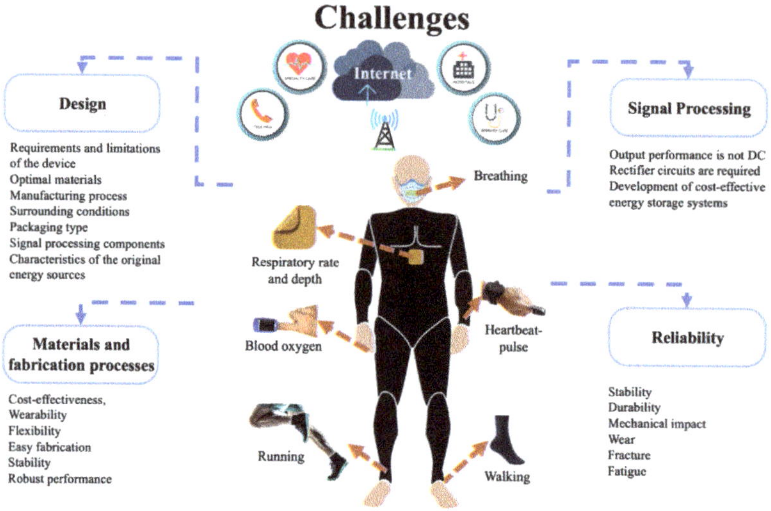

Figure 29. Summary of challenges of piezoelectric and triboelectric nanogenerators in IoMT electronic devices.

5. Conclusions

Recent progress on piezoelectric and triboelectric nanogenerators for self-powered monitoring healthcare devices was reviewed. The descriptions of the working principle, materials, and fabrication processes of a variety of these nanogenerators were reported. Discussions on the output electrical performance of piezoelectric and triboelectric nanogenerators are included. Furthermore, we presented the signal processing components and packaging types of different nanogenerators. The design, materials, and electromechanical behavior of different piezoelectric and triboelectric nanogenerators for biomechanical energy harvesting to power potential medical applications were considered. Also, we proposed some challenges and perspectives of these nanogenerator types, including the design, materials and manufacturing processes, signal processing, and reliability.

Author Contributions: E.D.-A., J.M.-C., L.Z.-P., J.A.G.-C., R.L.-E., M.W.A., S.T. and A.L.H.-M. wrote, coordinated, reviewed, and contributed to the scientific aspects of the article. All authors have read and agreed to the published version of the manuscript.

Funding: This research received no external funding.

Data Availability Statement: Not applicable.

Acknowledgments: Thanks to Universidad Veracruzana for the support and facilities for the publication of this article.

Conflicts of Interest: The authors declare no conflict of interest.

References

1. Silva, A.F.; Tavakoli, M. Domiciliary hospitalization through wearable biomonitoring patches: Recent advances, technical challenges, and the relation to COVID-19. *Sensors* **2020**, *20*, 6835. [CrossRef] [PubMed]
2. Remuzzi, A.; Remuzzi, G. COVID-19 and Italy: What next? *Lancet* **2020**, *395*, 1225–1228. [CrossRef] [PubMed]
3. Moes, S.L.; Depmann, M.; Lely, T.A.; Bekker, M.N. Telemonitoring for COVID-19 positive pregnant women; feasibility and user experience of SAFE@home Corona: Prospective pilot study. *BMC Pregnancy Childbirth* **2022**, *22*, 556. [CrossRef] [PubMed]
4. Chavda, E.; Guedon-Moreau, L.; Williatte, L.; Cordova, E. Post-emergency teleconsultations during COVID crisis: TELE-SCOPE tool's feedback and epidemiological analysis. *Digit. Health* **2022**, *8*, 2055–2076. [CrossRef] [PubMed]

5. Scherrenberg, M.; Storms, V.; Van der Velde, A.E.; Boyne, J.; Bruins, W.; Vranken, J.; Leenen, J.P.L.; Brunner-La Rocca, H.-P.; De Kluiver, E.P.; Dendale, P. A Home hospitalisation strategy for patients with an acute episode of heart failure using a digital health-supported platform: A multicentre feasibility study—A rationale and study design. *Cardiology* **2021**, *146*, 793–800. [CrossRef]
6. Tang, M.; Reddy, A. Telemedicine and its past, present, and future roles in providing palliative care to advanced cancer patients. *Cancers* **2022**, *14*, 1884. [CrossRef]
7. Disalvo, D.; Agar, M.; Caplan, G.; Murtagh, F.E.M.; Luckett, T.; Heneka, N.; Hickman, L.; Kinchin, I.; Trethewie, S.; Sheehan, C.; et al. Virtual models of care for people with palliative care needs living in their own home: A systematic meta-review and narrative synthesis. *Palliat. Med.* **2021**, *35*, 1385–1406. [CrossRef]
8. International Programs Center at the U.S., Census Bureau, Population Division, International Programs Center at the U.S. 2019. Available online: https://www.census.gov/data-tools/demo/idb/#/country?COUNTRY_YEAR=2022&COUNTRY_YR_ANIM=2022 (accessed on 6 June 2022).
9. Gelbman, B.; Reed, C. An integrated, multimodal, digital health solution for chronic obstructive pulmonary disease: Prospective observational pilot study. *JMIR Form Res* **2022**, *6*, e34758. Available online: https://formative.jmir.org/2022/3/e34758 (accessed on 6 June 2022). [CrossRef]
10. Nagase, F.I.; Stafinski, T.; Avdagovska, M.; Stickland, M.K.; Etruw, E.M.; Menon, D. Effectiveness of remote home monitoring for patients with chronic obstructive pulmonary disease (COPD): Systematic review. *BMC Health Serv. Res.* **2022**, *22*, 646. [CrossRef]
11. Mangla, C.; Rani, S.; Herencsar, N. An energy-efficient and secure framework for IoMT: An application of smart cities. *Sustain. Energy Technol. Assess.* **2022**, *53*, 102335. [CrossRef]
12. Sharma, G.; Joshi, A.M.; Pilli, E.S. DepML: An efficient machine learning-based MDD detection system in IoMT framework. *SN Comput. Sci.* **2022**, *3*, 394. [CrossRef]
13. Ahmed, S.T.; Kumar, V.V.; Singh, K.K.; Singh, A.; Muthukumaran, V.; Gupta, D. 6G enabled federated learning for secure IoMT resource recommendation and propagation analysis. *Comput. Electr. Eng.* **2022**, *102*, 108210. [CrossRef]
14. Idrees, A.K.; Idrees, S.K.; Couturier, R.; Ali-Yahiya, T. An edge-fog computing-enabled lossless EEG data compression with epileptic seizure detection in IoMT networks. *IEEE Internet Things J.* **2022**, *9*, 13327–13337. [CrossRef]
15. Demirel, B.U.; Bayoumy, I.A.; Faruque, M.A.A. Energy-efficient real-time heart monitoring on edge–fog–cloud internet of medical things. *IEEE Internet Things J.* **2022**, *9*, 12472–12481. [CrossRef]
16. Alzahrani, F.A.; Ahmad, M.; Ansari, M.T.J. Towards design and development of security assessment framework for internet of medical things. *Appl. Sci.* **2022**, *12*, 8148. [CrossRef]
17. Sadhu, P.K.; Yanambaka, V.P.; Abdelgawad, A.; Yelamarthi, K. Prospect of internet of medical things: A review on security requirements and solutions. *Sensors* **2022**, *22*, 5517. [CrossRef]
18. Chen, C.-M.; Chen, Z.; Kumari, S.; Lin, M.-C. LAP-IoHT: A lightweight authentication protocol for the internet of health things. *Sensors* **2022**, *22*, 5401. [CrossRef]
19. Kakhi, K.; Alizadehsani, R.; Kabir, H.M.D.; Khosravi, A.; Nahavandi, S.; Acharya, U.R. The internet of medical things and artificial intelligence: Trends, challenges, and opportunities. *Biocybern. Biomed. Eng.* **2022**, *42*, 749–771. [CrossRef]
20. Phan, D.T.; Nguyen, C.H.; Nguyen, T.D.P.; Tran, L.H.; Park, S.; Choi, J.; Lee, B.-I.; Oh, J. A Flexible, wearable, and wireless biosensor patch with internet of medical things applications. *Biosensors* **2022**, *12*, 139. [CrossRef]
21. Rizi, K.S. The smartphone biosensors for point-of-care detection of human infectious diseases: Overview and perspectives—A systematic review. *Curr. Opin. Electrochem.* **2022**, *32*, 100925. [CrossRef]
22. Scrugli, M.A.; Loi, D.; Raffo, L.; Meloni, P. An adaptive cognitive sensor node for ECG monitoring in the internet of medical things. *IEEE Access* **2022**, *10*, 1688–1705. [CrossRef]
23. Verma, D.; Singh, K.R.B.; Yadav, A.K.; Nayak, V.; Singh, J.; Pratima, R.; Solanki, P.R.; Singh, R.P. Internet of things (IoT) in nano-integrated wearable biosensor devices for healthcare applications. *Biosens. Bioelectron. X* **2022**, *11*, 100153. [CrossRef]
24. Hasan, M.K.; Ghazal, T.M.; Saeed, R.A.; Pandey, B.; Gohel, H.; Eshmawi, A.A.; Abdel-Khalek, S.; Alkhassawneh, H.M. A review on security threats, vulnerabilities, and counter measures of 5G enabled Internet-of-Medical-Things. *IET Commun.* **2022**, *16*, 421–432. [CrossRef]
25. Taimoor, N.; Rehman, S. Reliable and resilient AI and IoT-based personalised healthcare services: A survey. *IEEE Access* **2022**, *10*, 535–563. [CrossRef]
26. Vankdothu, R.; Hameed, M.A.; Ameen, A.; Unnisa, R. Brain image identification and classification on internet of medical things in healthcare system using support value based deep neural network. *Comput. Electr. Eng.* **2022**, *102*, 108196. [CrossRef]
27. Kwarteng, E.; Cebe, M. A survey on security issues in modern implantable devices: Solutions and future issues. *Smart Health* **2022**, *25*, 100295. [CrossRef]
28. Zhao, J.; Zhang, S.; Sun, Y.; Zhou, N.; Yu, H.; Zhang, H.; Jia, D. Wearable optical sensing in the medical internet of things (MIoT) for pervasive medicine: Opportunities and challenges. *ACS Photonics* **2022**, *9*, 2579–2599. [CrossRef]
29. Dwivedi, R.; Mehrotra, D.; Chandra, S. Potential of internet of medical things (IoMT) applications in building a smart healthcare system: A systematic review. *J. Oral Biol. Craniofac. Res.* **2022**, *12*, 302–318. [CrossRef]
30. Li, S.; Guo, H.; He, S.; Yang, H.; Liu, K.; Gaigai Duan, G.; Jiang, S. Advanced electrospun nanofibers as bifunctional electrocatalysts for flexible metal-air (O_2) batteries: Opportunities and challenges. *Mater. Des.* **2022**, *214*, 110406. [CrossRef]

31. Yang, H.; Lee, J.; Cheng, J.Y.; Wang, Y.; Duan, G.; Hou, H.; Jiang, S.; Kim, I.-D. Molecular engineering of carbonyl organic electrodes for rechargeable metal-ion batteries: Fundamentals, recent advances, and challenges. *Energy Environ. Sci.* **2021**, *14*, 4228–4267. [CrossRef]
32. Mainar, A.R.; Leonet, O.; Bengoechea, M.; Boyano, I.; Meatza, I.D.; Kvasha, A.; Guerfi, A.; Blázquez, J.A. Alkaline aqueous electrolytes for secondary zinc–air batteries: An overview. *Int. J. Energy Res.* **2016**, *40*, 1032–1049. [CrossRef]
33. Yang, L.; Guo, X.; Jin, Z.; Guo, W.; Duan, G.; Liu, X.; Li, Y. Emergence of melanin-inspired supercapacitors. *Nano Today* **2021**, *37*, 101075. [CrossRef]
34. Wang, Y.; Zhang, L.; Hou, H.; Xu, W.; Duan, G.; He, S.; Liu, K.; Jian, S. Recent progress in carbon-based materials for supercapacitor electrodes: A review. *J. Mater. Sci.* **2021**, *56*, 173–200. [CrossRef]
35. Poonam; Sharma, K.; Arora, A.; Tripathi, S.K. Review of supercapacitors: Materials and devices. *J. Energy Storage* **2019**, *21*, 801–825. [CrossRef]
36. Delgado-Alvarado, E.; Elvira-Hernández, E.A.; Hernández-Hernández, J.; Huerta-Chua, J.; Vázquez-Leal, H.; Martínez-Castillo, J.; García-Ramírez, P.J.; Herrera-May, A.L. Recent progress of nanogenerators for green energy harvesting: Performance, applications, and challenges. *Nanomaterials* **2022**, *12*, 2549. [CrossRef]
37. Mariello, M. Recent Advances on hybrid piezo-triboelectric bio-nanogenerators: Materials, architectures and circuitry. *Nanoenergy Adv.* **2022**, *2*, 64–109. [CrossRef]
38. Zhang, J.; He, Y.; Boyer, C.; Kalantar-Zadeh, K.; Peng, S.; Chu, D.; Wang, C.H. Recent developments of hybrid piezo–triboelectric nanogenerators for flexible sensors and energy harvesters. *Nanoscale Adv.* **2021**, *3*, 5465–5486. [CrossRef]
39. Han, S.A.; Lee, J.; Lin, J.; Kim, S.-W.; Kim, J.H. Piezo/triboelectric nanogenerators based on 2-dimensional layered structure materials. *Nano Energy* **2019**, *57*, 680–691. [CrossRef]
40. Mariello, M.; Fachechi, L.; Guido, F.; De, M. Conformal, ultra-thin skin-contact-actuated hybrid piezo/triboelectric wearable sensor based on AlN and parylene-encapsulated elastomeric blend. *Adv. Funct. Mater.* **2021**, *31*, 2101047. [CrossRef]
41. Zhou, L.; Zhu, L.; Yang, T.; Hou, X.; Du, Z.; Cao, S.; Wang, H.; Chou, K.-C.; Wang, Z.L. Ultra-stable and durable piezoelectric nanogenerator with all-weather service capability based on N doped 4H-SiC nanohole arrays. *Nano-Micro Lett.* **2022**, *14*, 30. [CrossRef]
42. Nguyen, Q.H.; Ta, Q.T.H.; Tran, N. Review on the transformation of biomechanical energy to green energy using triboelectric and piezoelectric based smart materials. *J. Clean. Prod.* **2022**, *371*, 133702. [CrossRef]
43. Hou, X.; Zhong, J.; Yang, C.; Yang, Y.; He, J.; Mu, J.; Geng, W.; Chou, X. A high-performance, single-electrode and stretchable piezo-triboelectric hybrid patch for omnidirectional biomechanical energy harvesting and motion monitoring. *J. Mater.* **2022**, *8*, 958–966. [CrossRef]
44. Park, G.G.; Lee, E.J.; Jung, S.; Jeong, S.; Kim, H.-S.; Choi, Y.; Lee, S.Y. Double nanocomposites-based piezoelectric nanogenerators for high-performance energy harvester. *ACS Appl. Energy Mater.* **2022**, *5*, 8835–8843. [CrossRef]
45. Fu, H.; Long, Z.; Lai, M.; Cao, J.; Zhou, R.; Gong, J.; Chen, Y. Quantum dot hybridization of piezoelectric polymer films for non-transfer integration of flexible biomechanical energy harvesters. *ACS Appl. Mater. Interfaces* **2022**, *14*, 29934–29944. [CrossRef]
46. Bouhamed, A.; Jöhrmann, N.; Naifar, S.; Böhm, B.; Hellwig, O.; Wunderle, B.; Kanoun, O. Collaborative filler network for enhancing the performance of $BaTiO_3$/PDMS flexible piezoelectric polymer composite nanogenerators. *Sensors* **2022**, *22*, 4181. [CrossRef]
47. Chen, L.; He, M.; Li, L.; Yuan, S.; Chen, A.; Chen, M.; Wang, Y.; Sun, L.; Wei, L.; Zhang, T.; et al. Hydrochromic CsPbBr3-KBr microcrystals for flexible anti-counterfeiting and wearable self-powered biomechanical monitoring. *Chem. Eng. J.* **2022**, *450*, 138279. [CrossRef]
48. Li, M.; Xu, B.; Li, Z.; Gao, Y.; Yang, Y.; Huang, X. Toward 3D double-electrode textile triboelectric nanogenerators for wearable biomechanical energy harvesting and sensing. *Chem. Eng. J.* **2022**, *450*, 137491. [CrossRef]
49. Qu, M.; Shen, L.; Wang, J.; Zhang, N.; Pang, Y.; Wu, Y.; Ge, J.; Peng, L.; Yang, J.; He, J. Superhydrophobic, humidity-resistant, and flexible triboelectric nanogenerators for biomechanical energy harvesting and wearable self-powered sensing. *ACS Appl. Nano Mater.* **2022**, *5*, 9840–9851. [CrossRef]
50. Yang, M.; Liu, J.; Liu, D.; Jiao, J.; Cui, N.; Liu, S.; Xu, Q.; Gu, L.; Qin, Y. A fully self-healing piezoelectric nanogenerator for self-powered pressure sensing electronic skin. *Research* **2021**, *2021*, 9793458. [CrossRef]
51. Wan, X.; Wang, Z.; Zhao, X.; Hu, Q.; Li, Z.; Wang, Z.L.; Li, L. Flexible and highly piezoelectric nanofibers with organic–inorganic coaxial structure for self-powered physiological multimodal sensing. *Chem. Eng. J.* **2022**, *451*, 139077. [CrossRef]
52. Zhou, J.; Gou, X.; Fan, D.; Wang, J.; Wan, Z. Polydimethylsiloxane/$BaTiO_3$ nanogenerators with a surface-assembled mosaic structure for enhanced piezoelectric sensing. *ACS Appl. Mater. Interfaces* **2022**, *14*, 38105–38115. [CrossRef]
53. Li, H.; Chang, T.; Gai, Y.; Liang, K.; Jiao, Y.; Li, D.; Jiang, X.; Wang, Y.; Huang, X.; Wu, H.; et al. Human joint enabled flexible self-sustainable sweat sensors. *Nano Energy* **2022**, *92*, 106786. [CrossRef]
54. Xu, Z.; Zhang, D.; Cai, H.; Yang, Y.; Zhang, H.; Du, C. Performance enhancement of triboelectric nanogenerators using contact-separation mode in conjunction with the sliding mode and multifunctional application for motion monitoring. *Nano Energy* **2022**, *102*, 107719. [CrossRef]
55. Liu, Y.; Wong, T.H.; Huang, X.; Yiu, C.K.; Gao, Y.; Zhao, L.; Zhou, J.; Park, W.; Zhao, Z.; Yao, K.; et al. Skin-integrated, stretchable, transparent triboelectric nanogenerators based on ion-conducting hydrogel for energy harvesting and tactile sensing. *Nano Energy* **2022**, *99*, 107442. [CrossRef]

56. Sahu, M.; Hajra, S.; Panda, S.; Rajaitha, M.; Panigrahi, B.K.; Rubahn, H.-G.; Mishra, Y.K.; Kim, H.J. Waste textiles as the versatile triboelectric energy-harvesting platform for self-powered applications in sports and athletics. *Nano Energy* **2022**, *97*, 107208. [CrossRef]
57. Yang, J.; An, J.; Sun, Y.; Zhang, J.; Zu, L.; Li, H.; Jiang, T.; Chen, B.; Wang, Z.L. Transparent self-powered triboelectric sensor based on PVA/PA hydrogel for promoting human-machine interaction in nursing and patient safety. *Nano Energy* **2022**, *97*, 107199. [CrossRef]
58. Sengupta, A.; Das, S.; Dasgupta, S.; Sengupta, P.; Datta, P. Flexible nanogenerator from electrospun PVDF–polycarbazole nanofiber membranes for human motion energy-harvesting device applications. *ACS Biomater. Sci. Eng.* **2021**, *7*, 1673–1685. [CrossRef] [PubMed]
59. Cao, S.; Zou, H.; Jiang, B.; Li, M.; Yuan, Q. Incorporation of ZnO encapsulated MoS_2 to fabricate flexible piezoelectric nanogenerator and sensor. *Nano Energy* **2022**, *102*, 107635. [CrossRef]
60. Li, Y.; Tan, J.; Liang, K.; Li, Y.; Sun, J.; Zhang, H.; Luo, C.; Li, P.; Xu, J.; Jiang, H.; et al. Enhanced piezoelectric performance of multi-layered flexible polyvinylidene fluoride–$BaTiO_3$–rGO films for monitoring human body motions. *J. Mater. Sci. Mater. Electron.* **2022**, *33*, 4291–4304. [CrossRef]
61. Bairagi, S.; Ghosh, S.; Ali, S.W. A fully sustainable, self-poled, bio-waste based piezoelectric nanogenerator: Electricity generation from pomelo fruit membrane. *Sci. Rep.* **2022**, *10*, 12121. [CrossRef] [PubMed]
62. Zhang, J.; Zhao, X.; Wang, Z.; Liu, Z.; Yao, S.; Li, L. Antibacterial, antifreezing, stretchable, and self-healing organohydrogel electrode based triboelectric nanogenerator for self-powered biomechanical sensing. *Adv. Mater. Interfaces* **2022**, *9*, 2200290. [CrossRef]
63. Yang, Y.; Hou, X.; Geng, W.; Mu, J.; Zhang, L.; Wang, X.; He, J.; Xiong, J.J.; Chou, X. Human movement monitoring and behavior recognition for intelligent sports using customizable and flexible triboelectric nanogenerator. *Sci. China Technol. Sci.* **2022**, *65*, 826–836. [CrossRef]
64. Wang, J.; Zhao, Z.; Zeng, X.; Liu, X.; Hu, Y. A Tubular flexible triboelectric nanogenerator with a superhydrophobic surface for human motion Detecting. *Sensors* **2021**, *21*, 3634. [CrossRef]
65. Li, R.; Wei, X.; Xu, J.; Chen, J.; Li, B.; Wu, Z.; Wang, Z.L. Smart wearable sensors based on triboelectric nanogenerator for personal healthcare monitoring. *Micromachines* **2021**, *12*, 352. [CrossRef]
66. Liu, J.; Cui, N.; Du, T.; Li, G.; Liu, S.; Xu, Q.; Wang, Z.; Gu, L.; Qin, Y. Coaxial double helix structured fiber-based triboelectric nanogenerator for effectively harvesting mechanical energy. *Nanoscale Adv.* **2020**, *2*, 4482–4490. [CrossRef]
67. Kaur, G.; Meena, J.S.; Jassal, M.; Agrawal, A.K. Synergistic Effect of Polyurethane in Polyurethane–Poly(vinylidene fluoride) Nanofiber-Based Stretchable Piezoelectric Nanogenerators (S-PENGs). *ACS Appl. Polym. Mater.* **2022**, *4*, 4751–4764. [CrossRef]
68. Waseem, A.; Bagal, I.V.; Abdullah, A.; Kulkarni, M.A.; Thaalbi, H.; Ha, J.-S.; Lee, J.K.; Ryu, S.-W. High performance, stable, and flexible piezoelectric nanogenerator based on GaN:Mg nanowires directly grown on tungsten foil. *Small* **2022**, *18*, 2200952. [CrossRef]
69. Yu, J.; Hou, X.; Cui, M.; Zhang, N.; Zhang, S.; He, J.; Chou, X. Skin-conformal $BaTiO_3$/ecoflex-based piezoelectric nanogenerator for self-powered human motion monitoring. *Mater. Lett.* **2020**, *269*, 127686. [CrossRef]
70. Ponnamma, D.; Parangusan, H.; Tanvir, A.; AlMa'adeed, M.A.A. Smart and robust electrospun fabrics of piezoelectric polymer nanocomposite for self-powering electronic textiles. *Mater. Des.* **2019**, *184*, 108176. [CrossRef]
71. Zhu, Y.; Xia, Y.; Wu, M.; Guo, W.; Jia, C.; Wang, X. Wearable, freezing-tolerant, and self-powered electroluminescence system for long-term cold-resistant displays. *Nano Energy* **2022**, *98*, 107309. [CrossRef]
72. Liu, J.; Li, S.; Yang, M.; Wang, Y.; Cui, N.; Gu, L. Coaxial spring-like stretchable triboelectric nanogenerator toward personal healthcare monitoring. *Front. Bioeng. Biotechnol.* **2022**, *10*, 889364. [CrossRef] [PubMed]
73. Yang, Y.; Chen, L.; He, J.; Hou, X.; Qiao, X.; Xiong, J.; Chou, X. Flexible and extendable honeycomb-shaped triboelectric nanogenerator for effective human motion energy harvesting and biomechanical sensing. *Adv. Mater. Technol.* **2022**, *7*, 2100702. [CrossRef]
74. Bai, Z.; Xu, Y.; Li, J.; Zhu, J.; Gao, C.; Zhang, Y.; Wang, J.; Guo, J. An eco-friendly porous nanocomposite fabric-based triboelectric nanogenerator for efficient energy harvesting and motion sensing. *ACS Appl. Mater. Interfaces* **2020**, *12*, 42880–42890. [CrossRef] [PubMed]
75. Venkatesan, M.; Chen, W.-H.; Cho, C.-H.; Veeramuthu, L.; Chen, L.-G.; Li, K.-Y.; Tsai, M.-L.; Lai, Y.-C.; Lee, W.-Y.; Chen, W.-C.; et al. Enhanced piezoelectric and photocatalytic performance of flexible energy harvester based on CsZn0.75Pb0.25I3/CNC–PVDF composite nanofibers. *Chem. Eng. J.* **2022**, *433*, 133620. [CrossRef]
76. Lo, W.C.; Chen, C.C.; Fuh, Y.K. 3D Stacked Near-Field Electrospun Nanoporous PVDF-TrFE Nanofibers as Self-Powered Smart Sensing in Gait Big Data Analytics. *Adv. Mater. Technol.* **2021**, *6*, 2000779. [CrossRef]
77. Lee, D.W.; Jeong, D.G.; Kim, J.H.; Kim, H.S.; Murillo, G.; Lee, G.-H.; Song, H.-C.; Jung, J.H. Polarization-controlled PVDF-based hybrid nanogenerator for an effective vibrational energy harvesting from human foot. *Nano Energy* **2020**, *76*, 105066. [CrossRef]
78. Deng, C.; Tang, W.; Liu, L.; Chen, B.; Li, M.C.; Wang, Z.L. Self-powered insole plantar pressure mapping system. *Adv. Funct. Mater.* **2018**, *28*, 1801606. [CrossRef]
79. Gulahmadov, O.; Muradov, M.B.; Kim, J. Gait analysis by using electric signals from a triboelectric nanogenerator. *Eng. Res. Express* **2022**, *4*, 035027. [CrossRef]

80. Rahman, M.T.; Rana, S.M.S.; Salauddin, M.; Zahed, M.A.; Lee, S.; Yoon, E.-S.; Park, J.Y. Silicone-incorporated nanoporous cobalt oxide and MXene nanocomposite-coated stretchable fabric for wearable triboelectric nanogenerator and self-powered sensing applications. *Nano Energy* **2022**, *100*, 107454. [CrossRef]
81. Shao, Y.; Zhou, F.; Wang, F. A Triboelectric sensor with a dual working unit for race walking motion monitoring. *J. Electron. Mater.* **2022**, *51*, 3569–3578. [CrossRef]
82. Park, D.; Hong, J.-H.; Choi, D.; Kim, D.; Jung, W.H.; Yoon, S.S.; Kim, K.H.; An, S. Biocompatible and mechanically-reinforced tribopositive nanofiber mat for wearable and antifungal human kinetic-energy harvester based on wood-derived natural product. *Nano Energy* **2022**, *96*, 107091. [CrossRef]
83. Kim, Y.W.; Lee, H.B.; Yoon, J.; Park, S.-H. 3D customized triboelectric nanogenerator with high performance achieved via charge-trapping effect and strain-mismatching friction. *Nano Energy* **2022**, *95*, 107051. [CrossRef]
84. Yang, P.; Shi, Y.; Li, S.; Tao, X.; Liu, Z.; Wang, X.; Wang, Z.L.; Chen, X. Monitoring the degree of comfort of shoes in-motion using triboelectric pressure sensors with an ultrawide detection range. *ACS Nano* **2022**, *16*, 4654–4665. [CrossRef]
85. Lackermair, K.; Fischer, F.; Manhart, J.; Scheurer, E.; Graw, M.; Boy, D.; Lenz, C.; Hartrampf, B.; Kellnar, A.; Sams, L.; et al. Determination of time of death by blinded post-mortem interrogation of cardiac implantable electrical devices. *Sci. Rep.* **2022**, *12*, 8199. [CrossRef]
86. Yang, E.; Suzuki, M.; Nazarian, S.; Halperin, H.R. Magnetic resonance imaging safety in patients with cardiac implantable electronic devices. *Trends Cardiovasc. Med.* **2022**, *32*, 440–447. [CrossRef]
87. Krucoff, M.O.; Wozny, T.A.; Lee, A.T.; Rao, V.R.; Chang, E.F. Operative technique and lessons learned from surgical implantation of the NeuroPace responsive Neurostimulation® System in 57 consecutive patients. *Oper. Neurosurg.* **2021**, *20*, E98–E109. [CrossRef]
88. Barri, K.; Zhang, Q.; Swink, I.; Aucie, Y.; Holmberg, K.; Sauber, R.; Altman, D.T.; Cheng, B.C.; Wang, Z.L.; Alavi, A.H. Patient-specific self-powered metamaterial implants for detecting bone healing progress. *Adv. Funct. Mater.* **2022**, *32*, 2203533. [CrossRef]
89. Lone, S.A.; Lim, K.C.; Kaswan, K.; Chatterjee, S.; Fan, K.-P.; Choi, D.; Lee, S.; Zhang, H.; Cheng, J.; Lin, Z.-H. Recent advancements for improving the performance of triboelectric nanogenerator devices. *Nano Energy* **2022**, *99*, 107318. [CrossRef]
90. Kabir, H.; Dehghan, H.K.; Mashayekhan, S.; Bagherzadeh, R.; Sorayani Bafqi, M.S. Hybrid fibrous (PVDF-$BaTiO_3$)/ PA-11 piezoelectric patch as an energy harvester for pacemakers. *J. Ind. Text.* **2022**, *51*, 4698S–4719S. [CrossRef]
91. Al-Suhaimi, E.A.; Aljafary, M.A.; Alfareed, T.M.; Alshuyeh, H.A.; Alhamid, G.M.; Sonbol, B.; Almofleh, A.; Alkulaifi, F.M.; Altwayan, R.K.; Alharbi, J.N.; et al. Nanogenerator-Based Sensors for Energy Harvesting from Cardiac Contraction. *Front. Energy Res.* **2022**, *10*, 900534. [CrossRef]
92. Torres, F.G.; Troncoso, O.P.; De-la-Torre, G.E. Hydrogel-based triboelectric nanogenerators: Properties, performance, and applications. *Int. J. Energy Res.* **2022**, *46*, 5603–5624. [CrossRef]
93. Ryu, H.; Park, H.m.; Kim, M.K.; Kim, B.; Myoung, H.S.; Kim, T.Y.; Yoon, H.-J.; Kwak, S.S.; Kim, J.; Hwang, T.H.; et al. Self-rechargeable cardiac pacemaker system with triboelectric nanogenerators. *Nat. Commun.* **2021**, *12*, 4374. [CrossRef] [PubMed]
94. Parandeh, S.; Etemadi, N.; Kharaziha, M.; Chen, G.; Nashalian, A.; Xiao, X.; Chen, J. Advances in triboelectric nanogenerators for self-powered regenerative medicine. *Adv. Funct. Mater.* **2021**, *31*, 2105169. [CrossRef]
95. Chen, P.; Wu, P.; Wan, X.; Wang, Q.; Xu, C.; Yang, M.; Feng, J.; Hu, B.; Luo, Z. Ultrasound-driven electrical stimulation of peripheral nerves based on implantable piezoelectric thin film nanogenerators. *Nano Energy* **2021**, *86*, 106123. [CrossRef]
96. Long, Y.; Li, J.; Yang, F.; Wang, J.; Wang, X. Wearable and Implantable Electroceuticals for Therapeutic Electrostimulations. *Adv. Sci.* **2021**, *8*, 2004023. [CrossRef]
97. Zhang, S.; Bick, M.; Xiao, X.; Chen, G.; Nashalian, A.; Chen, J. Leveraging triboelectric nanogenerators for bioengineering. *Matter* **2021**, *4*, 845–887. [CrossRef]
98. Mathew, A.A.; Chandrasekhar, A.; Vivekanandan, S. A review on real-time implantable and wearable health monitoring sensors based on triboelectric nanogenerator approach. *Nano Energy* **2021**, *80*, 105566. [CrossRef]
99. Owida, H.A.; Al-Nabulsi, J.I.; Turab, N.M.; Alnaimat, F.; Rababah, H.; Shakour, M.I. Autocharging Techniques for Implantable Medical Applications. *Int. J. Biomater.* **2021**, *2021*, 6074657. [CrossRef]
100. Khandelwal, G.; Joseph Raj, N.P.M.; Kim, S.-J. Triboelectric nanogenerator for healthcare and biomedical applications. *Nano Today* **2020**, *33*, 100882. [CrossRef]
101. Yoon, H.-J.; Kim, S.-W. Nanogenerators to Power Implantable Medical Systems. *Joule* **2020**, *4*, 1398–1407. [CrossRef]
102. Sun, J.; Yang, A.; Zhao, C.; Liu, F.; Li, Z. Recent progress of nanogenerators acting as biomedical sensors in vivo. *Sci. Bull.* **2019**, *64*, 1336–1347. [CrossRef]
103. Chen, X.; Song, Y.; Su, Z.; Chen, H.; Cheng, X.; Zhang, J.; Han, M.; Zhang, H. Flexible fiber-based hybrid nanogenerator for biomechanical energy harvesting and physiological monitoring. *Nano Energy* **2017**, *38*, 43–50. [CrossRef]
104. Sun, J.-G.; Yang, T.-N.; Wang, C.-Y.; Chen, L.-J. A flexible transparent one-structure tribo-piezo-pyroelectric hybrid energy generator based on bio-inspired silver nanowires network for biomechanical energy harvesting and physiological monitoring. *Nano Energy* **2018**, *48*, 383–390. [CrossRef]
105. Zhu, M.; Shi, Q.; He, T.; Yi, Z.; Ma, Y.; Yang, B.; Chen, T.; Lee, C. Self-powered and self-functional cotton sock using piezoelectric and triboelectric hybrid mechanism for healthcare and sports monitoring. *ACS Nano* **2019**, *13*, 1940–1952. [CrossRef]
106. Li, M.; Jie, Y.; Shao, L.H.; Guo, Y.; Cao, X.; Wang, N.; Wang, Z.L. All-in-one cellulose based hybrid tribo/piezoelectric nanogenerator. *Nano Res.* **2019**, *12*, 1831–1835. [CrossRef]

107. Syu, M.H.; Guan, Y.J.; Lo, W.C.; Fuh, Y.K. Biomimetic and porous nanofiber-based hybrid sensor for multifunctional pressure sensing and human gesture identification via deep learning method. *Nano Energy* **2020**, *76*, 105029. [CrossRef]
108. Huang, T.; Zhang, Y.; He, P.; Wang, G.; Xia, X.; Ding, G.Q.; Tao, T.H. Self-matched tribo/piezoelectric nanogenerators using vapor-induced phase-separated poly (vinylidene fluoride) and recombinant spider silk. *Adv. Mater.* **2020**, *32*, 1907336. [CrossRef]
109. Du, M.; Cao, Y.; Qu, X.; Xue, J.; Zhang, W.; Pu, X.; Shi, B.; Li, Z. Hybrid nanogenerator for biomechanical energy harvesting, motion state detection, and pulse sensing. *Adv. Mater. Technol.* **2022**, *7*, 2101332. [CrossRef]
110. Ouyang, H.; Liu, Z.; Li, N.; Shi, B.; Zou, Y.; Xie, F.; Ma, Y.; Li, Z.; Li, H.; Zheng, Q.; et al. Symbiotic cardiac pacemaker. *Nat. Commun.* **2019**, *10*, 1821. [CrossRef]
111. Ouyang, H.; Tian, J.J.; Sun, G.L.; Zou, Y.; Liu, Z.; Li, H.; Zhao, L.M.; Shi, B.J.; Fan, Y.B.; Fan, Y.F.; et al. Self-powered pulse sensor for antidiastole of cardiovascular disease. *Adv. Mater.* **2017**, *29*, 1703456. [CrossRef]
112. Chu, Y.; Zhong, J.; Liu, H.; Ma, Y.; Liu, N.; Song, Y.; Liang, J.; Shao, Z.; Sun, Y.; Dong, Y.; et al. Human pulse diagnosis for medical assessments using a wearable piezoelectret sensing system. *Adv. Funct. Mater.* **2018**, *28*, 1803413. [CrossRef]
113. Liu, Z.; Xu, L.; Zheng, Q.; Kang, Y.; Shi, B.; Jiang, D.; Li, H.; Qu, X.; Fan, Y.; Wang, Z.L.; et al. Human motion driven self-powered photodynamic system for long-term autonomous cancer therapy. *ACS Nano* **2020**, *14*, 8074–8083. [CrossRef] [PubMed]
114. Zhang, J.; Hu, Y.; Lin, X.; Qian, X.; Zhang, L.; Zhou, J.; Lu, A. High-performance triboelectric nanogenerator based on chitin for mechanical-energy harvesting and self-powered sensing. *Carbohydr. Polym.* **2022**, *291*, 119586. [CrossRef] [PubMed]
115. Zhang, Y.; Zhou, Z.; Sun, L.; Liu, Z.; Xia, X.; Tao, T.H. Genetically engineered biofunctional triboelectric nanogenerators using recombinant spider silk. *Adv. Mater.* **2018**, *30*, 1805722. [CrossRef] [PubMed]
116. Saqib, Q.M.; Chougale, M.Y.; Khan, M.U.; Shaukat, R.A.; Kim, J.; Bhat, K.S.; Bae, J. Triboelectric nanogenerator based on lignocellulosic waste fruit shell tribopositive material: Comparative analysis. *Mater. Today Sustain.* **2022**, *18*, 100146. [CrossRef]
117. Khandelwal, G.; Minocha, T.; Yadav, S.K.; Chandrasekhar, A.; Joseph Raj, N.P.M.; Gupta, S.C.; Kim, S.-J. All edible materials derived biocompatible and biodegradable triboelectric nanogenerator. *Nano Energy* **2019**, *65*, 104016. [CrossRef]
118. Chi, Y.; Xia, K.; Zhu, Z.; Fu, J.; Zhang, H.; Du, C.; Xu, Z. Rice paper-based biodegradable triboelectric nanogenerator. *Microelectron. Eng.* **2019**, *216*, 111059. [CrossRef]
119. Han, Y.; Han, Y.; Zhang, X.; Lin, L.; Zhang, C.; Liu, J.; Lu, G.; Yu, H.-D.; Huang, W. Fish gelatin based triboelectric nanogenerator for harvesting biomechanical energy and self-powered sensing of human physiological signals. *ACS Appl. Mater. Interfaces* **2020**, *12*, 16442–16450. [CrossRef]
120. Oliveira, G.S.; Candido, I.C.M.; Lima, R.M.A.P.; Oliveira, H.P. All-in-one energy harvesting/storage integrated systems based on eggshell membranes. *ACS Appl. Electron. Mater.* **2022**, *4*, 4708–4718. [CrossRef]
121. Yan, S.; Zhang, Z.; Shi, X.; Xu, Y.; Li, Y.; Wang, X.; Li, Q.; Turng, L.-S. Eggshell membrane and expanded polytetrafluoroethylene piezoelectric-enhanced triboelectric bio-nanogenerators for energy harvesting. *Int. J. Energy Res.* **2021**, *45*, 11053–11064. [CrossRef]
122. Ghaffarinejad, A.; Hasani, J.Y.; Hinchet, R.; Lu, Y.; Zhang, H.; Karami, A.; Galayko, D.; Kim, S.-W.; Basset, P. Aconditioning circuit with exponential enhancement of output energy for triboelectric nanogenerator. *Nano Energy* **2018**, *51*, 173–184. [CrossRef]
123. Morel, A.; Brenes, A.; Gibus, D.; Lefeuvre, E.; Gasnier, P.; Pillonnet, G.; Badel, A. A comparative study of electrical interfaces for tunable piezoelectric vibration energy harvesting. *Smart Mater. Struct.* **2022**, *31*, 045016. [CrossRef]
124. Rincón-Mora, G.A.; Yang, S. Tiny piezoelectric harvesters: Principles, constraints, and power conversion. *IEEE Trans. Circuits Syst. I Regul. Pap.* **2016**, *63*, 639–649. [CrossRef]
125. Long, Z.; Li, P.; Chen, J.; Chung, H.S.-H.; Yang, Z. Self-Powered Single-Inductor Rectifier-Less SSHI Array Interface with the MPPT Technique for Piezoelectric Energy Harvesting. *IEEE Trans. Ind. Electron.* **2022**, *69*, 10172–10181. [CrossRef]
126. Wang, J.; Chen, Z.; Li, Z.; Jiang, J.; Liang, J.; Zeng, X. Piezoelectric energy harvesters: An overview on design strategies and topologies. *IEEE Trans. Circuits Syst. II Express Briefs* **2022**, *69*, 3057–3063. [CrossRef]

MDPI
St. Alban-Anlage 66
4052 Basel
Switzerland
www.mdpi.com

Nanomaterials Editorial Office
E-mail: nanomaterials@mdpi.com
www.mdpi.com/journal/nanomaterials

www.ingramcontent.com/pod-product-compliance
Lightning Source LLC
Chambersburg PA
CBHW061605210926
51850CB00009B/2221